UNSATURATED FATTY ACIDS

UNSATURATED FATTY ACIDS
Nutritional and physiological significance

The Report of the British Nutrition Foundation's Task Force

The British Nutrition Foundation

SPRINGER-SCIENCE+BUSINESS MEDIA, B.V.

First edition 1992

© Springer Science+Business Media Dordrecht 1992
Originally published by The British Nutrition Foundation in 1992
Softcover reprint of the hardcover 1st edition 1992
Typeset in 10 on 12pt Helvetica by Columns of Reading Ltd.

ISBN 978-0-442-31621-1   ISBN 978-1-4899-4429-0 (eBook)
DOI 10.1007/978-1-4899-4429-0

A catalogue record for this book is available from the British Library

Library of Congress Cataloging-in-Publication data

British Nutrition Foundation
      Unsaturated fatty acids : nutritional and physiological
significance : the report of the British Nutrition Foundation's task
force.
          p.   cm.
      Includes bibliographical references and index.
      1. Unsaturated fatty acids in human nutrition.   2. Unsaturated
fatty acids—Pathophysiology.   I. Title.
QP752.F35B75   1992
612.3'97—dc20
                                              92–11590
                                                    CIP

CONTENTS

BNF TASK FORCE ON UNSATURATED FATTY ACIDS: NUTRITIONAL AND PHYSIOLOGICAL SIGNIFICANCE

TERMS OF REFERENCE

The Task Force was invited by the Council of the British Nutrition Foundation to:

1. Review the present state of knowledge of:
   (a) the occurrence of unsaturated fatty acids in foods used for human consumption;
   (b) past and present sources and intakes of unsaturated fatty acids;
   (c) the metabolic and physiological functions of unsaturated fatty acids;
   (d) the nutritional requirements for, and desirable intakes of, unsaturated fatty acids.
2. Prepare a report and, should it see fit, to draw conclusions, make recommendations and identify areas for future research.

FOREWORD

The British Nutrition Foundation organises independent 'Task Forces' to review, analyse and report in depth upon specific areas of interest and importance in the field of human nutrition.

These expert committees consist of acknowledged specialists and operate independently of the Foundation.

The Unsaturated Fatty Acids Task Force has reviewed and discussed much published information. This report summarises the deliberations and findings of the Task Force and gives its conclusions and recommendations.

I am most grateful to the members of the Task Force who have contributed their time and expertise so generously. My sincere thanks also go to the Secretariat for their excellent support.

Dr Alan Garton FRS
Chairman of the Task Force

A TRIBUTE TO PRESCIENCE

This Task Force Report would be incomplete if recognition were not accorded to the profound influence of the late Dr Hugh Sinclair (1910–1990) on the development of our understanding of the nutritional and physiological significance of unsaturated fatty acids. In his student days, Sinclair became convinced that many 'diseases of civilization' were, at least in part, due to a chronic relative dietary deficiency of essential fatty acids. At the time, and for many years thereafter, the study of the possible biological functions of lipids and fatty acids was, to say the least, 'unfashionable', but Hugh Sinclair consistently confirmed his conviction and undertook experiments in attempts to validate aspects of his thesis. †he scientific and medical 'establishments' of the day took no notice of him and, even when they did, his views were often dismissed as the ramblings of an eccentric Oxford don.

Nevertheless, Sinclair doggedly persisted and, in due course, he was vindicated when the crucial role which dietary acids play in the maintenance of cellular integrity and in the regulation of metabolic processes became firmly established. In retrospect, a remarkably prophetic letter which Hugh Sinclair wrote to the Editor of *The Lancet* in 1956 probably marked the turning point in attitudes and resulted in stimulating world-wide research. In the letter he elaborated on his contention that degenerative diseases, such as atherosclerosis, are associated with deficiency of unsaturated fatty acids, notably arachidonic acid. It is noteworthy that the letter, the longest ever to appear in *The Lancet*, was selected by *Current Contents* as a 'Citation Classic'.

Hugh Sinclair, Emeritus Fellow of Magdalen College, Oxford, was a memnber of the Council of The British Nutrition Foundation from its establishment in 1967 until his death. He had agreed to advise this Task Force and to comment on the report as it was prepared, but unfortunately his wise counsel was to be denied to us. We trust he would have approved of our conclusions and recommendations in the knowledge that the topics to which he gave so much thought and effort are continuing to attract ever-increasing attention.

Alan Garton

ABBREVIATIONS USED IN TEXT

ADP	adenosine diphosphate		LT	leukotriene
ATP	adenosine triphosphate		MCT	medium chain triacylglycerol(s)
BP	blood pressure		MI	myocardial infarction
BSA	bovine serum albumin		MS	multiple sclerosis
cAMP	cyclic adenosine monophosphate		MUFA	monounsaturated fatty acid(s)
CDAGP	cytidyl diacyl glycerophosphate		NADH	nicotinamide adenine dinucleotide
CHD	coronary heart disease		NADPH	nicotinamide adenine dinucleotide phosphate
CHO	carbohydrate			
Cho	choline		NFS	National Food Survey
CM	chylomicrons		NIDDM	non-insulin dependent diabetes mellitus
CMP	cytidyl monophosphate			
CoA	coenzyme A		NSAID	non-steroidal anti-inflammatory drugs
DAG	diacylglycerol		PAF	platelet activating factor
DGLA	dihomo gamma linolenic acid		PCTA	percutaneous transluminal coronary angioplasty
DHA	docosahexaenoic acid			
DMBA	dimethylbenz-[a] anthracene		PDGF	platelet derived growth factor
DPA	docosapentaenoic acid		PG	prostaglandin
DRV	dietary reference value(s)		PGI	prostacyclin
EAR	estimated average requirement		PIP_1	phosphatidyl inositol phosphate
EC	European Community		PIP_2	phosphatidyl inositol 4,5 bisphosphate
ECG	electro-cardiogram			
EDRF	endothelium-derived relaxing factor		PKC	protein kinase C
EFA	essential fatty acids		PLA	phospholipase A
EPA	eicosapentaenoic acid		PLC	phospholipase C
EPSO	evening primrose seed oil		PLG	plasminogen
Etn	ethanolamine		Ptd	phosphatidyl
FAD	flavin adenine dinucleotide		PUFA	polyunsaturated fatty acid(s)
FFA	free fatty acid		RNI	reference nutrient intake
FLAP	5-lipoxygenase activity protein		Ser	serine
GLA	gamma linolenic acid		SFA	saturated fatty acid(s)
GLC	gas liquid chromatography		SLE	systemic lupus erythematosus
GP	glycoproteins		SPE	sucrose polyesters
HDL	high density lipoproteins		SRS-A	slow reacting substances A
HMWK	high molecular weight kininogen		TBARM	thiobarbituric acid reactive material
5-HPETE	5-hydroxyperoxyeicosatetraenoic acid		TNF	tumour necrosis factor
IDDM	insulin-dependent diabetes mellitus		TDS	Total Diet Study
IDL	intermediate density lipoproteins		tPA	tissue plasminogen activator
Ins	inositol		TPN	total parenteral nutrition
IP_3	inositol 1,4,5 trisphosphate		TX	thromboxane
IRR	initial rate of response		UFA	unsaturated fatty acid
LCAT	lecithin-cholesterol acyl transferase		VLC PUFA	very long chain polyunsaturated fatty acid
LCP	long chain polyunsaturated fatty acid			
LDL	low density lipoproteins		VLDL	very low density lipoproteins
LPL	lipoprotein lipase(s)		WHHL	Watanabe hereditary hyperlipidaemia

INTRODUCTION

Lipids containing saturated and unsaturated fatty acids are present in the diet in many forms. They can be overtly present, as in salad oils, butter and margarines or they can be concealed in cooked and processed foods such as biscuits, cakes and confectionery. In whatever form lipids are eaten, they comprise about 40% of the energy value of the average diet – a value which, according to the UK Government's Committee on Medical Aspects of Food Policy (COMA) (Department of Health, 1984), is too high because of the risk of inducing unduly high levels of blood cholesterol thereby increasing the possibility that coronary heart disease may ensue. That Committee concurred with earlier conclusions that high blood cholesterol levels were associated with the intakes of saturated fatty acids and accordingly suggested that the public should be encouraged to eat less fat (less than 35% of energy intake from food) and that not more than 15% should be derived from saturated fatty acids. At the same time, COMA indicated that polyunsaturated fatty acids (P) could, with benefit, be 'substituted' in the diet for saturated fatty acids (S), thereby increasing the 'P/S ratio'. The nature of the polyunsaturated acids was not clearly stated.

Whereas some progress has been made to encourage people to eat less fat and to consume margarines and spreads with a high content of polyunsaturated fatty acids, the public is still confused by much of the dietary information and advice purveyed by the media. Unsaturated fatty acids are sometimes discussed as if they were a group of substances with exactly equivalent nutritional and physiological properties and even the injunction to eat more foods rich in unsaturated fatty acids is called into question in the popular press from time to time because of misunderstandings about increased cancer risk.

The apparent virtues of supplementing one's diet with fish oil and/or certain unusual vegetable oils have been widely canvassed, accompanied by somewhat extravagant claims for their potential to cure, alleviate or prevent all manner of conditions ranging from atopic eczema to rheumatoid arthritis. It is not only the public which is confused; many members of the medical profession, nutrition scientists, dietitians and food scientists have voiced their concern over the lack of information on which to base dietary guidance which they would like to be able to provide.

The British Nutrition Foundation commissioned this Task Force to review objectively the present state of knowledge of unsaturated fatty acids. For consistency and convenience, the Report has considered unsaturated fatty acids in three classes: monounsaturated fatty acids, and polyunsaturated fatty acids which are sub-divided into two families, commonly referred to as n-6 polyunsaturates and n-3 polyunsaturates. The monounsaturates can be synthesised in the body, and thus there is no dietary essentiality for them. The parent acids of the two polyunsaturated families cannot be synthesised in the body and must be provided in the diet. The term 'essential fatty acids' has become widely known, prompting the question 'essential for what and for whom?' The Report addresses this question in some detail.

The Report starts by outlining the structure of the different unsaturated fatty acids (Chapter 1). It specifies in what foods they are found, which foods contribute most to intakes (Chapter 2), how much of them are eaten and how this has changed in recent years (Chapter 3). It then discusses how they are digested, absorbed and transported in the body (Chapters 4–6), and what they do (Chapters 7–9). Finally, it examines the role of unsaturated fatty acids in a variety of different disease states (Chapters 10–20) and considers the implications for nutrition labelling (Chapter 21). On the basis of these considerations, the Task Force has put forward its recommendations for intakes of unsaturated fatty acids in the diet (Chapter 22) which complement and extend the conclusions of the COMA 1991 report (DH, 1991) in respect of lipids.

The Task Force hopes that its discussions and conclusions will go some way to create a better general understanding of the nature, nutritional and physiological significance of unsaturated fatty acids. In addition, the Task Force has put forward suggestions for future progress which, if implemented, should shed further light on the metabolic importance of all unsaturated fatty acids and should enable dietary recommendations to be met.

1

CHEMISTRY OF UNSATURATED FATTY ACIDS

1.1 STRUCTURE AND NOMENCLATURE OF FATTY ACIDS

A fatty acid is made up of a hydrocarbon chain, from which the properties of lipid solubility derive, and a terminal carboxyl group, giving acidic properties. Fatty acids with chain lengths from 2 to over 30 carbon atoms are known, but the commonest are in the range of 12 to 22 carbon atoms (C12–C22). When each of the carbon atoms in the chain, except the two terminal ones, is bonded to two hydrogen atoms, the acids are said to be saturated, since all the bonding capacity of the carbons is saturated with hydrogen. When each of two adjacent carbon atoms is bonded to only one hydrogen, there is an ethylenic double bond between the pair of carbons and the fatty acid is said to be unsaturated. If the chain contains only one double bond, it is a monounsaturated fatty acid (MUFA) and if the chain contains more than one double bond, it is a polyunsaturated fatty acid (PUFA) (Figure 1.1).

Saturated fatty acids are given systematic chemical names that denote the number of carbon atoms (see Table 1.1). Thus the eighteen-carbon, straight chain saturated fatty acid is octadecanoic acid. The presence of a double bond is indicated by the change of the suffix from -anoic to -enoic. Thus monounsaturated and polyunsaturated fatty acids can also be called monoenoic or polyenoic fatty acids (sometimes shortened to monoenes and polyenes). In this Report, we will restrict ourselves to the most familiar terms, saturated and unsaturated. An eighteen-carbon acid with one double bond is octadecenoic acid and with two double bonds, octadecadienoic acid. Table 1.1 lists the commonest saturated and unsaturated fatty acids.

Over the years, a system of trivial names has grown up beside those of the more rigorous chemical nomenclature. Thus the most widespread form of an octadecenoic acid is also known as oleic acid, the most common form of octadecadienoic acid is known as linoleic acid and the most abundant octadecatrienoic acid is alpha linolenic acid. Some fatty acids (DGLA, EPA, DPA and DHA) do not have trivial names and are usually referred to using the abbreviated form of the systematic name.

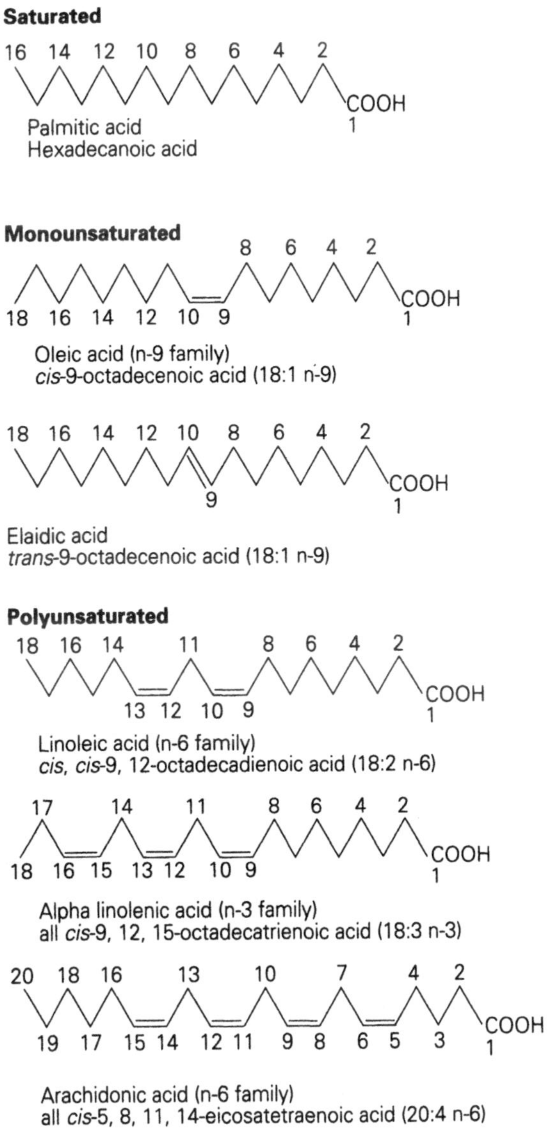

Figure 1.1 Chemistry of unsaturated fatty acids.

Table 1.1 Fatty acid nomenclature

Systematic name	Trivial name/abbreviation	Shorthand notation
Saturated		
dodecanoic	lauric	(12:0)
tetradecanoic	myristic	(14:0)
hexadecanoic	palmitic	(16:0)
octadecanoic	stearic	(18:0)
Unsaturated		
cis-9-hexadecenoic	palmitoleic	(16:1 n-7)
cis-9-octadecenoic	oleic	(18:1 n-9)
trans-9-octadecenoic	elaidic	(18:1 n-9)
cis-11-eicosaenoic	gadoleic	(20:1 n-9)
cis-13-docasaenoic	erucic	(22:1 n-9)
cis-11-docasaenoic	cetoleic	(22:1 n-11)
cis-15-tetracosaenoic	nervonic	(24:1 n-9)
cis,cis,9,12-octadecadienoic	linoleic	(18:2 n-6)
trans-5,*cis*-9,*cis*-12-octadecatrienoic	columbinic	(18:3 n-6)
all *cis*,9,12,15-octadecatrienoic	alpha linolenic	(18:3 n-3)
all *cis*,6,9,12-octadecatrienoic	gamma linolenic	(18:3 n-6)
all *cis*,6,9,12,15-octadecatetraenoic	stearidonic	(18:4 n-3)
all *cis*,11,14,17-eicosatrienoic	Mead	(20:3 n-9)
all *cis*,8,11,14-eicosatrienoic	DGLA	(20:3 n-6)
all *cis*,8,11,14,17-eicosatetraenoic	ETA	(20:4 n-3)
all *cis*,5,8,11,14-eicosatetraenoic	arachidonic	(20:4 n-6)
all *cis*,5,8,11,14,17-eicosapentaenoic	EPA	(20:5 n-3)
all *cis*,7,10,13,16-docosatetraenoic	adrenic	(22:4 n-6)
all *cis*,7,10,13,16,19-docosapentaenoic		(22:5 n-3)
all *cis*,4,7,10,13,16-docosapentaenoic	DPA	(22:5 n-6)
all *cis*,4,7,10,13,16,19-docosahexaenoic	DHA	(22:6 n-3)

1.2 STRUCTURAL FEATURES OF UNSATURATED FATTY ACIDS

1.2.1 The chemistry of the ethylenic double bond

The electronic structure of the ethylenic bond, with its so-called *pi*-electrons, gives the double bond much greater chemical reactivity than the single bond. Thus, it can be the site of addition of a number of chemical constituents. Addition of two hydrogen atoms, for example, converts the acid into a saturated fatty acid and this is important in some types of industrial processing ('hardening') of fats and oils. The elements of water can be added to produce hydroxy fatty acids and oils. Most important of all, the double bond is particularly susceptible to attack by oxygen to produce a variety of oxygenated fatty acids, some of which are of considerable biological significance. The presence of double bonds also influences the shape of the molecules, as discussed below.

1.2.2 Geometrical isomerism

The ethylenic double bond can adopt two distinct geometrical configurations, denoted by chemists as the Z and E configurations. An older and more familiar system is the *cis/trans* nomenclature which will be used in this Report. A projection of the *cis* double bond would have the two hydrogens on the same side of the molecule, whilst in the *trans* configuration they are on opposite sides. This has implications for the shape of the molecule. Fatty acids can adopt a large number of configurations or shapes but while it is possible for saturated chains and those containing the *trans* double bond to adopt a straight configuration, the *cis* forms always have a kink in the chain. This influences the physical properties of fatty acids as discussed in Section 1.3.

Cis-double bonds, rather than *trans*-double bonds, are more commonly found in natural lipids. *Trans* bonds do occur sometimes, as intermediates in the biosynthesis of fatty acids, in ruminant fats, in plant leaf lipids and in some seed oils (see Chapter 2).

1.2.3 Positional isomerism

Irrespective of whether double bonds are *cis* or *trans*, they can be located at different positions in the hydrocarbon chain. During biosynthesis, the double bond is introduced between carbon atoms 9 and 10 in the most commonly occurring monounsaturated fatty acid, oleic acid (see Chapter 7).

However, small amounts of monounsaturated fatty acids with double bonds in other positions are known. When a second double bond is introduced in the biosynthetic process, it is usually separated from the first double bond by a methylene (CH_2) group and this pattern is continued with the introduction of subsequent double bonds (Table 1.1). These structures are known as 'methylene-interrupted' polyunsaturated fatty acids. Far less common in natural oils are conjugated, or widely separated, double-bond systems.

During chemical reactions, such as catalytic hydrogenation or oxidation, the double bonds may be shifted along the chain and/or geometrically isomerised to yield a much wider variety of structural isomers than normally occurs in natural oils (see Chapter 7).

The position of double bonds in the hydrocarbon chain of a fatty acid is denoted by a prefix indicating the number of the first of the pair of carbon atoms forming the double bond, starting to number from the carboxyl carbon as carbon atom number one (Figure 1.1 and Table 1.1). Thus, oleic acid is *cis*-9-octadecenoic acid and linoleic acid is *cis*, *cis*-9, 12-octadecadienoic acid. A useful short-hand nomenclature consists of two numbers separated by a colon. The number before the colon gives the carbon chain length while the figure after denotes the number of double bonds. Thus stearic acid is 18:0, oleic acid is 18:1 and linoleic acid is 18:2. When a fatty acid chain is elongated, two carbon atoms are added at the carboxyl end of the chain. Thus the numbering of the double bonds in the systematic name changes each time the chain is elongated (compare arachidonic acid with gamma linolenic acid in Table 1.1).

Sometimes, to emphasise metabolic relationships, it is helpful to use a shorthand notation which numbers the first double bond from the methyl terminal carbon atom rather than from the carboxyl end. Under this system, the shorthand notation for oleic acid is 18:1 n-9 and that for linoleic acid is 18:2 n-6. Arachidonic acid becomes 20:4 n-6 and gamma linolenic acid is 18:3 n-6 and it is easier to see that the former has been formed from the latter by the addition of two carbon atoms and one double bond.

1.3 TRIACYLGLYCEROLS

Triacylglycerols were previously called triglycerides. Triacylglycerol molecules consist of three fatty acids esterified to a glycerol moiety. In most natural oils and fats, the three constituent fatty acids are not randomly distributed (see Table 1.2). There is a tendency for specific fatty acids to be located at particular positions on the three glycerol carbons

(i.e. positions 1, 2 and 3). In cows' milk fat, the characteristic short chain fatty acids are found at position 3, whilst in human milk fat, the saturated acid, palmitic acid is found at position 2 and unsaturated acids at position 1. Animal depot fats have saturated fatty acids mainly at position 1, unsaturated and short chain fatty acids at position 2. Position 3 seems to have a more random population, although polyunsaturated acids tend to accumulate at position 3 in most mammals. Phosphoglycerides usually have a saturated fatty acid at position 1 and an unsaturated fatty acid at position 2.

Table 1.2 Characteristic positioning of fatty acids in triacylglycerol and phosphoglyceride molecules

	Cows' milk fats	Human milk fats	Animal depot fats	Phospho-glycerides
Position 1	Random	Unsaturated fatty acids	Saturated fatty acids	Saturated fatty acids
Position 2	Random	Saturated fatty acids	Unsaturated fatty acids and short chain fatty acids	Unsaturated fatty acids
Position 3	Short chain fatty acids		Random	Phosphate group

These specific distributions are important both in terms of the physico-chemical properties of the lipids and of their metabolic activities, as discussed in Chapters 7 and 8.

1.4 PHOSPHOGLYCERIDES

In mammals, the lipids involved in membrane structures are mainly the phosphoglycerides and unesterified (free) cholesterol (Figure 1.2). Phosphoglycerides belong to a more general class of phosphorus-containing lipids, the phospholipids. Most of the phospholipids considered in this report are phosphoglycerides and the terms may be used interchangeably. The only common phospholipids which are not phosphoglycerides are sphingo-myelins and the cerebrosides which are mainly found in nervous tissues (see Figure 1.2). These phospholipids are based on the amino alcohol sphingosine rather than glycerol.

Phospholipids have important properties in membranes and foods because of their amphiphilic nature: i.e they possess chemical groupings that associate with water (hydrophilic groups) in juxtaposition with hydrophobic moieties (amphiphilic is derived from the Greek for 'liking both'). In contrast,

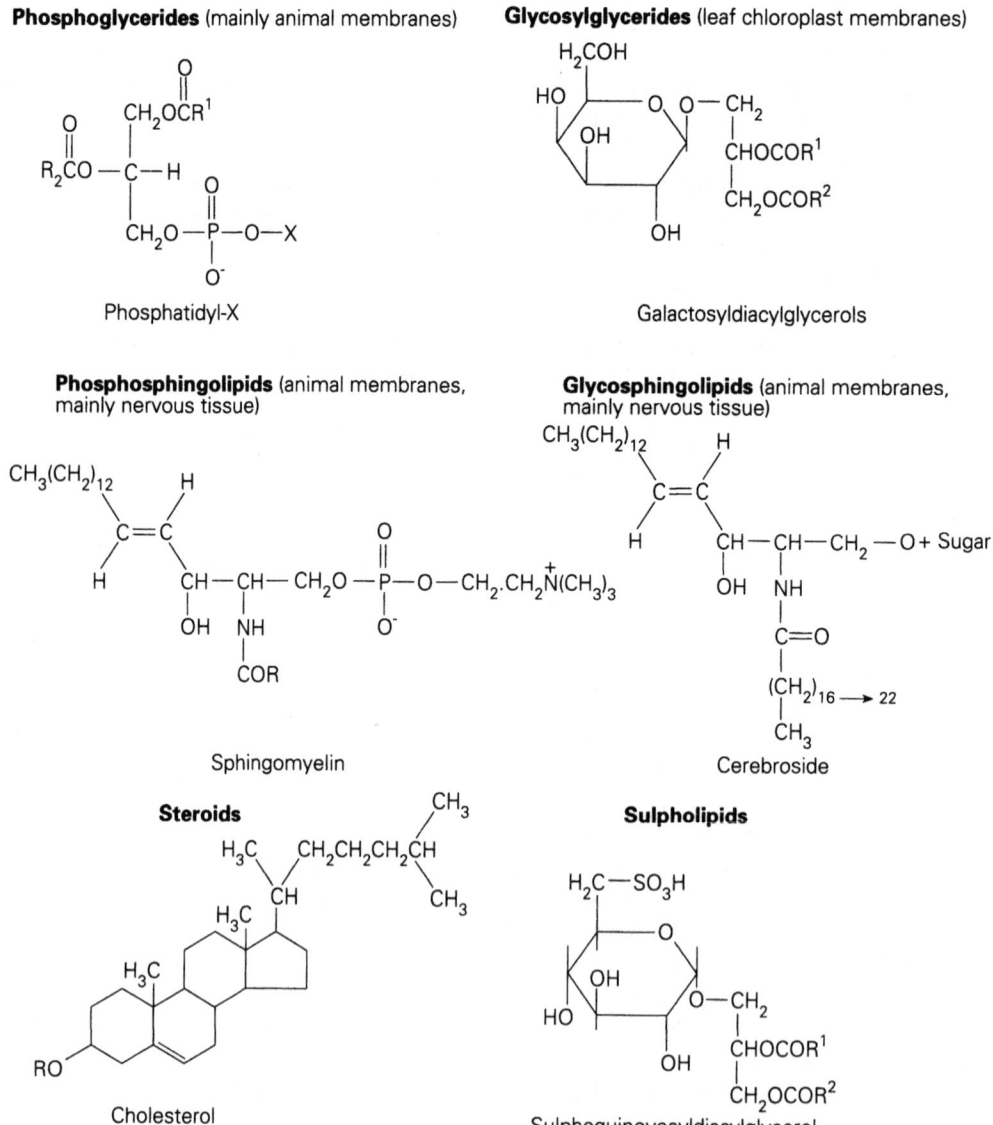

Figure 1.2 Structural lipids in plant and animal membranes.

hydrophobic fats, without polar groups, such as triacylglycerols, wax esters and sterols are often called neutral, apolar or non-polar lipids, but these are imprecise terms.

Phosphoglycerides are amphiphilic lipids in which the polar moiety is an ester of phosphorus with an organic base, such as choline in phosphatidylcholine (PtdCho), ethanolamine in phosphatidylethanolamine (PtdEth), serine in phosphatidylserine (PtdSer) or inositol in phosphatidlyinositol (PtdIns).

1.5 PHYSICAL PROPERTIES OF FATS AND OILS AND THEIR CONSTITUENT FATTY ACIDS

The physical properties of food fats (triacylglycerols) are influenced largely by the nature of their constituent fatty acids, although the presence of minor components such as sterols and phospholipids can also be important. Table 1.3 shows that in general, the higher the chain length, the higher the melting point (compare 12:0 and 16:0); the greater the number of double bonds, the lower the melting point (compare 18:1, 18:2 and 18:3). However, because fatty acids with *trans* unsaturation have melting points higher than the corresponding acids with *cis*-unsaturation (compare *trans* 18:1 with *cis* 18:1), the total degree of unsaturation can sometimes be an unreliable guide to physical properties. As discussed above, the physical properties of triacylglycerols are also dependent on the distribution of fatty acids on the three positions of the glycerol backbone, so that interesterification can also modify physical properties. In most foods, fats and oils are not present in isolation from other components. Physical properties of fats and oils can be markedly influenced by the formation of water-in-oil or oil-in-water emulsions exemplified by margarine and mayonnaise respectively.

Table 1.3 The melting points of some fatty acids

Chain length	Fatty acid	Melting point (°C)
12	12:0	44.2
16	16:0	62.7
18	18:0	69.6
18	c-18:1	13.2
18	t-18:1	44.0
18	c,c-18:2	−5.0
18	t,t-18:2	28.5
18	c,c,c-18:3	−11.0
18	t,t,t-18:3	29.5

2

SOURCES OF UNSATURATED FATTY ACIDS IN THE DIET

The raw materials from which food fats are derived are mainly the storage fats of land and marine mammals and fish and the seed oils of plants. To convert them into a suitable condition for food use, various extraction and purification processes are used that may modify the composition to some degree. More extensive compositional changes are brought about by other specific processes such as industrial hydrogenation which affects particularly the unsaturated fatty acids. These manufacturing processes are described in Section 2.4.

2.1 NATURALLY OCCURRING STORAGE FATS FOR FOOD USE

2.1.1 Storage lipids of land animals

2.1.1.1 Pigs and poultry

The composition of the adipose tissue of simple-stomached animals, of which pigs and poultry are economically the most important, is markedly affected by the fat in the diet. When these species

are given low fat, cereal-based diets, as in traditional farming practice (columns A and D in Table 2.1), the adipose tissue fatty acids are synthesised in the body from dietary carbohydrates and the storage fat is composed mainly of saturated and monounsaturated fatty acids (MUFA). However, cereals contain structural lipids that influence storage fat composition to some extent, depending on the type of cereal.

Inclusion of vegetable oils, such as soya bean oil, in the diet (columns B and E) results in higher proportions of linoleic acid and lower proportions of oleic acid than the inclusion of tallow (column C) which tends to give a storage fat similar in composition to that of animals fed on cereals (column A).

2.1.1.2 Ruminants

The adipose tissue of ruminants is less variable than that of simple-stomached animals because about 90% of the unsaturated fatty acids originally present in the animals' diet are hydrogenated (i.e. converted into relatively more saturated fatty acids)

Table 2.1 The fatty acid composition of some animal storage fats used in human foods showing the influence of different feeding practices

Fatty acid (g/100g total fatty acids)		Pig (lard)			Poultry		Beef (suet)		Lamb	
		A	B	C	D	E	F	G	H	I
Myristic	(14:0)	1	1	1	1	1	3	3	3	4
Palmitic	(16:0)	29	21	21	27	22	26	20	21	19
Stearic	(18:0)	15	12	17	7	6	8	10	20	16
Palmitoleic	(16:1)	3	3	4	9	5	9	4	4	6
Oleic	(18:1)	43	46	54	45	27	45	33	41	37
Linoleic	(18:2)	9	16	3	11	35	2	23	5	12
Long chain MUFA	(20:1, 22:1)	0	0	0	0	0	0	0	0	0
Long chain PUFA	(20:5, 22:5, 22:6)	0	0	0	0	0	0	0	0	0
Others		0	1	0	0	4	7	7	6	6

A, pig fed low fat cereal-based diet; B, pig fed high fat diet containing soyabean oil; C, pig fed high fat diet containing beef tallow; D, poultry fed low fat cereal-based diet; E, poultry fed high fat diet containing soyabean oil; F, cattle fed diet based on hay; G, cattle fed diet containing 'protected' safflower oil; H, lambs fed cereal-based concentrate diet; I, lamb fed diet containing 'protected' safflower oil (Adapted from Gurr, 1984) by bacteria and other microorganisms in the rumen before they reach the adipose tissue. The adipose tissue of ruminants contains a larger proportion of saturated and MUFA and a lower proportion of PUFA than the adipose tissue of simple-stomached animals (see columns F and H). It also contains more *trans* and branched-chain fatty acids derived from bacteria. The amounts and types of these more unusual fatty acids depend on the ruminant species and the way in which the diet affects the activities of the microorganisms.

In traditional agricultural practice, ruminants are given diets in which fats comprise no more than about 2–5% of digestible energy. Under these circumstances, most of the body fats (except those that are dietary essentials) are synthesised within the body from the products of microbial fermentation of dietary carbohydrates in the rumen. A recent trend to improve the efficiency of animal production is to give ruminants, especially dairy cows, feeds which contain fat supplements or whole oil seeds such as those from safflower oil. In this way, modest changes in the composition of the carcass fat can be effected.

More extensive changes can be brought about by feeding so-called 'protected' fat. In this process, fat particles are coated with protein which is then crosslinked by treatment with formaldehyde, making the fat resistant to degradation in the rumen. On reaching the acidic environment of the true stomach, the protein is digested and the fat passes unchanged into the small intestine where it can be digested and absorbed as in a simple-stomached animal. In this way, large increases in the proportion of linoleic acid can be induced in the storage fats of ruminants (see columns G and I).

2.1.2 Fats in milk

The fatty acid composition of cows', sheeps' and goats' milks is characterised by relatively high proportions of short and medium chain fatty acids and small proportions of PUFA (Table 2.2). Ruminant milks also contain small quantities of a wide variety of branched and odd chain fatty acids, the nature and amounts depending on the nature of the feed. The proportions of unsaturated to saturated fatty acids in cow's milk change during the season. 'Summer milk' contains somewhat more saturated fatty acids than 'winter milk'.

Feeding fat supplements influences the degree of unsaturation of milk fats as it does with storage fat (see Table 2.1). Large increases in linoleic acid can only be produced by giving 'protected' fat (compare columns A and B). However, the more economically feasible method of giving animals unprotected supplements, either in the form of

Table 2.2 Fatty acid composition of some milk fats

Fatty acid (g/100g total fatty acids)	Cow[a] A	Cow[a] B	Goat	Sheep	Human
Saturated 4:0	3		2	4	0
6:0	2		2	3	0
8:0	1	10	3	3	0
10:0	3		9	9	1
12:0	4		5	5	5
14:0	12	8	11	12	7
16:0	26	17	27	25	27
18:0	11	8	10	9	10
	62	43	69	70	50
Monounsaturated 14:1	1	0	1	1	1
16:1	3	0	2	3	4
18:1	28	21	26	20	35
	32	21	29	24	40
Polyunsaturated 18:2	2[d]	35[e]	2	2	7
18:3[b]	1	1	0	1	1
20:4	0	0	0	0	tr
	3	36	2	3	8
Others[c]	3	0	0	3	2

[a] A, cow grazing normally; B, cow fed supplement of 'protected' safflower oil.
[b] Mainly alpha linolenic acid (18:3 n-3); human milk may contain a trace of gamma linolenic acid (18:3 n-6).
[c] Includes odd-chain and branched chain fatty acids in the case of ruminant milks and very long chain polyunsaturated fatty acids (22:4 n-6, 22:5 n-6; 20:5 n-3; 22:5 n-3; 22:6 n-3) in the case of human milk.
[d] Contains linoleic acid (18:2 n-6) and other positional and geometric 18:2 isomers.
[e] Mainly linoleic acid (18:2 n-6).
tr = trace
(Adapted from Gurr, 1981)

crushed oilseeds, or varieties of oats with a high fat content, can increase the MUFA content of milk to over 40%.

2.1.3 Fats in eggs

The egg of the domestic fowl provides a significant source of fat in many human diets. The triacylglycerols are present in the form of lipoproteins, and the fatty acids are predominantly saturated and monounsaturated (Table 2.3). However, the phospholipid fraction contributes significant amounts of polyunsaturated fatty acids.

2.1.4 Fish oils

Fish can be classified broadly into lean fish that store their reserve fats as triacylglycerols in the

Table 2.3 Fatty acid composition of egg yolk lipids

Fatty acid (g/100g total fatty acids)		Egg yolk	Low density lipoprotein fraction (LDL)
Palmitic	(16:0)	29	32
Stearic	(18:0)	9	8
Palmitoleic	(16:1 n-9)	4	8
Oleic	(18:1 n-9)	43	45
Linoleic	(18:2 n-6)	11	7
Others		4	–

(Adapted from Johnson and Davenport, 1971)

liver (e.g. cod) and 'fatty' (or 'oily') fish whose triacylglycerols are stored in the flesh (e.g. mackerel, herring). Fish have no adipose tissue though some (e.g. the orange roughy, found in the Pacific) store reserve oil as wax esters that have poor nutritive value for humans.

Although fish oils differ in composition between species and according to diet and the season when the fish are caught, the oils nevertheless have some common characteristics that distinguish them from mammalian and avian fats. They have a high content of fatty acids with 20 or more carbon atoms and are normally rich in PUFA of the n-3 family, although in some species there is a predominance of long chain monounsaturated fatty acids (Table 2.4). Most fish oils used in food have been heavily hydrogenated and are therefore not a source of these long chain PUFA.

2.1.5 Seed oils

Seed oils are a form of energy storage for many plants just as the adipose tissue is the main form of energy storage for mammals. In species that are of commercial importance in food manufacture, the seed oil is composed predominantly of triacylglycerols that are stored mainly in the fleshy fruit exocarp (e.g. olive) or in the seed endosperm (e.g. rape). Some, like palm, contain triacylglycerols in both the exocarp (palm oil) and the endosperm (palm kernel oil).

Seed oils vary widely in fatty acid composition. One fatty acid often predominates, is frequently of unusual structure, and is characteristic of a particular plant family. The commercially important seed oils (Table 2.5) are, however, generally those in which the predominant fatty acids are the common ones: palmitic, stearic, oleic and linoleic acids. Exceptions are coconut and palm kernel oils, which are unusual in containing a high proportion of saturated fatty acids of medium chain length.

2.2 NATURALLY OCCURRING STRUCTURAL LIPIDS FOR FOOD USE

Structural lipids are those that are present in the cell membranes of plant and animal tissues. They are mainly phospholipids and glycolipids and their fatty acid compositions differ markedly from those of storage lipids. In general, they contain a much higher proportion of PUFA than animal storage lipids, with the exception of fish oils.

2.2.1 Muscle and offal fats

Because of their role in membranes, the fatty acid composition of lean meats tends to be less variable and less susceptible to dietary influence than that of the storage fats. There is a high proportion of

Table 2.4 Fatty acid composition of some fish oils

Fatty acid (g/100g total fatty acids)	Herring	Menhaden	Cod liver	Trout liver	Fish oil concentrate[a]
14:0	9	8	1	9	5
16:0	15	22	19	18	6
16:1	7	11	4	14	7
18:0	1	3	5	6	3
18:1 n-9	16	21	15	14	10
18:2 n-6	1	2	2	3	4
20:1 n-9	16	2	10	2	8
20:4 n-6	tr	2	1	1	tr
20:5 n-3	3	14	6	16	15
22:0	tr	2	4	2	tr
22:1 n-9	23	2	2	3	7
22:6 n-3	3	10	27	7	13
Others	6	1	4	5	22

tr = trace; [a] MaxEPA
(Adapted from Gunstone et al., 1990)

Table 2.5 Fatty acid composition of some seed oils for human consumption

Fatty acid (g/100g total fatty acids)	Coconut	Maize	Olive	Palm	Palm kernel	Peanut	Rape Low erucic	Rape High erucic	Soyabean	Sunflower
Saturated										
8:0	8	0	0	0	4	0	0	0	0	0
10:0	7	0	0	0	4	0	0	0	0	0
12:0	48	0	0	tr	45	tr	0	0	tr	tr
14:0	16	1	tr	1	18	1	tr	tr	tr	tr
16:0	9	14	12	42	9	11	4	4	10	6
18:0	2	2	2	4	3	3	1	1	4	6
20:0	1	tr	tr	tr	0	1	1	1	tr	tr
22:0	0	tr	0	0	0	3	tr	tr	tr	tr
Monounsaturated										
16:1	tr	tr	1	tr	0	tr	2	tr	tr	tr
18:1	7	30	72	43	15	49	54	24	25	33
Polyunsaturated										
18:2	2	50	11	8	2	29	23	16	52	52
18:3	0	2	1	tr	0	1	10	11	7	tr
Others	0	1	1	2	0	2	5	10[a] 33[b]	2	3

tr = trace; [a] (20:1); [b] (22:1)
(Adapted from Gurr, 1984)

PUFA, even in ruminant meats, and these foods are a major source of dietary arachidonic acid (20:4 n-6) (see columns A, C and E in Table 2.6). Brain, though not important in the UK diet, contains a high proportion of structural fat (see column D in Table 2.6).

The overall fatty acid composition of the food itself will also depend on the content of non-structural fat (triacylglycerols in the 'marbling' fat of muscle meats or in fatty liver) which will tend to be richer in saturated and monounsaturated fatty acids than the structural fats.

2.2.2 Plant leaves

The fatty acid composition of lipids in plant membranes is very simple and varies little between different types of leaves. Five fatty acids generally account for over 90% of the total: palmitic, palmitoleic, oleic, linoleic and alpha linolenic (see column H in Table 2.6). Of these, alpha linolenic acid is quantitatively the most important and green vegetables provide much of the intake of this essential fatty acid in the human diet.

Table 2.6 Fatty acid composition of some structural lipids occurring in foods for human consumption

Fatty acids (g/100g total fatty acids)		Beef A	Beef B	Lamb C	Lamb D	Chicken E	Chicken F	Pork G	Green leaves H
Palmitic	(16:0)	16	14	22	22	23	25	19	13
Stearic	(18:0)	11	14	13	18	12	17	12	tr
Palmitoleic	(16:1)	2	2	2	1	6	3	2	3
Oleic	(18:1)	20	5	30	28	33	26	19	7
Linoleic	(18:2)	26	47	18	1	18	15	26	16
Alpha linolenic	(18:3)	1	1	4	0	1	1	0	56
Arachidonic	(20:4)	13	11	7	4	6	6	8	0
Long chain PUFA	(20:5, 22:5, 22:6)	0	0	0	14	0	6	tr	0
Others		11	6	4	12	1	1	14	5

A, muscle, cattle fed low fat diet; B, muscle, cattle fed 'protected' safflower oil diet; C, muscle; D, brain; E, muscle; F, liver; G, muscle; H, membranes
(Adapted from Gurr, 1984)

2.3 UNUSUAL FATS AND OILS CONTAINING UNSATURATED FATTY ACIDS

Of the several hundred varieties of plants known to have oil-bearing seeds, only twelve are important commercially and of these, two are used almost entirely for industrial purposes other than as edible oils. Thus linseed oil, with a high proportion of alpha linolenic acid, is more important in paints and varnishes. Castor oil is unusual in having a high proportion of ricinoleic acid (12-hydroxy-cis-9-octadecenoic) and is used as a laxative. Older varieties of rapeseed oil contained nearly 60% of erucic acid (cis-13-docosaenoic acid) but due to its reported toxicity, new varieties in which the erucic acid has been bred out and replaced by oleic acid are now almost always used for foods.

Demands for food fats with particular fatty acid compositions, or at a lower price, could stimulate the agricultural production of oil-seed crops that have not so far been commercially exploited (e.g. lupin or evening primrose). The oil of the latter contains about 8% of gamma linolenic acid, a fatty acid normally found only in very low quantities. Seed breeding (by conventional or biotechnological methods) may lead to the development of new varieties of traditional oil-seed crops which have novel fatty acid composition (e.g. high oleic sunflower and linseed varieties). In addition, new oil-seed crops such as plants of the family Cuphea containing seed oils rich in medium chain fatty acids may be developed.

2.4 EFFECTS OF INDUSTRIAL PROCESSING

2.4.1 Extraction

Plant oils are extracted from seeds by pressure or by extraction with solvents. The main objective is to obtain a clean product that is composed mainly of triacylglycerols. 'Degumming' removes a variety of substances such as phospholipids and polysaccharide gums that would separate out from the oil on storage. Lipases are inactivated by brief exposure to high temperature as the presence of significant amounts of non-esterified fatty acids could cause rancidity and spoil the quality of the final oil. These processes are not such as to alter significantly the fatty acid composition of the lipids however. The same is true of the 'rendering' process used to extract animal carcass fats which is somewhat milder than the treatment needed for seed-oil extraction.

2.4.2 Heating

In some processes, such as frying, which may be used either in the industrial preparation of foods or in cooking in the home, fats may be subjected to high temperatures. During heating, some changes may occur in the structure of the fats, the nature of which depends on several factors like temperature, the length of time for which the fat is heated and the amount of air to which it is exposed: e.g. the temperature involved in industrial deep-fat frying could be about 180°C. At this temperature, the production of steam carries volatile compounds out of the fat ('steam-stripping'). The formation of a layer of volatile compounds and steam above the fat may act as a barrier to prevent exposure of the oil to air.

In the presence of air, the first products that are formed at lower temperature are the hydroperoxides of unsaturated fatty acids which break down at higher temperatures to give a variety of oxygenated fatty acids. Further heating leads to the formation of polymeric and other products. This can result in significant reductions in the concentration of PUFA over a period of time.

2.4.3 Hydrogenation

Whether a fat is liquid or solid at ambient temperatures is largely determined by the proportion of unsaturated fatty acids in its constituent triacylglycerols (see Chapter 1). The highly unsaturated oils found in many seeds and in fish are unsuitable for most food uses because they have very low melting points and also because they are more susceptible to oxidative deterioration. The aim of hydrogenation is to reduce the degree of unsaturation, thereby increasing the melting point of the oil. By careful choice of catalyst and temperature, the oil can be selectively hydrogenated so as to achieve a product with the desired characteristics. Indeed the hydrogenation process is seldom taken to completion, since completely saturated fats, especially those derived from the long chain fatty acids of fish oils, would have melting points that were too high.

Hydrogenation is carried out in the presence of a finely powdered catalyst (usually nickel) at temperatures up to 180°C, after which all traces of catalyst are removed by filtration. In the course of hydrogenation, some of the cis double bonds in natural oils are isomerised to trans and there is a migration of double bonds along the fatty acid chain.

2.4.4 Interesterification

Natural fats tend to have an asymmetrical distribution of fatty acids on the three positions of the triacylglycerol molecule (see Chapter 1). Interesterification is a means of altering the physical properties of the fat by randomizing the positions of the fatty acids. Heating the triacylglycerols in the presence of a catalyst can cause fatty acids of the same triacylglycerol molecule to exchange positions or can cause exchanges between fatty acids of different molecules. More precise interesterification can be achieved enzymically.

2.5 POST-PROCESSING DETERIORATION

The chemical mechanism involved in the oxidation of unsaturated fatty acids is discussed in Chapter 7. This process occurs at varying rates throughout the life of the oil: during storage and distribution, in food preparation and storage of the final food product. In general terms, the rate of oxidation is proportional to the degree of unsaturation of the oil and its temperature although other factors, such as the presence of antioxidants, may also affect the rate. Many of the compounds formed during oxidation are unstable themselves particularly at higher temperatures.

Many oils and fats are used for frying foods or are subjected to other forms of heating during the preparation of food. In the presence of air and water vapour, this leads to the formation of many breakdown products including hydroperoxides and conjugated ketodienes. In addition, cyclic monomeric, dimeric and polymeric fatty acids/triacylglycerols are produced. The possible toxic effects of these breakdown products have been reviewed by Gurr (1988) who concluded from the available evidence that the risk was negligible at the levels consumed on the basis that they are poorly absorbed and that flavour deterioration occurs before they are formed in significant amounts.

2.6 UNSATURATED FATTY ACID ANALYSIS AND DOCUMENTATION

2.6.1 Methodology

The methods for the fatty acid analysis of oils and fats have developed considerably in recent years with the advent of capillary gas liquid chromatographic (GLC) techniques. The fatty acid composition of many vegetable oils which have not been hydrogenated is relatively simple with ten or fewer fatty acids present. In these cases, simple packed column GLC can give the necessary resolution of the fatty acids for their identification.

The hydrogenation of oils leads to the formation of both geometric and positional isomers which are much more difficult to analyse and require the use of a capillary column GLC to ensure correct identification. This is particularly the case for partially hydrogenated fish oils which may contain very many isomers which can only be identified by repeated analysis on two or more different columns. The skill of the analyst is still required to interpret the complex chromatograms produced.

2.6.2 Deficiencies of food tables and other available sources of fatty acid data

A number of food tables exist which list the fatty acid composition of a wide range of foods. In the UK, McCance and Widdowson's *The Composition of Foods* is the most widely used (Holland *et al.*, 1991). The data given in such compositional tables are based on a large number of sources and sometimes may not reflect changes which have occurred since their publication in the formulation of the products. In the case of cooked foods, the different ways of cooking food including the type of cooking oil which may be used cannot always be included. In addition, techniques are constantly improving allowing more accurate analytical data to be produced.

3

INTAKES OF UNSATURATED FATTY ACIDS IN THE POPULATION

3.1 INTRODUCTION

This chapter describes the current methods used for estimating dietary intakes, with particular reference to unsaturated fatty acids. The reliability and relevance of the results obtained from these surveys are critically examined.

Several Government surveys have been undertaken in the UK to estimate the intake of nutrients in the diet. The Ministry of Agriculture, Fisheries and Food (MAFF) has traditionally conducted the National Food Survey (NFS) and the Total Diet Study (TDS). However, the survey that will be the focus of this chapter is the recently published OPCS/MAFF/DH survey, entitled 'The Dietary and Nutritional Survey of British Adults' (Gregory et al., 1990) which will be referred to as the Adult Survey. This is the most recent and comprehensive survey available. The Government has also undertaken a national survey of the nutrient intake of infants, which will also be discussed (MAFF, unpublished). In addition to these Government surveys, other organisations have conducted studies to measure nutrient intake. The findings of these other non-central Government surveys will be discussed where relevant.

3.2 NATURE AND SCOPE OF SURVEYS OF FOOD INTAKE

3.2.1 Adult Survey

The Adult Survey was carried out between October 1986 and August 1987. It was the first survey of its kind to be conducted in Great Britain. The survey covered a nationally representative sample of 2197 adults aged 16 to 64 years living in private households in England, Wales and Scotland. Pregnant women were excluded from the sample. The sample was recruited using a multi-stage random probability design. The survey included a seven-day weighed dietary record in which inform-

ants kept a record of weighed intakes of all food and drink consumed both in and out of the home. For each item of food consumed, informants recorded a description of the food (including the cooking method), and its brand if relevant, the weight served, and the weight and description of any 'leftovers'. When an item could not easily be weighed (i.e. if it was consumed outside the home), an accurate description was recorded and the weight was estimated later. Each food was allocated a code number which enabled it to be linked to the relevant nutrient information on a computerised databank. This allowed an accurate estimate of nutrient intake including the kind and amount of fat as well as the content of saturated fatty acids (SFA), trans fatty acids, cis n-3 and n-6 polyunsaturated fatty acids (PUFA) and monounsaturated fatty acids (MUFA) for several population groups.

3.2.2 The National Food Survey (NFS)

The NFS is a survey of about 7000 households. The amounts, cost and nutritional value of food brought into homes in Great Britain are recorded. The survey was started in 1940, and until 1950 it covered only urban working class households. Since 1950, the NFS has covered the whole of Great Britain. The survey determines what households, rather than individuals, consume. The NFS sample is selected by means of a three-stage stratified random-sampling scheme. A major limitation of the NFS is that it excludes alcoholic and soft drinks, confectionery and food eaten outside the home which results in an underestimate of the total intake of nutrients. However, because the survey is on-going, trends in food type and nutrient intake can be measured. The NFS routinely provides the intakes of total fat, total SFA, total MUFA and total PUFA.

3.2.3   Non-central Government surveys

The results of four non-central Government surveys are used for comparative purposes in this Chapter. These surveys have each used the seven-day weighed record for dietary assessment.

3.2.3.1   The Caerphilly and Speedwell studies (C and S)

These were carried out in 1980–83 and 1986 respectively. The C and S dietary studies are long-term studies based on representative samples of adults (predominantly men) in Caerphilly, a small town in South Wales; and Speedwell, a district of Bristol (England). The C and S studies have many aims, one of which is a detailed examination of dietary factors of possible relevance to coronary heart disease. Dietary surveys were completed by 807 adults. Nutrient intakes were calculated from food composition tables, together with data for some additional foods (see Fehily et al., 1984; Fehily et al., 1987).

3.2.3.2   The Edinburgh/Fife study

This was carried out between 1980 and 1982. 164 men aged 45–54 participated in the dietary survey. The subjects were from North Edinburgh and west Fife in Scotland (Thomson et al., 1985).

3.2.3.3   Edinburgh study

This was carried out in 1976. 176 forty year old men from Edinburgh were asked to participate in the survey; 97 successfully completed the dietary record (Thomson et al., 1982).

3.2.3.4   Northern Ireland diet, lifestyle and health study

This was carried out between 1986 and 1987. The survey was based on a random sample of 616 men and women (aged 16–64) (Barker et al., 1989). Fehily et al. (1990) analysed the diets of 55 men aged 45–64 years in the Northern Ireland study; the results are presented in this Chapter.

3.3   UNSATURATED FATTY ACID INTAKE BY ADULTS

3.3.1   Total fatty acid intake from different dietary surveys

The average daily intakes of fatty acids reported in the six surveys described in Section 3.2 are summarised in Tables 3.1 and 3.2. Detailed comparative analysis has not been attempted because different approaches were used. Fehily et al. (1990), however, have recently compared the

Table 3.1 Average daily intakes of energy, fat and fatty acids

		Adult Survey		NFS	Caerphilly		Speedwell	Edinburgh/ Fife	Edinburgh	Northern Ireland
		1986–1987		1988	1980–83		1986	1980–1982	1976	1986–1987
		1087 men	1110 women	2197 both sexes	7320 house-holds	665 men aged 45–63	49 women aged 40–59	93 men aged 45–63	164 men aged 45–54	97 men aged 40	55 men aged 45–64
		aged 16–64									
Energy intake (kcal)		2450	1680	2060	1998[d]	2395	1600	2318	2700	2895	2342
Energy intake (MJ)		10.3	7.1	8.6	8.4[d]	10.0	6.7	9.7	11.2	12.1	9.8
Total fat (g)		102	74	88	93	99	73	99	114	121	102
Saturates (g)		42.0	31.1	36.5	38.3	46	33	46	49	53	44
Trans fatty acids (g)		5.6	4.0	4.8	–	–	–	–	–	–	–
Monounsaturates[a] (g)		31.4	22.1	26.7	33.8	38	28	38	41.5	42.9	37
Polyunsaturates[b] of which											
Linoleic	(18:2)	13.8	9.6	11.7	–	9.6	7.3	10.2	10.4	9.3	–
Alpha linolenic	(18:3)	2.0	1.4	1.6	–	1.3	1.1	1.3	1.4	1.5	–
Arachidonic	(20:4)	–	–	–	–	1.2	1.0	1.3	–	0.8	–
EPA	(20:5)	–	–	–	–	0.1	–	0.1	–	–	–
Total PUFA (g)		15.8	11.0	13.3	14.2	13	10	14	13.7	12.2	16
P/S ratio[c]		0.40	0.38	0.39	0.37	0.31	0.31	0.32	0.30	0.23	0.38

[a] Adult Survey MUFA values include cis-MUFA only. The trans-MUFA are included in the value for trans fatty acids
[b] Adult Survey PUFA are classified as n-3 or n-6 in the report. It has been assumed for this table that all n-6 PUFA is 18:2 and all n-3 PUFA is 18:3
[c] Adult Survey P/S ratios are calculated using cis-PUFA only. Trans-PUFA are excluded
[d] Alcohol is not included

Table 3.2 Average daily intakes of fatty acids (g) and fatty acids as a percentage of total fatty acids

		Adult Survey 1986–1987			NFS 1988	Caerphilly 1980–83		Speedwell 1986	Edinburgh/ Fife 1980–1982	Edinburgh 1976	Northern Ireland 1986–1987
		1087 men	1110 women	2197 both sexes	7320 house- holds	665 men aged 45–63	49 women aged 40–59	93 men aged 45–63	164 men aged 45–54	97 men aged 40	55 men aged 45–64
		aged 16–64									
Total fatty acids	(g)	94.8	68.2	81.3	86.3	97	71	98	104.2	108.1	97
Saturates	(g)	42.0	31.1	36.5	38.3	46	33	46	49	53	44
	(%)	(44)	(46)	(45)	(44)	(47)	(47)	(47)	(47)	(49)	(45)
Trans fatty acids	(g)	5.6	4.0	4.8	–	–	–	–	–	–	–
	(%)	(6)	(6)	(6)							
Monounsaturates[a]	(g)	31.4	22.1	26.7	33.8	38	28	38	41.5	42.9	37
	(%)	(33)	(32)	(33)	(39)	(39)	(39)	(39)	(40)	(40)	(38)
Polyunsaturates[b] of which											
Linoleic (18:2)	(g)	13.8	9.6	11.7	–	9.6	7.3	10.2	10.4	9.3	–
	(%)	(15)	(14)	(14)		(10)	(10)	(10)	(10)	(9)	
Alpha linolenic (18:3)	(g)	2.0	1.4	1.6	–	1.3	1.1	1.3	1.4	1.5	–
	(%)	(2.1)	(2.1)	(2.0)		(1.3)	(1.5)	(1.3)	(1.3)	(1.4)	–
Total PUFA	(g)	15.8	11.0	13.3	14.2	13	10	14	13.7	12.2	16
	(%)	(17)	(16)	(16)	(17)	(13)	(14)	(14)	(13)	(11)	(17)

[a] Adult Survey MUFA values include cis-MUFA only. The trans-MUFA are included in the value for trans fatty acids
[b] Adult Survey PUFA are classified as n-3 or n-6 in the report. It has been assumed for this table that all n-6 PUFA is 18:2 and all n-3 PUFA is 18:3

nutrient intakes of the four non-central Government surveys, to assess if the differences in nutrient intake of men parallel the differences in coronary heart disease mortality rates.

Of particular interest to this Task Force is the comparison between fatty acid intakes. PUFA intakes are highest in Northern Ireland (both in absolute terms and as a percentage contribution to energy intake). There is also a marked difference in the P/S ratio, with Northern Ireland having a significantly higher ratio than any other region. It is also possible that fatty acid intakes are changing with time, and therefore that these differences are time-related rather than being actual regional differences. Fehily et al. (1990) suggest that this difference can in part be explained by the higher consumption of polyunsaturated margarine in Northern Ireland. (Polyunsaturated margarines constitute 33% of the total 'spreading fats' purchased in Northern Ireland, compared with 24% in Wales, 20% in Scotland and 20% in South West England.)

Table 3.2 shows the percentage contribution the different fatty acids make to the intake of total fatty acids. The intake of SFA accounts for nearly half of the total fatty acid intake (an average of 45% in the Adult Survey).

MUFA contribute more than PUFA to the dietary intake of unsaturated fatty acids. MUFA account for around 40% of the total fatty acid intake (or 33% if only the cis-MUFA are included). The Adult Survey MUFA values are lower than those reported in the other surveys, because only the cis-MUFA are

included. The trans fatty acid values are given separately. Trans-MUFA fatty acids are more abundant in the diet than trans-PUFA. By summing the trans fatty acid values and cis-MUFA values in the Adult Survey, an estimation of total cis- and trans-MUFA is obtained. The value is a slight over-estimate, because trans-PUFA values will also be included.

The trans fatty acids contribute 6% of the total fatty acid intake. This result is in agreement with an earlier BNF report (British Nutrition Foundation, 1987) which found that the average dietary intake of trans fatty acids was 7 g per person per day (range 5–27 g per day).

PUFA contribute approximately 16% of the intake of fatty acids. Linoleic acid (18:2, n-6) is the major contributor to the intake of PUFA (14% of total fatty acid intake). By contrast, alpha linolenic acid (18:3, n-3) contributes only 2% of total fatty acid intake (or around 1–2 g per day).

3.3.2 Fatty acid intake as a percentage of energy intake

Table 3.3 shows the results from the various surveys of fatty acid intakes as percentages of energy intake – both total energy intake and food energy intake (i.e total energy intake excluding energy from alcoholic beverages). Dietary guidelines and recommendations for fat intakes are usually given in these terms and Table 3.3 will allow the

Table 3.3 Average daily intakes of fat and fatty acids as a percentage of total energy and food energy

	Adult Survey 1986–1987			NFS 1988	Caerphilly 1980–83		Speedwell 1986	Edinburgh/ Fife 1980–1982	Edinburgh 1976	Northern Ireland 1986–1987
	1087 men	1110 women	2197 both sexes	7320 house-holds	665 men aged 45–63	49 women aged 40–59	93 men aged 45–63	164 men aged 45–54	97 men aged 40	55 men aged 45–64
	aged 16–64									
Total energy intake[a] (kcal)	2450	1680	2080	–	2395	1600	2318	2700	2895	2342
Fat (% total energy)	37.6	39.2	38.4	–	41.3	45.6	42.7	42.2	41.7	43.5
Food energy intake (kcal)	2279	1641	1957	1998	2254	1585	2222	2508	2621	2290
Fat (% food energy)	40.4	40.3	40.3	41.9	39.5	41.5	40.1	40.9	41.7	40.1
Saturates (% food energy)	16.5	17.0	16.8	17.3	18.4	18.7	18.6	17.6	18.2	17.3
Monounsaturates[b] (% food energy)	14.6	14.3	14.5	15.2	15.2	15.9	15.4	14.9	14.7	14.5
Polyunsaturates (% food energy)	6.2	6.0	6.1	6.4	5.2	5.7	5.7	4.9	4.2	6.3

[a] includes alcohol
[b] includes *trans* fatty acids

reader to compare current intakes with those in the recommendations which are discussed in Chapter 22.

As can be seen from Tables 3.1, 3.2 and 3.3, the non-central Government survey results are in broad agreement with the Adult Survey data, and it is therefore reasonable to focus on the Adult Survey data in this Chapter because of its wider scope, greater detail and the relatively recent sampling period.

3.3.3   Effect of sex and age on adult intake of fatty acids

The Adult Survey provides a detailed breakdown of fatty acid intake by age and sex. These intakes are summarised in Table 3.4, in terms of absolute amounts (g per day) and as a percentage of total energy and food energy.

Men consumed significantly more saturated fatty acids than women (42 g vs 31.1 g). Intakes of saturated fatty acids did not vary by age for either sex. However, on average, saturated fatty acids contributed a significantly greater proportion of the

Table 3.4 Average daily intakes of fatty acids by age and sex of adults

	Average daily intake (g)					Percentage of total energy					Percentage of food energy				
	SFA	*Trans* fatty acids	MUFA *cis*	PUFA n-6	n-3	SFA	*Trans* fatty acids	MUFA *cis*	PUFA n-6	n-3	SFA	*Trans* fatty acids	MUFA *cis*	PUFA n-6	n-3
Men															
16–24	41.6	5.9	32.4	14.2	2.0	15.2	2.1	11.9	5.2	0.7	16.1	2.3	12.6	5.5	0.8
25–34	41.7	5.5	32.0	14.3	2.0	15.3	2.0	11.8	5.3	0.8	16.5	2.2	12.8	5.7	0.8
35–49	42.0	5.7	31.5	14.3	2.0	15.1	2.1	11.3	5.2	0.7	16.3	2.2	12.3	5.6	0.8
50–64	42.7	5.3	29.9	12.1	1.8	16.1	2.0	11.3	4.6	0.7	17.2	2.1	12.1	4.9	0.7
All men, 16–64	42.0	5.6	31.4	13.8	2.0	15.4	2.0	11.6	5.1	0.7	16.5	2.2	12.4	5.4	0.8
Women															
16–24	30.4	4.0	22.7	9.9	1.4	16.0	2.1	12.0	5.2	0.8	16.4	2.2	12.3	5.3	0.8
25–34	30.9	4.0	22.2	9.9	1.4	16.4	2.1	12.0	5.3	0.7	16.9	2.2	12.4	5.5	0.8
35–49	31.7	4.1	22.6	9.9	1.4	16.4	2.1	11.8	5.1	0.7	16.9	2.2	12.2	5.3	0.7
50–64	30.8	3.8	21.0	8.8	1.3	17.1	2.1	11.7	4.9	0.7	17.5	2.1	12.0	5.0	0.7
All women, 16–64	31.1	4.0	22.1	9.6	1.4	16.5	2.1	11.8	5.1	0.7	17.0	2.2	12.2	5.3	0.8

(From *Gregory et al.*, 1990)

total energy intake of women (16.5%) than of men (15.4%).

Men had significantly greater average intakes of *trans* fatty acids than women (5.6 g compared with 4.0 g). Older men (aged 50–64) had significantly lower intakes of *trans* fatty acids than younger men (aged 16 to 24). All these significant differences disappear when values are expressed in terms of energy intake.

The average daily intake of *cis*-MUFA recorded for men (31.4 g) was significantly higher than that for women (22.1 g). Among men, intake did not vary significantly with age, but women aged 50 years or over had a significantly lower intake than younger women. Again, these differences are due to differences in food energy intake.

The average daily intakes of n-6 PUFA were significantly higher than n-3 PUFA for both men and women. The average daily intakes of n-6 PUFA were significantly lower among men and women in the 50–64 age group compared with the average value for the whole sample. These intakes were also lower when expressed as a percentage of food energy. n-3 PUFA intakes were significantly lower in the oldest men and women in the survey.

3.3.4 Effect of geographical region on adult intake of fatty acids

The Adult Survey embraced four broad regions: Scotland; Northern England; Central England, South

West England and Wales; and London and South East England. In spite of its small population, Scotland was separately identified because of the region's particularly high rate of mortality from heart disease, although the data must be interpreted with some caution because of the small number of people studied (9% of total sample).

The results from all regions are summarised in Table 3.5. Compared with average values for all men in the sample, men in Scotland consumed significantly less total fat and all types of fatty acids except for n-3 PUFA than did men in other regions. However, men in Scotland, and men in London and the South East, derived a greater proportion of food energy from saturated fatty acids than did men in other regions; proportions of other fatty acids did not differ significantly. Thus, although men in Scotland consumed less saturated fatty acids and PUFA than men in other regions, these differences were not apparent when expressed in terms of percentage of food energy. There were no significant differences between regions for intakes of fatty acids by women.

3.3.5 Effect of racial groups on fatty acid intakes by adults

3.3.5.1 Total fat

The Afro-Caribbeans and Asian Indians have fat intakes that differ from the rest of the general

Table 3.5 Average daily intakes of fatty acids by adults – effect of geographical region

	Average daily intake (g)					Percentage of total energy					Percentage of food energy				
	SFA	Trans fatty acids	MUFA cis	PUFA n-6	n-3	SFA	Trans fatty acids	MUFA cis	PUFA n-6	n-3	SFA	Trans fatty acids	MUFA cis	PUFA n-6	n-3
Men															
Scotland	38.5	5.2	29.1	12.0	1.8	15.5	2.0	11.8	4.8	0.7	16.8	2.2	12.8	5.2	0.8
Northern	40.8	5.7	31.2	13.6	1.9	14.9	2.1	11.5	5.0	0.7	16.2	2.3	12.4	5.4	0.8
Central, South-West and Wales	42.6	5.7	31.6	14.2	1.9	15.3	2.0	11.4	5.1	0.7	16.3	2.2	12.2	5.5	0.8
London and South East	43.4	5.6	31.8	14.0	2.0	15.8	2.0	11.7	5.1	0.7	17.0	2.2	12.6	5.5	0.8
All men	42.0	5.6	31.4	13.8	2.0	15.4	2.0	11.6	5.1	0.7	16.5	2.2	12.4	5.4	0.8
Women															
Scotland	30.6	3.9	22.3	9.5	1.4	16.5	2.1	12.1	5.0	0.8	16.9	2.1	12.4	5.2	0.8
Northern	30.6	4.0	21.9	9.4	1.4	16.4	2.1	11.9	5.0	0.7	16.9	2.2	12.2	5.2	0.8
Central, South-West and Wales	31.4	4.1	22.2	9.7	1.3	16.4	2.1	11.7	5.1	0.7	16.9	2.2	12.0	5.2	0.7
London and South East	31.2	3.9	22.2	9.9	1.4	16.6	2.1	11.9	5.3	0.8	17.1	2.1	12.3	5.4	0.8
All women	31.1	4.0	22.1	9.6	1.4	16.5	2.1	11.8	5.1	0.7	17.0	2.2	12.2	5.3	0.8

(From Gregory *et al.*, 1990)

population. The Indian population can be divided into Hindu, Sikh and Moslem groups. Both the Sikh and Moslems consume meat and fish whereas the Hindus are predominantly vegetarians. As with other vegetarian groups, their diet is devoid of long chain (C20–22) n-3 polyunsaturated fatty acids. There are limited data published by Government on the fatty acid composition of ethnic foods. However, a number of analyses have been undertaken recently by various research groups.

Maize oil is the preferred cooking oil among the Gujarati population and ghee is also used. The composition of ethnic take-away foods is largely determined by the type of fat used in cooking and this depends upon the composition of blended cooking oil. Total fat intakes tend to be lower among the Indian population but so do total energy intakes. The proportion of energy derived from fat therefore tends to be similar to that of the general population. The Afro-Caribbean population, on the other hand, tend to have a lower intake of fat as a proportion of the total dietary energy.

3.3.5.2 P/S ratio

The P/S ratio in the diet of Asian Indians has been estimated at 0.85 by McKeigue *et al.* (1985) using a household inventory method and at 0.52 for Hindus using a weighed dietary intake survey (Miller *et al.*, 1988). A lower P/S ratio of 0.46 was reported in a Bangladeshi community in East London (McKeigue *et al.*, 1988). Nutrient intakes in vegetarians and non-vegetarian households have been estimated (McKeigue *et al.*, 1989). Linoleic acid intakes were estimated to be 28 g/day in vegetarians and 24 g/day in non-vegetarians. The P/S ratios were 0.93 and 0.7 respectively.

3.3.5.3 Unsaturated fatty acids

Linoleic acid intakes tend to be higher among the Asian Indians than the general population, with those of the Afro-Caribbean population intermediate between the two. This high intake of linoleic acid is reflected by higher levels of linoleic acid in plasma phospholipids and in breast milk. However, Indians have low levels of docosahexaenoic acid (DHA) and higher than normal levels of arachidonic acid in their plasma phospholipids (McKeigue *et al.*, 1985; Miller *et al.*, 1988). These levels are similar to those found in Caucasian vegetarians (Sanders *et al.*, 1978). Higher arachidonic acid levels have also been found in the cord blood of Indian vegetarians (Stammers *et al.*, 1989).

Due to the paucity of data on fatty acid composition of ethnic foods, the studies published so far have probably underestimated linoleic acid intakes. Comparisons that have been made between British and Indian vegetarians show the

levels of linoleic acid in plasma phospholipids to be similar. In a total diet study, British vegans were found to have intakes of linoleic acid in the range of about 26 g/day (Sanders and Roshanai, 1983). No studies have measured levels of linoleic acid in adipose tissue of the Asian Indians or vegetarians.

3.4 INTAKES OF FATTY ACIDS IN INFANTS

MAFF conducted a national survey of the food consumption patterns and nutrient intakes of British infants aged 6–12 months in November 1986 (MAFF, unpublished). 488 mothers completed a quantitative record of all food consumed by their infant over a seven-day period. The survey will be referred to here as the Infant Survey.

Table 3.6 Average daily intake of fat and fatty acids by infants

	6–9 months (n = 258)	9–12 months (n = 230)	6–12 months (n = 488)
Energy (kcal)	815	928	868
Total fat (g)	33	39	36
Saturates (g)	15.9	19.5	17.6
Monounsaturates (g)	9.8	11.9	10.8
Polyunsaturates (g)	3.2	3.8	3.5
% energy from fat	36.3	37.1	36.7
% energy from SFA	17.5	18.8	18.1
% energy from PUFA	3.6	3.7	3.6
P/S ratio	0.20	0.20	0.20

(From MAFF, unpublished)

The fatty acid intakes are summarised in Table 3.6. The infants in the older age group (9–12 months) had greater intakes of all fatty acids than those in the younger age group (6–9 months). PUFA contributed 3.6% to the total energy intake across the whole age group. The P/S ratio was 0.20. This ratio is considerably less than the P/S ratio for adults (see Table 3.1). The difference is due to the infants' greater dependency on milk which is comparatively rich in saturated fatty acids (DHSS, 1984).

3.5 CONTRIBUTIONS OF FOOD GROUPS TO INTAKES OF FATTY ACIDS

Table 3.7 shows the contribution that different types of food make to the intakes of fatty acids provided in the diet. The table summarises the values from the Adult Survey which reflects food as consumed. (Results from the 1988 NFS, which reflects food as purchased, are broadly similar.)

Table 3.7 Contribution of food groups to daily intakes of fat and fatty acids by adults

	Total fat	Average daily intake in g (% of total intake) Fatty acids				
		Saturated	Trans	MUFA	PUFA n-6	n-3
Milk and milk products	13.4(15)	8.5(23)	0.5(10)	3.3(12)	0.2 (2)	0.1 (6)
Fat spreads (including butter and margarine)	14.1(16)	6.2(17)	1.4(30)	3.0(11)	2.4(20)	0.3(15)
Meat and meat products	21.5(24)	8.3(23)	0.9(18)	8.3(31)	2.0(17)	0.3(19)
Fish and fish dishes	2.5 (3)	0.6 (2)	0.1 (1)	0.9 (3)	0.5 (4)	0.2(14)
Eggs and egg dishes	3.6 (4)	1.1 (3)	0.1 (2)	1.4 (5)	0.5 (4)	0.0 (2)
Cereal products	16.9(19)	6.6(18)	1.3(27)	4.7(18)	2.6(22)	0.3(17)
Vegetables (including roast and fried)	9.7(11)	2.4 (6)	0.3 (6)	3.2(12)	2.8(24)	0.4(22)
Fruit and nuts	0.7 (1)	0.1 (0)	0.0 (0)	0.3 (1)	0.2 (2)	0.0 (1)
Sugar, confectionery and preserves	2.3 (3)	1.3 (4)	0.2 (3)	0.6 (2)	0.1 (1)	0.0 (1)
Beverages	0.2 (0)	0.2 (0)	0.0 (0)	0.0 (0)	0.0 (0)	0.0 (0)
Other	3.1 (3)	1.2 (3)	0.1 (2)	0.9 (3)	0.5 (5)	0.0 (2)
Total	88	36.5	4.8	26.7	11.7	1.6

(From Gregory et al., 1990)

Almost one-quarter (24%) of total fat intake was derived from meat and meat products. Almost one-fifth (19%) of the total fat intake was derived from the fat added to cereal products, particularly breakfast cereals, cakes and puddings. Milk and milk products and fat spreads also contributed significantly to the fat intake (15% and 16%, respectively). The contribution from vegetables is artificially high (11%) because this category includes roast and fried vegetables. In fact, nearly half (5%) was from roast and fried potatoes (including chips).

The foods which contributed most to the intake of saturated fatty acids were broadly similar to those contributing to total fat intake. Milk and milk products were, however, of greater significance and vegetables of lesser importance, than they were for total fat intake.

Trans fatty acids were derived chiefly from the hydrogenated fat used in fat spreads (30%) and cereal products (27%). Meat and meat products were also an important source (18%), followed by milk and milk products (10%). *Trans* fatty acids are present in milk and milk products and also occur in meat of ruminant origin.

The major sources of *cis*-MUFA were meat and meat products (31%). Cereal products (18%), milk and milk products (12%), fat spreads (11%), and vegetables (12%) were also significant sources.

PUFA were derived chiefly from margarine (especially soft margarine) and other fats and oils. This is seen more clearly in data from the NFS and TDS (not shown), where margarine and other fats and oils are identified separately and can be seen to provide the greatest contribution of n-6 and n-3 PUFA.

Vegetables contributed 22% and 24% of the total intake of n-3 and n-6 PUFA respectively, but about half of this intake from vegetables was derived from roast and fried vegetables. In addition to vegetables, the other sources of n-6 PUFA were cereal products and fat spreads, which each provided about one-fifth of the total intake. Meat and meat products contributed 17% of the total.

n-3 PUFA were further derived from meat, cereal products and fat spreads, which each contributed between 15 and 20% of total intake. Fish provided 14% of the total, with oily fish contributing about half of this (i.e. 0.1 g of marine n-3 PUFA). This result is in agreement with the intake of long chain n-3 PUFA observed in the Caerphilly and Speedwell Studies (Table 3.1).

3.6 LONGITUDINAL TRENDS IN FAT AND FATTY ACID INTAKE

The results of the National Food Survey (NFS), which uses consistent methodology from year to year, allow the trends in fat and fatty acid intake to be examined. These results are presented in Table 3.8.

The consumption of fat has fallen from 110 g per person per day in 1959, to 106 g in 1979, to 86 g in 1990. The 1991 COMA Report (Committee on Medical Aspects of Food Policy) recommended that the average fat intake for the population should be 35% of food energy (DH, 1991). In fact, it has remained unchanged over the past 20 years at about 42% because as fat intake decreased, so has energy intake.

The consumption of saturated fatty acids as a

Table 3.8 Total fat and fatty acids (g/person/day and as % of food energy[a]) and ratio of polyunsaturated/saturated fatty acids (P/S ratio) in the average household diet 1959–1990

	Total fat		Saturates		Monounsaturates		Polyunsaturates		P/S ratio
	g	%E	g	%E	g	%E	g	%E	
1959	110	38.3	53.0	18.5	43.0	15.0	9.2	3.2	0.17
1969	120	42.0	56.7	19.9	46.5	16.3	11.0	3.9	0.19
1972	112	41.5	52.0	19.3	42.9	15.9	11.5	4.3	0.22
1973	111	42.0	51.5	19.3	41.9	15.7	11.5	4.3	0.22
1974[b]	110	41.9	51.4	19.6	41.2	15.7	10.8	4.1	0.21
1974	106	41.3	50.7	19.7	39.8	15.4	10.6	4.1	0.20
1975	107	42.2	51.7	20.3	39.8	15.6	10.1	4.0	0.19
1976	105	41.7	50.1	19.8	39.7	15.7	10.5	4.1	0.20
1977	105	41.9	47.5	18.9	39.0	15.5	10.4	4.1	0.21
1978	106	42.0	47.2	18.8	39.3	15.7	10.6	4.2	0.22
1979	106	42.4	47.8	19.1	39.7	15.9	10.7	4.3	0.22
1980	106	42.6	46.8	18.9	39.6	16.0	11.3	4.6	0.24
1981	104	42.2	45.6	18.6	38.9	15.8	11.4	4.6	0.25
1982	103	42.6	44.4	18.3	38.7	16.0	12.1	5.0	0.27
1983	101	42.5	44.4	18.7	37.0	15.6	12.8	5.4	0.29
1984	97	42.4	41.9	18.3	35.1	15.3	12.7	5.5	0.30
1985	96	42.8	40.6	18.1	34.7	15.5	13.1	5.8	0.32
1986	98	42.6	40.6	17.7	35.8	15.6	14.3	6.2	0.35
1987	96	42.3	39.4	17.4	34.8	15.3	14.5	6.4	0.37
1988	93	41.9	38.3	17.2	33.8	15.2	14.2	6.4	0.37
1989	90	41.7	36.9	17.1	33.1	15.3	13.6	6.3	0.37
1990	86	41.6	34.6	16.6	31.8	15.3	13.9	6.7	0.40

[a] Energy excluding energy from alcoholic beverages
[b] In 1974, changes in methods resulted in two estimates
(From National Food Survey, MAFF)

percentage of food energy increased from 18.5% in 1959 to 20.3% in 1975. From then, it has declined to 16.6% in 1990. The recent COMA report (DH, 1991) set a target for saturated fatty acids to provide, on average, 11% of food energy so there is still scope for decreased intake.

The P/S ratio has been increasing from 0.17 in 1959 to 0.22 in 1979 and to 0.40 in 1990. The 1984 COMA report recommended a P/S ratio of 0.45. This change in P/S ratio has been brought about largely by a change in consumption from butter to margarine, which is reflected in the increased intake of PUFA over the years. The recent COMA report has not recommended a P/S ratio.

Trends estimated from NFS data cover food eaten inside the home only. It is possible that these trends are being offset by food eaten outside the home, especially foods with a high content of fat. The Adult Survey will be repeated in about eight years' time, which will enable a more accurate measure of trends in fat consumption of the diet as a whole to be given. In addition, the inclusion of foods eaten outside the home may be included in the NFS if a feasibility study conducted in 1991 proves successful.

4

DIGESTION OF LIPIDS TO UNSATURATED FATTY ACIDS

4.1 DEVELOPMENTAL ASPECTS OF LIPID DIGESTION

4.1.1 Adults

Most dietary fat is provided by triacylglycerols, which must be extensively hydrolysed to their constituent fatty acids before they can be assimilated by the body. In most adults, the process of fat digestion is very efficient and the hydrolysis of triacylglycerols is accomplished almost entirely in the small intestine by the lipase secreted from the pancreas.

4.1.2 Neonates

At birth, the newborn baby has to adapt to the relatively high fat-content of breast milk after relying mainly on glucose as an energy substrate in fetal life. Fat digestion presents two major problems which are even more acute in the premature infant:

(i) The pancreatic secretion of lipase is rather low
(ii) The immature liver is unable to provide sufficient bile salts to solubilize the digested lipids.

The newborn baby can, however, digest fat, albeit less efficiently than the older child or adult. This is now attributed primarily to the activity of a lipase secreted from the serous glands of the tongue (lingual lipase). This is carried into the stomach where hydrolysis occurs without the need for bile salts, at a pH of around 4.5–5.5. The secretion is probably stimulated both by the action of sucking and the presence of fat in the mouth, although the evidence for this was obtained from experiments with rat pups rather than human babies. The products of digestion are mainly 2-monoacylglycerols, diacylglycerols and free fatty acids, the latter being relatively richer in medium chain length fatty acids than the original triacylglycerols.

There is also evidence that a lipase present in human breast milk contributes to fat digestion in the newborn baby. The milk fat of most mammals is relatively rich in medium chain-length fatty acids rather than the usual fatty acids with chain lengths of 16–20 carbon atoms found in most storage fats. The relative ease with which lipids containing medium chain fatty acids can be absorbed certainly helps newborn babies to acquire lipids.

4.1.3 Post-weaning

As the baby is weaned on to solid food, the major site of fat digestion moves from the stomach to the duodenum. The stomach still has a role to play since its churning action creates a coarse oil-in-water emulsion, stabilized by phospholipids. Proteolytic digestion in the stomach releases lipids from the food particles where they are generally associated with proteins as lipoprotein complexes. The fat emulsion that enters the intestine from the stomach is modified by mixing with bile from the gallbladder and enzymes from the pancreas. Bile supplies bile salts (which are mainly the glycine and taurine conjugates of tri- and dihydroxycholic acids), formed from cholesterol in the liver. Much of the intestinal phospholipid in man comes from the bile and is estimated at between 7 and 22 g/day compared with a dietary contribution of 4–8 g/day. The secretion of bile is enhanced as the amount of fat in the diet increases.

4.2 MECHANISMS OF HYDROLYSIS OF LIPIDS TO FATTY ACIDS

4.2.1 Modification of emulsion particles

Pancreatic lipase attacks triacylglycerol molecules at the surface of the large emulsion particles (the oil–water interface). Before lipolysis can occur, the surface tension of the emulsion particles must be lowered and the enzyme must be modified to allow interaction to take place. First, bile salt molecules accumulate on the surface of the lipid droplet, displacing other surface active constituents. As 'amphiphilic molecules' they are uniquely designed for this task since one side of the rigid planar structure of the steroid nucleus is hydrophobic and can essentially dissolve in the oil surface. The other face contains hydrophilic groups that interact with the aqueous phase.

The presence of the bile salts gives a negative charge to the oil droplets, which attracts a protein, colipase, to the surface. Colipase has a molecular mass of 10,000 Daltons. Its function is to attract, and anchor, the pancreatic lipase to the surface of the droplets. Thus, bile salts, colipase and pancreatic lipase interact in a ternary complex requiring calcium ions for full lipolytic activity.

As digestion progresses, the large emulsion particles, which may be about 1,000 nm in diameter, decrease in size as the digestion products pass into large molecular aggregates called 'mixed micelles'. The components of these mixed micelles have been classified either as 'insoluble swelling amphiphiles' (Type II) or 'soluble amphiphiles' (Type III) according to their physical properties and the way in which they interact with water (Borgstrom, 1980). The main components are monoacyl-glycerols, lysophospholipids and fatty acids which leave the surface of the lipid particles to be incorporated into the micelles.

Fatty acids are in the form of soluble amphiphiles since the pH in the proximal part of the small intestine, where digestion takes place, has risen from the acid values in the stomach to around pH 5.8–6.5. The partition of long chain fatty acids into the micellar phase is favoured by the gradual increase in pH that occurs in the luminal contents as they pass into the more distal parts of the small intestine. Bile salts are also incorporated into micelles. The presence of these 'soluble amphiphiles' helps to incorporate very insoluble non-polar molecules like cholesterol and the fat-soluble vitamins into the micelles and aid their absorption. The picture of the luminal lipid digestion contents being distributed between coarse triacylglycerol emulsion particles and small mixed micelles is probably a grossly oversimplified one. Carey et al. (1983) have described a 'hierarchy' of lipid particles, in which non-micellar liposomes with diameters larger than the mixed micelles may also play an important role.

4.2.2 Hydrolysis of triacylglycerols

Pancreatic lipase catalyses the hydrolysis of fatty acids from positions 1 and 3 of triacylglycerols to yield 2-monoacylglycerols; there is very little hydrolysis of the fatty acid in position 2 and isomerization to the 1-monoacylglycerols is very limited.

4.2.2.1 Influence of unsaturation on digestive efficiency

The fate of dietary fats during the various stages of digestion depends on their constituent fatty acids – the early assumption of physiological equality is not justified.

The rate at which dietary fats are absorbed and metabolised may theoretically be influenced by differences in the rate of lipolysis of triacylglycerols, as well as by differences in the site and rate of transport of particular fatty acids across the intestinal mucosa (see Chapter 5). Both of these mechanisms are known to influence the absorption of dietary fat containing unsaturated fatty acids, although the practical nutritional significance of this is not entirely clear.

4.2.2.2 Evidence from simulated digestion studies

Some years ago, during the course of studies on the structure of marine mammal oils, it was observed that after hydrolysis of triacylglycerols by pancreatic lipase in vitro, the concentration of 20:5 and 22:6 unsaturated fatty acids became enriched in the diacylglycerol and triacylglycerol fractions and depleted in the free fatty acid and monoacylglycerol fractions (Brockerhoff et al, 1966; Bottino et al., 1967). These observations indicated that eicosapentaenoic acid (EPA) and docosahexaenoic acid (DHA) occupied the 1 and 3 positions of the triacylglycerols but were resistant to lipolysis by mammalian pancreatic enzymes. Recent increased interest in the metabolic effects of n-3 polyunsaturated fatty acids (PUFA) has focused attention on this problem. There has also been a search for purified sources of EPA which are chemically stable and suitable for oral administration in clinical trials. Synthetic ethyl esters have been widely used for this purpose, but there is increasing evidence that the absorption of EPA derived from this source is less efficient than that from natural sources. The rate of hydrolysis of primary n-alcohol esters depends both upon the fatty acid constituents and the chain length of the alkyl group. Ethyl esters are hydrolysed relatively slowly in vitro, perhaps because of adverse molecular orientation at the oil-water interface (Mattson and Volpenheim, 1969). Although the behaviour of fats during hydrolysis in vitro is not necessarily an accurate guide to digestion in vivo, subsequent studies with both experimental animals and man are consistent with these observations.

4.2.2.3 Evidence from studies with experimental animals

To investigate the uptake of n-3 unsaturated fatty acids from triacylglycerols to the intact intestine, Chen et al. (1987) introduced aqueous emulsions of oleic acid, maize oil, menhaden oil, or a fish oil concentrate directly into the duodenum of rats. The absorption of fatty acids was estimated from recovery of the substrate in thoracic duct lymph. Under these conditions, the total recovery of fatty acids from menhaden oil and fish oil was little more than 50% of that from maize oil. In earlier work, the authors had established that free EPA and arachidonic acid were absorbed as efficiently as oleic acid over a 24 hour period (Chen et al., 1985). They therefore attributed the poor absorption of menhaden oil and fish oil concentrate to a failure of lipolysis rather than to limited absorption of fatty acids. Fatty acid ethyl esters have also been found to be poorly absorbed in the rat (Lawson et al., 1985) but ethyl esters of EPA are known to be available for incorporation into plasma lipids (Hamazaki et al., 1987).

4.2.2.4 Evidence from experimental studies in man

There are relatively few studies in which the absorption of unsaturated fatty acids has been studied in man. The observations are, by necessity, indirect, and it is often difficult to establish the underlying physiological mechanisms. Harris and Connor (1980) fed a test meal containing salmon oil, which is high in n-3 fatty acids, to ten healthy subjects. The post-prandial rise in plasma triacylglycerols was both slower and less pronounced than that following a meal containing no polyunsaturated fatty acids. This may indicate slow lipolysis of the salmon oil, although the authors themselves attributed their results to a faster clearance of chylomicrons containing n-3 fatty acids from the plasma, rather than to a slower rate of intestinal absorption.

El Boustani et al. (1987) have recently reported that ethyl esters of EPA are less readily absorbed and incorporated than the free fatty acid or other esterified forms. Lawson and Hughes (1988) have recently investigated the appearance of fish oil fatty acids in plasma triacylglycerols of adult men, after oral administration in the form of free fatty acids, triacylglycerols, or ethyl esters. The authors calculated that about 95% of the free fatty acids were absorbed, whereas the absorption of EPA and DHA was less than 70% from the triacylglycerols and only about 20% from ethyl esters. Differences of this magnitude in the rate and extent of absorption of PUFA make interpretation of different studies very difficult (Ackman, 1988), and are likely to have metabolic consequences which require

investigation. In particular, further research is needed to determine the most effective vehicle for EPA administration in man.

Infants may be particularly vulnerable to the metabolic consequences of differences in the rate of fat digestion because their ability to absorb fat is significantly low during the first few months of life. For example, in newborn infants fed on formula diets, stearic and palmitic acids are less completely absorbed from random triacylglycerols than from natural lard containing a high proportion of these fatty acids in the 2-position (Filer et al., 1969).

4.2.3 Hydrolysis of phospholipids and cholesterol esters

Phospholipase A_2 hydrolyses the fatty acid in position 2 of phospholipids, the most abundant being phosphatidylcholine. The enzyme is present as an inactive proenzyme in pancreatic juice and is activated by the tryptic hydrolysis of a heptapeptide from the N-terminus. The major digestion products that accumulate in intestinal contents are lysophospholipids. Any cholesterol ester entering the small intestine is hydrolysed by a pancreatic cholesterol-ester hydrolase.

4.2.4 Microbial hydrolysis

In ruminant animals, the complex population of microorganisms contains lipases that split triacylglycerols completely to glycerol and free fatty acids. Some of the unsaturated fatty acids undergo several metabolic transformations catalysed by enzymes in the rumen microorganisms before passing into the small intestine where they are absorbed. Principal among these is hydrogenation in which double bonds are reduced by a process that is strictly anaerobic (see Chapter 2). During hydrogenation, the double bonds are isomerized from the *cis* to the *trans* geometrical configuration. The result is a complex mixture of fatty acids, generally less unsaturated than the fatty acids in the ruminant's diet and containing a wide spectrum of positional and geometrical isomers. Dairy products and beef are therefore significant sources of *trans* fatty acids in a human diet (see Chapter 3).

4.3 RESISTANCE TO DIGESTION BY PANCREATIC LIPASE – THE POTENTIAL FOR SUCROSE POLYESTER?

Resistance to pancreatic lipase is a property which has been put to practical use in the development of sucrose polyester (SPE), which is a generic term for mixtures of hexa-, hepta- and octa- fatty acid esters of sucrose. The sucrose molecule has eight hydroxyl groups available for esterification with a variety of fatty acids of different chain lengths and degrees of unsaturation. SPE can therefore be synthesised to display a considerable range of physical and chemical properties, including those of culinary fats and cooking oils. The compositional criteria for one such product includes a requirement to contain not less than 70% octa-ester and not more than 2% penta- or lower esters. The total unsaturated fatty acid content is not more than 70%, with chain lengths primarily in the C16–C18 range.

Sucrose polyesters containing six or more ester groups are completely resistant to hydrolysis by pancreatic lipase (Mattson and Volpenheim, 1972) and also by microbial enzymes, and they are therefore not absorbed from the alimentary tract in detectable quantities (Mattson and Nolen, 1972a, b). The functional properties of the compounds can be manipulated by varying the characteristics of conventional fats, and this has led to their proposed use as low, or zero-energy, fat substitutes for incorporation into manufactured foods. No such product has yet been approved for commercial application.

The significance of SPE in the present context stems from their ability to form a non-absorbed oil phase in the gut lumen, into which lipid-soluble nutrients can be partitioned during digestion. This serves the useful purpose of reducing the absorption of both dietary and biliary cholesterol, thereby increasing the faecal excretion of neutral sterols (Jandacek et al., 1980). This mechanism has been shown to reduce blood cholesterol levels in hypercholesterolaemic patients (Mellies et al., 1983) but the availability of essential nutrients may also be impaired. In rats, dietary supplementation with SPE leads to a significant reduction in hepatic vitamin A

stores (Mattson and Hollenbach, 1979). In man, plasma levels of vitamin A and E have been shown to fall in hypercholesterolaemic patients given SPE at levels sufficient to reduce plasma cholesterol, although vitamin levels remained within the normal range (Glueck et al., 1980).

Should commercial preparations of SPE ever be approved for use in foods, they may be supplemented with vitamins A and E to overcome the problem of reduced availability (Jandacek, 1977). However, in view of the probable importance of these vitamins in the prevention of chronic diseases, and the present uncertainty about optimum intake of these nutrients, vigilance will be necessary to ensure that high levels of consumption of SPE do not lead to low status of these vitamins in potentially vulnerable groups such as children.

4.4 DEFECTS IN DIGESTION

Maldigestion leading to a failure to assimilate dietary lipids into the body can occur because of incomplete lipolysis. Thus pancreatic insufficiency, which may result from pancreatitis, pancreatic tumour or in states of malnutrition such as kwashiorkor, can lead to a failure to secrete enough lipase or the production of lipase with reduced activity. Alternatively, the lipase may be fully functional, but a failure to produce bile (generally arising from hepatic insufficiency) may result in an inability to effect micellar solubilization of lipolysis products. This, in turn, can cause inhibition of lipolysis. Gastric disturbances that result in abnormal acid secretion also inhibit pancreatic lipase and, furthermore, gastric problems may cause poor initial emulsification of the lipid in the stomach, further reducing the efficiency of digestion. Thus maldigestion seems to arise from defects in a variety of organs contributing to different aspects of the digestive process.

5

ABSORPTION OF UNSATURATED FATTY ACIDS

5.1 TRANSPORT ACROSS THE INTESTINAL MUCOSA

Lipid absorption in man occurs largely from the jejunum. The principal molecular species passing across the brush-border membrane of the enterocyte are the monoacylglycerols and free fatty acids. The bile salts themselves are not absorbed in the jejunum or proximal small intestine but pass on to the ileum where they are absorbed and recirculated via the portal blood to the liver and then to the bile for re-entry into the duodenum.

The digestion products encounter two main barriers to their absorption:

(i) the layer of 'unstirred' water at the surface of the microvillus membrane which is thought to be the rate-limiting step in the absorption process;
(ii) the microvillus membrane (brush-border membrane) itself. This membrane has a particularly robust structure. After many years of research, it is still not entirely clear how the lipid digestion products cross this barrier.

5.1.1 Differences in the rate of uptake of fatty acids

There are important differences in the mechanisms by which different free fatty acids are transported across the small intestinal mucosa. As with other lipids, at high luminal concentrations, the transfer of polyunsaturated fatty acids (PUFA) across the intestinal brush-border membrane appears to be essentially a passive diffusion process. However, the transport of linoleic acid and arachidonic acid shows structural specificity at concentrations lower than 1.0 mM/l in the rat, which implies that a carrier-mediated mechanism may also exist (Chow and Hollander, 1978, 1979).

For fatty acids of equal chain length, an increasing degree of unsaturation appears to favour more complete absorption in the rat. Thus, in balance studies, Feldman et al. (1979) observed that the absorption of dietary fatty acids from tristearin by rats was about 25% less than that of fatty acids from triolein or safflower oil. However, this was assumed to be due to incomplete lipolysis rather than limited absorption of the saturated fatty acids.

5.2 INTRACELLULAR PHASE

5.2.1 Maintenance of diffusion gradient

For efficient absorption into the enterocytes to occur, it is essential that an inward diffusion gradient of lipolysis products is maintained. Two cellular events ensure that this occurs.

(i) On entering the cells the fatty acids bind to a fatty acid binding protein (or 'Z protein') of molecular mass 12,000 Daltons. The protein binds long chain unsaturated fatty acids in preference to long, medium or short chain saturated fatty acids and this may explain why oleic acid is absorbed more rapidly than stearic acid. Until this point, the absorption process is not dependent on a source of energy.
(ii) The next phase which removes free fatty acids, thereby maintaining the gradient, is the energy-dependent re-esterification of the absorbed fatty acids into triacylglycerols and phospholipids.

5.2.2 Re-esterification of fatty acids

5.2.2.1 Products of triacylglycerol digestion

The first step in re-esterification is the activation of fatty acids to their acyl-CoA thiolesters. Again, the preferred substrates are the long-chain fatty acids. In man and other simple-stomached animals, the major acceptors for esterification of acyl-CoAs are the 2-monoacylglycerols, which together with the free fatty acids are the major forms of absorbed lipids. Resynthesis of triacylglycerols, therefore, occurs mainly via the monoacylglycerol pathway. In ruminant animals, the major absorbed products of lipid digestion are glycerol and free fatty acids and resynthesis occurs via the glycerol phosphate pathway after phosphorylation of glycerol catalysed by glycerol kinase. However, lipids usually form a minor part of ruminant diets, so that glucose is probably the major precursor of glycerol phosphate in the ruminant enterocyte.

5.2.2.2 Products of phospholipid digestion

The main absorbed product of phospholipid digestion is monoacyl-phosphatidylcholine (lysophosphatidylcholine). A fatty acid is re-esterified to position 1 to form phosphatidylcholine by an acyl transferase located in the villus tips of the intestinal brush border. The function of this phospholipid is to stabilise the triacylglycerol-rich particles, or chylomicrons, exported from the cell as described later. It is probable that the phosphatidylcholine used for the synthesis and repair of membranes in the enterocytes (cells with a rapid turnover) is synthesized by the CDP-choline pathway in cells at the villus crypts.

5.2.2.3 Cholesterol

The absorption of cholesterol is slower and less complete than that of the other lipids, about half of the absorbed sterol being lost during desquamation of cells. Most of the cholesterol which is absorbed is esterified either by reversal of cholesterol esterase or via acyl-CoA:cholesterol acyl transferase. The latter enzyme is induced by high concentrations of cholesterol in the digesta.

5.2.3 Export from the enterocyte

Once absorbed, the enterocyte resynthesises the lipids in a form that is stabilised for transport in the aqueous environment of the blood. Within minutes of the products of absorption entering the enterocyte, fat droplets are present within the cysternae of the smooth endoplasmic reticulum, where the enzymes of the monoacylglycerol pathway are located. The rough endoplasmic reticulum is the site of the synthesis of phospholipids and apolipoproteins which provide the coat that stabilises the lipid droplets. These gradually increase in size and are pinched off from the endoplasmic reticulum to form lipid vesicles and fuse with the Golgi apparatus. The Golgi apparatus also provides carbohydrate moieties for the apolipoproteins and acts as a vehicle for the transport of the particles (fully

formed chylomicrons) to the lateral surface of the enterocyte. The final phase of export from the cells involves fusion with the membrane and secretion into the intercellular space by a process known as exocytosis.

In general, short chain fatty acids, having carbon chains of between 8 and 12 units, are absorbed directly into the portal blood stream, whilst those with 14 to 24 carbons are absorbed primarily via the lymphatic route (Vallot et al., 1985).

5.2.4 The efficiency of lipid absorption

There are relatively few relevant studies in man, but lipid absorption is a very efficient process in healthy people. Differences in relative rates of uptake of fatty acids may be of metabolic significance but they do not seem to lead to significant malabsorption of conventional dietary fatty acids. Bonanome and Grundy (1988) have recently demonstrated complete absorption of stearic acid from a liquid formula diet to which it contributed about 17% of total calories. However, Jones et al. (1985) reported that purified stearic acid is less completely absorbed than either oleic acid or linoleic acid in human subjects. Moreover the absorbed stearic acid was less completely oxidised during the immediate post-prandial period. It was suggested that the transport step was rate-limiting for hepatic metabolism, and whereas stearic acid was absorbed almost entirely through the lymphatic pathway after incorporation into chylomicrons, there might be substantial transfer of the unsaturated fatty acids directly to the liver via the portal venous route. Some experimental evidence exists for this hypothesis (Surawicz et al., 1981).

5.2.5 Absorption of fatty acid peroxides

Unsaturated fatty acids are susceptible to peroxidation during cooking or after prolonged storage. The products are highly reactive and potentially damaging to tissues, and there has long been concern that lipid peroxidation products in foods may be absorbed and could be toxic to man. Bergan and Draper (1970) studied the absorption and metabolism of 1-^{14}C-methyl linoleate hydroperoxide which had been administered orally to rats as a suspension in triolein. A large proportion of the dose was retained in the gastrointestinal tissues, and the remainder was metabolised to non-toxic compounds. Although this particular compound may not be absorbed in a toxic form, there is evidence that it may damage the small intestine and interfere with the absorption of nutrients, including other lipids (Cutler and Hayward, 1974).

Further studies are therefore needed to assess the absorption of a wider range of lipid peroxides, and to determine the significance of their interaction with gastrointestinal tissues.

5.3 GASTROINTESTINAL ADAPTATION

The mammalian alimentary tract is capable of important adaptive changes in response to diet and other physiological stimuli. However, much of the evidence for such effects has been obtained from studies with experimental animals and, in general, their significance for human nutrition is not understood.

5.3.1 Adaptive responses to dietary fat

The small intestine of the rat responds to a high fat diet by increasing its capacity to absorb fat. For example, Singh et al. (1972) showed that when rats were given a diet containing 20% lipid, most of which was in the form of lard, there was a decrease in faecal fat excretion. The transport of fat across isolated small intestine segments was increased, and there was enhanced activity of mucosal esterifying enzymes. More recent studies in rats have confirmed that a diet providing 67% of calories from fat in the form of triolein leads to enhanced pancreatic and mucosal enzyme activity, as well as a twofold increase in the rate of fat absorption from the small intestinal lumen in vivo (Flores et al., 1990).

Apart from total dietary fat, there is also evidence to show that the proportion of unsaturated fatty acids in the diet influences small intestinal function. It has been reported that moderate dietary intakes of EPA lead to a reduction in the capacity of isolated rat jejunum to transport glucose and cholesterol in vitro (Thomson et al., 1989a). However, this effect was shown to be dependent, in a complex way, on the source and composition of other dietary lipids. Moreover, the observations with respect to cholesterol have not been confirmed by other workers (Reynier et al., 1989). The subject therefore requires further research, preferably using techniques in vivo.

5.3.2 Gastrointestinal membrane fluidity

Apart from the effect of total fat intake on enzyme systems involved in fat digestion and absorption, the dietary supply of unsaturated fatty acids may influence nutrient transport via a direct effect upon the structure of intestinal mucosal cell membranes. Brasitus et al. (1985) compared the composition and properties of brush border and basolateral cellular membranes from the small intestine of rats which had been given diets rich in linoleic acid or butter fat. After six weeks there were significant increases in the unsaturated fatty acid content and lipid fluidity of brush border and basolateral membranes of the mucosal cells from rats given linoleic acid. Although there was evidence of an adaptive increase in the membrane cholesterol/phospholipid ratio, which tended to counteract the increased fluidity, there were significant changes in sodium-pump activity which may have been due to increased passive sodium influx (see review by Thomson et al., 1989b).

5.4 DEFECTS IN ABSORPTION

Failure to assimilate lipids of dietary origin into the body may arise from defects in digestion (maldigestion) (see Chapter 4) or absorption (malabsorption).

Malabsorption may occur, even when digestion is functioning normally, due to defects in the small intestine which affect the absorptive surfaces. There may be a variety of causes, some common ones being bacterial invasion of the gut or sensitization of the gut to dietary components such as gluten, as in coeliac disease. Malabsorption syndromes (often called 'sprue') are characterized by dramatic changes in the morphology of the intestinal mucosa. The epithelium is flattened and irregular, and atrophy of the villi reduces the absorbing surface. 'Tropical sprue' is a prevalent disease in many countries of Africa and Asia.

A common feature of all fat-malabsorption syndromes is a massively increased excretion of fat in the faeces (steatorrhoea) which arises not only from unabsorbed dietary material but also from the breakdown of cells and from bacteria which proliferate in the gut. The bacteria undoubtedly affect the composition of the excreted fat. For example, a major component of faecal fat but not of dietary fat, is 10-hydroxystearic acid which is formed by bacteria from stearic acid. This normal component of faecal fat is found in particularly high concentration in the faeces of patients with steatorrhoea.

Malabsorption of fat can also occur in a number of inherited disorders in which the biosynthesis of different apoproteins in the enterocytes is impaired. Without the apoproteins, the stabilisation of the lipid droplets cannot occur and the fat cannot be transported out of the cell. Triacylglycerols then begin to accumulate in the enterocyte.

Patients with poor fat absorption are very much at risk from deficiencies of energy, fat-soluble vitamins and of essential fatty acids. The clinical management of fat malabsorption is facilitated by replacing normal dietary fats by medium chain triacylglycerols (MCT). This product is produced from the refining of coconut oil and is available as a cooking oil or as a fat spread. It is composed largely of triacylglycerols with C8 and C10 saturated fatty acids. These medium-chain fatty acids are rapidly hydrolysed and efficiently absorbed into the portal blood, thereby bypassing the normal absorptive route of long-chain fatty acids and chylomicron formation.

6

TRANSPORT OF UNSATURATED FATTY ACIDS

6.1 THE IMPORTANCE OF LIPOPROTEINS
(see review by Segrest and Albers, 1986)

Once the fatty acids from the diet are absorbed from the intestine and re-assembled into triacylglycerols, they face the biological problem of how water-immiscible lipids can be transported in the predominantly aqueous environment of the blood. This problem is overcome by stabilising the lipid particles with a coat of amphiphilic compounds: phospholipids and proteins. The resulting particles are lipoproteins (see Figure 6.1). They are not molecules in the normal sense. They are aggregates of individual lipid and protein molecules with a degree of structural organisation. Their mass should correctly be termed 'particle mass' rather than 'molecular mass'.

There are several types of lipoproteins with differing chemical compositions, physical properties and metabolic functions (Table 6.1) but their common role is to transport lipids from one tissue to another to supply the lipid needs of different cells. The different types may be classified in a number of ways depending on their origins, their major functions, their composition, physical properties or method of isolation. Lipoproteins differ according to the ratio of lipid to protein within the particle as well as having different proportions of lipids: triacylglycerols, esterified and non-esterified cholesterol and phospholipids.

These compositional differences influence the density of the particles and there is a strong relationship between biological function and the broad density classes into which they fall. It is convenient, therefore, to use density to separate and isolate lipoproteins by ultracentrifugation and to classify plasma lipoproteins into different density classes. From lowest to highest density, these are: chylomicrons (CM), very low density lipoproteins (VLDL), low density lipoproteins (LDL) and high density lipoproteins (HDL). As density decreases, particle size increases and so does the ratio of lipid to protein and the ratio of triacylglycerols to phospholipids and cholesterol (Table 6.1). The

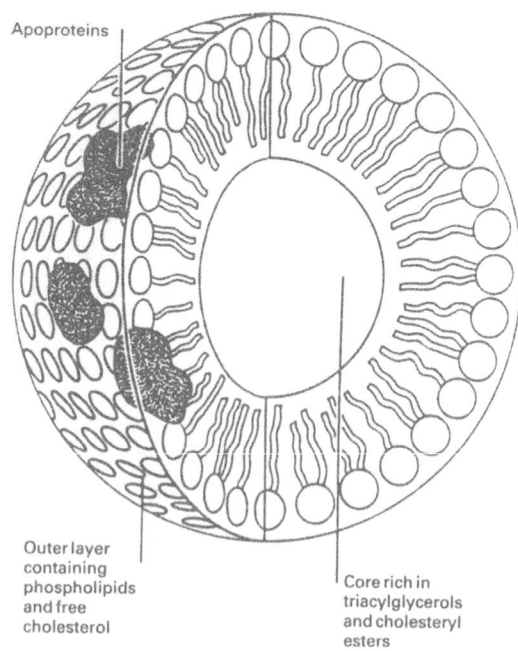

Figure 6.1 A typical lipoprotein. Lipoprotein particles are probably spherical structures with the least soluble lipids inside the core. This is surrounded by lipids capable of interacting with water, as well as by the apoproteins. Above is a schematic design of the probable structure of a high density lipoprotein (HDL) particle. (From Walker and Wood, 1991)

classes are not homogeneous: there is a wide variety of particle sizes and chemical compositions within each class and, therefore, a certain amount of overlap between them: see Gurr and Harwood (1991) for a more detailed discussion of this subject.

6.2 APOLIPOPROTEINS
(see review by Brewer et al., 1988)

The protein moieties of lipoproteins are called apolipoproteins. They fulfil two main functions (Table 6.2):

Table 6.1 Composition and characteristics of human plasma lipoproteins

	Chylomicrons	VLDL	LDL	HDL
Protein (% particle mass)	2	7	20	50
Triacylglycerols (% particle mass)	83	50	10	8
Cholesterol (% particle mass) (free + esterified)	8	22	48	20
Phospholipids (% particle mass)	7	20	22	22
Particle mass (10^6 Daltons)	0.4–30	10–100	2–3.5	0.175–0.36
Density range (g/ml)	<0.95	0.95–1.006	1.019–1.063	1.063–1.210
Diameter (nm)	>70	30–90	18–22	5–12
Major apolipoproteins	A_1, B–48 C_1, C_2, C_3	B–100, C	B–100	A_1, A_2, C
Trace apolipoproteins	E	A, E		E, D
Site of synthesis	Gut	Gut, liver	Capillaries of peripheral tissues, liver	Gut, liver
Major function(s)	Transport of dietary fat	Transport of endogenous fat	Transport of cholesterol to peripheral tissues	Reverse transport of cholesterol to liver

(i) They provide a means of solubilising the lipid particles and maintaining their structural integrity.

(ii) They confer specificity on the lipoprotein particle and thus direct its metabolism in specific ways.

The letters A–E are used to identify apolipoproteins but most of them can be divided further into several sub-classes (Table 6.2). These are usually referred to in abbreviated form, thus: ApoA$_1$, ApoC$_3$, etc.

The complete amino-acid sequences of apoA$_1$, A$_2$, A$_3$, C$_1$, C$_2$, C$_3$ and E are now known. Specific regions of these proteins, containing some of the structural and functional determinants of the lipoprotein particles, have been synthesised and tested for biological activity. Knowledge of the detailed structure of the polypeptides has recently been extended using the monoclonal antibody technique.

Much interest is centred on the properties of apolipoprotein B (apoB). It is common to the triacylglycerol-rich VLDL and cholesterol-rich LDL. It is important in the recognition of lipoproteins by cell-surface receptors and it is now being used as a marker for vascular disease. Its characterisation has been hampered by its insolubility, its susceptibility to degradation and its propensity for aggregation. The peptides can be solubilised by trypsin and detergent treatment. The amino-terminal residue is glutamic acid and the carboxyl-terminal residue is serine. ApoB has considerable beta-structure in addition to regions of random coil and alpha-helix. The content of beta-structure depends on the lipid content and the temperature at which the observations are made. ApoB is a glycoprotein linked to glucosamine through

Table 6.2 Characteristics of the human apolipoproteins

Shorthand name	Molecular mass (Daltons)	Amino acid residues	Function	Major site(s) of synthesis
A_1	28,000	243	Activates LCAT[a]	Liver Intestine
A_2	17,000	154	Inhibits LCAT? Activates hepatic lipase	Liver
B–48	–	–	Cholesterol clearance	Liver
B–100	350–550,000		Cholesterol clearance	Liver
C_1	6,605	57	Activates LCAT?	Liver
C_2	8,824	79	Activates LPL[b]	Liver
C_3	8,750	79	Inhibits LPL? Activates LCAT?	Liver
E	34,000	279	Cholesterol clearance	Liver

[a] LCAT: Lecithin–cholesterol acyltransferase
[b] LPL: Lipoprotein lipase
–: not available

asparagine residues; about one third to one half of the carbohydrate is released by trypsin treatment and, therefore, probably not buried in the lipid core.

ApoB exists as two variants of different molecular masses, (approximately 100 and 48 Daltons) which are designated as apoB-100 and apoB-48 respectively. Human VLDL are believed to contain, almost exclusively, the heavy variant: apoB-100.

Naturally occurring variants of apolipoproteins have been identified that result in specific metabolic disorders (apolipoproteinopathies).

6.3 ORIGINS OF LIPOPROTEINS

6.3.1 Chylomicrons
(see review by Redgrave, 1983)

Chylomicrons (CM) are the largest and least dense of the lipoproteins because they contain a high proportion of lipid relative to protein. Their function is to transport lipids of dietary origin. Their size depends on factors such as the rate of lipid absorption and the type of dietary fatty acids that predominate. Thus larger chylomicrons are produced after the consumption of large amounts of fat, at the peak of absorption or when apolipoprotein synthesis is limiting. When the fatty acids are largely unsaturated, the chylomicrons tend to be larger than those in which saturated fatty acids predominate.

Because of their role in transporting absorbed dietary fat, the principal components are triacylglycerols, with small amounts of phospholipids and proteins, sufficient to cover the surface (Table 6.1). The 'core' lipid also contains some cholesterol esters and minor fat-soluble substances absorbed along with the dietary fats: fat-soluble vitamins, carotenoids and possibly environmental contaminants.

Several apolipoproteins are present in the surface layer of chylomicrons: $apoA_1$ (15–35%); $apoA_4$ (10%); apoB (10%); the apoC group (45–50%) and apoE (5%). The A group apolipoproteins are synthesised on the endoplasmic reticulum of the intestinal epithelial cells, whereas apoC and apoE are acquired from other lipoproteins once the chylomicrons have entered the blood.

6.3.2 Very low density lipoproteins

Very low density lipoproteins (VLDL), like chylomicrons, contain predominantly triacylglycerols (Table 6.1). Their function is to transport triacylglycerols of endogenous origin, synthesised mainly in the liver or intestine. VLDL are spherical particles with a core consisting mainly of triacylglycerols and cholesterol esters with cholesterol, phospholipids and protein mainly on the surface.

VLDL are most conveniently isolated for compositional, structural and metabolic studies by flotation in an ultracentrifuge. The amount of apoB per VLDL particle is independent of the particle mass and is the same as that in LDL particles. In contrast, the amounts of $apoC_1$, C_2, C_3 and apoE are variable and decrease relative to apoB as particle density increases.

The major site of synthesis of VLDL is in the liver, although some VLDL are produced in the intestinal enterocytes. The source of carbon is mainly glucose derived from dietary carbohydrate and converted into the lipid precursors, glycerol-3-phosphate via the glycolytic pathway and long chain fatty acids via the malonyl-CoA pathway (see Chapter 7). Nevertheless, some VLDL fatty acids are derived from circulating albumin-bound free fatty acids (FFA), which in turn may originate from lipolysis of adipose tissue triacylglycerols, intravascular lipolysis of triacylglycerol-rich lipoproteins or directly absorbed medium chain fatty acids. As in chylomicron biosynthesis in the intestine, nascent VLDL particles can originate in the smooth endoplasmic reticulum and acquire phospholipids and apoproteins from the rough endoplasmic reticulum. The resulting vesicles move to the Golgi apparatus where some of the apoproteins are glycosylated. The Golgi vesicles then migrate to the cell surface where VLDL are exported by exocytosis.

In contrast to nascent chylomicrons, which acquire apoC and apoB only after they reach the plasma, VLDL receive their full complement of apoproteins in the hepatocyte or enterocyte.

6.3.3 Low density lipoproteins

The role of the low density lipoproteins (LDL) is to transport cholesterol to tissues where it may be required for membrane structure or conversion into various metabolites such as steroid hormones. LDL are the major carriers of plasma cholesterol in man, although this is not so in all mammals.

LDL are normally isolated from plasma by ultracentrifugation at salt densities between 1.019 and 1.063 g/ml. Each lipoprotein particle contains the same mass of their main protein, apoprotein B, but each differs with respect to the amount of bound lipid. An average composition is shown in Table 6.1.

LDL are largely derived from VLDL by a series of degradative steps that remove triacylglycerols, resulting in a series of particles that contain a progressively lower proportion of triacylglycerols and correspondingly richer proportion of cholesterol and phospholipids. These intermediate particles are called intermediate density lipoproteins (IDL) and the above reactions take place, first in the blood capillaries associated with adipose tissue and then in the liver. During the transformations, the apoB component remains with the LDL particles and the apoC and apoE components are progressively lost. ApoB has an important role in the recognition of LDL by cells since it must interact with specific cell-surface receptors before the LDL particle can be taken up and metabolised by the cell. Other receptors (e.g. on macrophages) recognise modified LDL and are responsible for the degradation of LDL particles that cannot be recognised by normal cell surface LDL receptors (see Chapter 12). ApoE plays a role in receptor binding while the C group apolipoproteins are involved in reactions by which the particles are sequentially degraded by lipases.

6.3.4 High density lipoproteins
(see review by Patsch and Gotto, 1987)

High density lipoproteins (HDL) carry cholesterol from peripheral cells to the liver, a process generally called reverse cholesterol transport (see Section 6.4.4). Table 6.1 summarises their composition and properties. HDL are usually divided into two subclasses, HDL_2 and HDL_3, because rate zonal centrifugation gives rise to a bimodal distribution in the HDL region, whereas other lipoprotein classes form a continuum. This distinction between HDL_2 and HDL_3 may have metabolic significance since HDL_2 appear to have a stronger inverse relationship with cardiovascular disease than HDL_3 (see Chapter 12). The major apolipoproteins of HDL are $apoA_1$ and $apoA_2$ but the surface coat also contains some apoC, apoE and a protein which is unique to HDL: apoD.

Unlike triacylglycerol-rich lipoproteins, nascent HDL have not been identified within subcellular compartments. At the early stage in the development of nascent HDL, they are disc-shaped rather than spherical and consist of a bilayer composed mainly of phosphatidylcholine with $apoA_1$ and apoE at the margins of the disc. Similar particles are also found in intestinal lymph. After delivery to the plasma, HDL acquire additional surface components: phospholipids, cholesterol and apolipoproteins by transfer from chylomicrons, and VLDL during their catabolism by lipoprotein lipase.

ApoA, much of which is synthesised in the intestine, and some apoC are transferred to HDL during the breakdown of chylomicrons. Further apoC is transferred from VLDL breakdown products. In plasma, a subfraction of HDL, containing $apoA_1$ and apoD, becomes associated specifically with an enzyme lecithin-cholesterol acyltransferase (LCAT), which is synthesised in the liver and exported into plasma. The function of this enzyme in reverse cholesterol transport is described in Section 6.4.4.

6.3.5 Plasma free fatty acid–albumin complexes

Fatty acids released after lipolysis from tissues such as adipose tissue are carried in the plasma as albumin complexes. Albumin has three types of binding sites that bind 2, 5 and 20 molecules of fatty acids. These sites also have a decreasing affinity for the fatty acids in the above order and there are striking differences in the binding constants of different fatty acids at each site. Tissue fatty acid binding proteins are also important in the intracellular transport of fatty acids and these, too, have markedly different affinities for particular fatty acids.

6.4 METABOLISM OF LIPOPROTEINS

6.4.1 Uptake of lipoprotein fatty acids into tissues by the lipoprotein lipase reaction

6.4.1.1 Chylomicrons (see review by Redgrave, 1983)

After the consumption of a meal containing an appreciable amount of fat, the absorbed, resynthesised triacylglycerols are transported in the blood as chylomicrons. (A small amount of lipid-carrying particles also circulate when no fat is being absorbed from the diet. Under these conditions, the chylomicron fatty acids are derived from biliary phospholipids or the lipids of cells shed from the gut mucosa and the VLDL fatty acids from FFA transported to the liver from adipose tissue.)

The chylomicrons initially bind to the enzyme lipoprotein lipase (LPL) on the endothelial surfaces of blood capillaries in muscle and other organs, but primarily in adipose tissue. This enzyme catalyses the rapid hydrolysis of chylomicron triacylglycerols, releasing fatty acids which are then re-esterified inside the tissue (Figure 6.2). $ApoC_2$ plays a key role in increasing the rate of lipolysis, probably by modifying the interaction between enzyme and substrate. At this stage, the A group apolipoproteins and the remaining apoC are transferred to HDL, as are the phospholipids. The remaining particle, although retaining the same basic structure, contains fewer triacylglycerols, is enriched in cholesterol esters and is known as a chylomicron remnant. These particles are no longer able to compete effectively for lipoprotein lipase and circu-

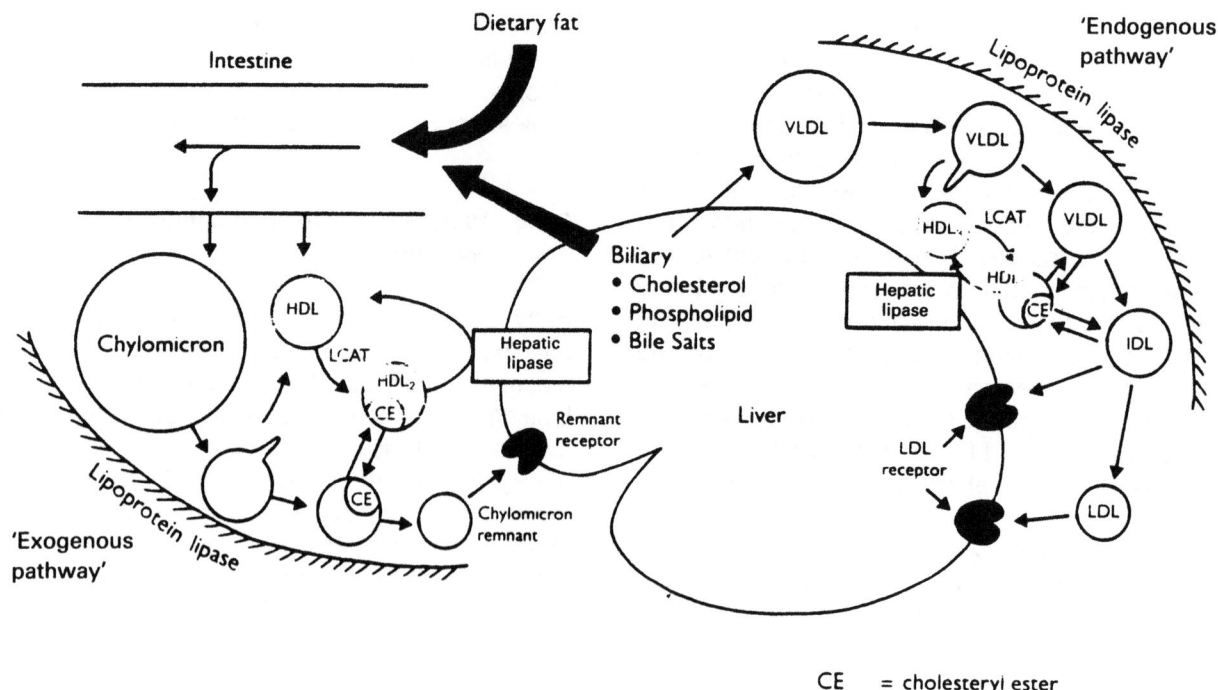

CE   = cholesteryl ester
LCAT = lecithin cholesterol acyltransferase

Figure 6.2 The metabolism of lipoproteins. (From Gurr, 1991)

late in the plasma to be taken up by liver cells by a receptor-mediated process (see below).

The regulation of LPL itself is crucial to the control of lipoprotein metabolism in different tissues in the body (see review by Quinn et al., 1982). The enzyme is synthesised in the parenchymal cells of the tissues and secreted into the capillary endothelium where it is bound to the cell surface by heparin sulphate. The activity of the enzyme is regulated by diet and hormones, of which the most important is insulin. After consuming a meal, when the supply of energy may exceed the body's immediate needs, the secretion of insulin ensures that the adipose tissue enzyme is active and the muscle enzyme suppressed. In a state of fasting, the adipose tissue LPL activity is suppressed and the hormone-sensitive lipase is 'switched on' allowing the mobilisation of FFA. At the same time the muscle LPL activity is elevated so that fatty acids from circulating lipoproteins can be used for energy. In lactation, the synthesis of LPL seems to be regulated by prolactin which promotes the utilisation of chylomicron triacylglycerol fatty acids for milk-fat synthesis.

In man, up to 300 g of chylomicron triacylglycerols can be hydrolysed by LPL each day, although less than 1% of the lipids may be found in the blood at any given time.

6.4.1.2 Very low density lipoproteins

Although arising primarily from endogenous syn-

thesis in the liver, rather than from the diet, the catabolism of VLDL is basically the same as for the chylomicrons (see Figure 6.2). They bind to endothelial lipoprotein lipases and the triacylglycerols are hydrolysed to free fatty acids. The larger the particles (and therefore the more apoC they contain) the greater the rate of hydrolysis. Phospholipids and remaining apolipoproteins are transferred to HDL while apoB and apoE are retained. The resulting particles, depleted of triacylglycerols and richer in cholesterol esters, are called VLDL remnants or more usually intermediate density lipoproteins (IDL), since further degradation results in the formation of LDL. In many species, most of the IDL are taken up by the liver by receptor-mediated endocytosis in a manner analogous to chylomicron remnants, but in man almost half are normally processed further to yield LDL.

Lipoproteins are not discrete particles of fixed size but form a distribution of sizes; thus the sizes and densities of IDL or remnant particles overlap those of the precursor VLDL or the product LDL but can be distinguished by their mobility on electrophoresis, presumably because of their different apolipoprotein composition.

The final processing step of the triacylglycerol-rich, apoB-containing lipoproteins is the further loss of triacylglycerols and phospholipids (catalysed by hepatic lipases) and of apoE, to form LDL which retain only apoB-100.

6.4.2   Uptake of lipoprotein fatty acids into tissues by the receptor mechanism

6.4.2.1   Chylomicrons and remnants

Liver cells have specific receptors on their surfaces that recognise and bind to the apoE component of chylomicron remnants. The whole receptor-remnant complex is then 'pinched off' from the membrane and is taken up into the cell (internalised). Here it is degraded by lysosomal enzymes which catalyse virtually the complete hydrolysis of the lipid and protein components (see Figure 6.2). The uptake of remnants is inhibited by the C group apolipoproteins, which thereby prevent the premature uptake of small unhydrolysed chylomicrons by the remnant receptor. A similar mechanism exists for removal of IDL.

6.4.2.2   Low density lipoproteins

LDL are less efficiently removed by hepatic receptors (although this can occur) and are principally removed by specific extra-hepatic LDL receptors (Brown and Goldstein, 1986).

The distribution of LDL to various tissues may depend on the rate of trans-capillary transport as well as the activities of the LDL receptors on the cell surfaces. Adipose tissue and muscle have few LDL receptors and take up LDL only slowly, whereas adrenal gland (important in the synthesis of steroid hormones derived from cholesterol) has a highly fenestrated epithelium with abundant receptors and avidly takes up LDL.

The human LDL receptor is a trans-membrane protein of 839 amino-acid residues which is synthesised in the rough endoplasmic reticulum and inserted at random in the plasma membrane. It then migrates laterally in the plane of the membrane until it reaches a pit that is coated with the protein clathrin. About 80% of the LDL receptors are concentrated in these coated pits which cover about 2% of the surface of the cell.

Both apoE (found in VLDL) and apoB (found in LDL) contain a recognition site for the LDL receptor. Once bound, the receptor-LDL complex is internalised by endocytosis and its component parts degraded by lysosomal enzymes. LDL receptors are recycled within minutes of endocytosis and are re-utilised many times before they are eventually catabolised. The regulation of LDL receptors is determined by the amount of incoming LDL. Normally, a high concentration of plasma LDL will suppress the number of specific cell-surface receptors ('down-regulation') which are increased under conditions in which tissue LDL are depleted.

Cholesterol esters, carried by LDL, are hydrolysed inside the cell by cholesterol-ester hydrolase. Incorporation of cholesterol into the endoplasmic reticulum membranes serves to inhibit hydroxy-methyl-glutaryl-CoA reductase (HMGCoA-reductase), the rate-limiting enzyme in cholesterol biosynthesis. An abundant supply of cholesterol in the plasma is therefore able to suppress its own endogenous biosynthesis and ensure that excessive amounts do not accumulate.

About 65% of LDL uptake is receptor-mediated. Some LDL in liver and other tissues is taken up by endocytosis that does not involve receptors, although this is less efficient.

6.4.3   Receptors for modified lipoproteins

Lipoproteins may be modified in various ways to probe the structural features needed for receptor uptake. This has involved reacting the lipoproteins with a number of chemical groups: acetyl, acetoacetyl, malonyl, succinyl and others (Mahley et al., 1979). Acetyl and some other groups convert the e-amino groups of lysine in apoB-100 to a neutral or negatively charged moiety that results in its inability to be recognised by the appropriate receptor. Many modified lipoproteins are, however, recognised by a so-called 'scavenger receptor' on some cells such as macrophages (Brown and Goldstein, 1986) (see Chapter 12).

6.4.4   Reverse cholesterol transport

A major function of HDL is to remove unesterified cholesterol which may have accumulated in cell membranes and plasma lipoproteins and transport it to the liver where it can be degraded and utilised for the synthesis of the bile acids. This process is now referred to as reverse cholesterol transport.

A key step in this process is catalysed by the enzyme, lecithin: cholesterol acyltransferase (LCAT). The enzyme catalyses the transfer of a fatty acid from phosphatidylcholine to cholesterol to form a cholesterol ester (Figure 6.2). In human plasma, LCAT is associated with a subfraction of HDL that contains apoA$_1$ and apoD. The phospholipid substrate for the reaction is present in the HDL particle, having been transferred from chylomicron remnants or IDL during the degradation of chylomicrons and VLDL respectively (Figure 6.2). The cholesterol substrate is derived from the surfaces of the plasma lipoproteins or the plasma membranes of cells. Molecules of the substrate cholesterol and the product, cholesterol ester, exchange readily between plasma lipoproteins and between lipoproteins and cell membranes. This exchange can be mediated by specific transfer proteins.

LCAT, by 'consuming' cholesterol, promotes its

net transfer from cells into plasma and from other lipoproteins to the site of esterification. One product, cholesterol ester, is redistributed among plasma lipoproteins. The other, lysophosphatidylcholine, is transferred to albumin from which it is rapidly removed from the blood and recycled. Molecules of cholesterol ester transferred to lipoproteins containing apoB-100 or apoE are taken up by the liver, thereby completing the process of reverse cholesterol transport.

7

METABOLISM OF UNSATURATED FATTY ACIDS

7.1 FATTY ACID BIOSYNTHESIS

7.1.1 Synthesis of saturated fatty acids

All mammals can synthesise saturated fatty acids *de novo* from simple precursors such as glucose or amino acids using a fundamentally similar pathway (Figure 7.1). Liver and adipose tissue are the most important organs for fatty acid biosynthesis; there are differences between species concerning the

relative importance of these two organs. The common precursor is acetyl CoA which is activated by acetyl CoA carboxylase in the presence of carbon dioxide to give malonyl CoA.

In the first cycle, malonyl CoA is condensed with acetyl CoA and the beta keto compound produced is reduced, dehydrated and finally reduced again to form a saturated 4-carbon acyl CoA. During subsequent cycles, sequential additions of two carbon units from malonyl CoA are made and the process is repeated. These reactions are catalysed

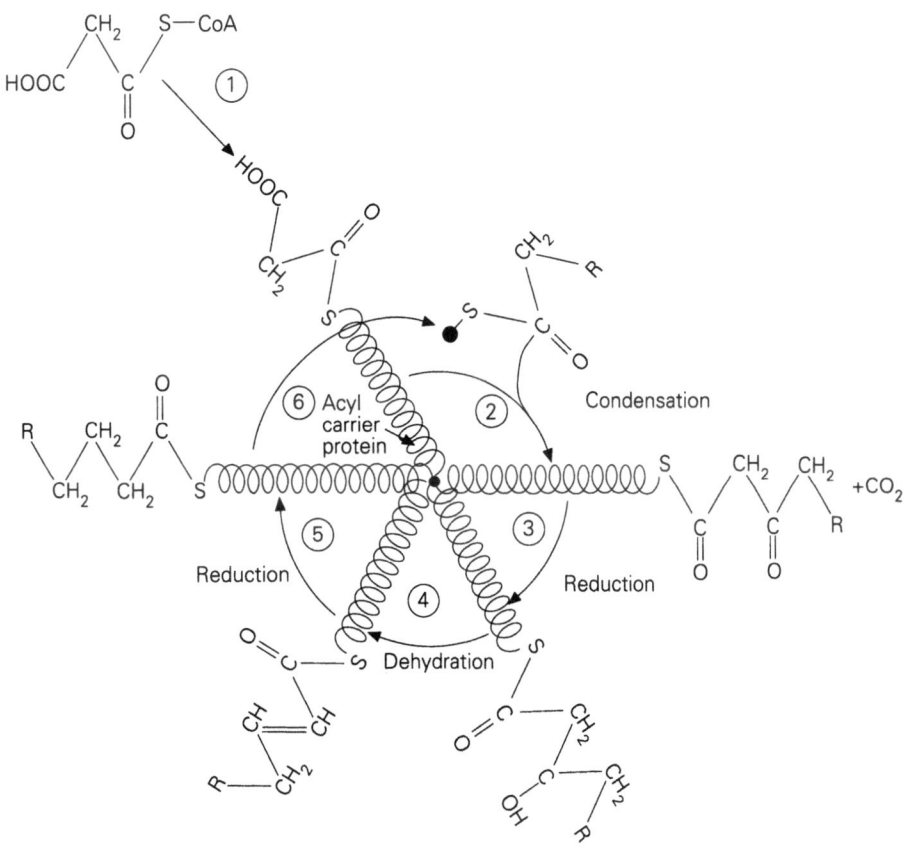

Figure 7.1 Fatty acid biosynthesis.
First cycle, R = H; second cycle, R = CH$_2$CH$_3$;
third cycle, R = CH$_2$CH$_2$CH$_2$CH$_3$.

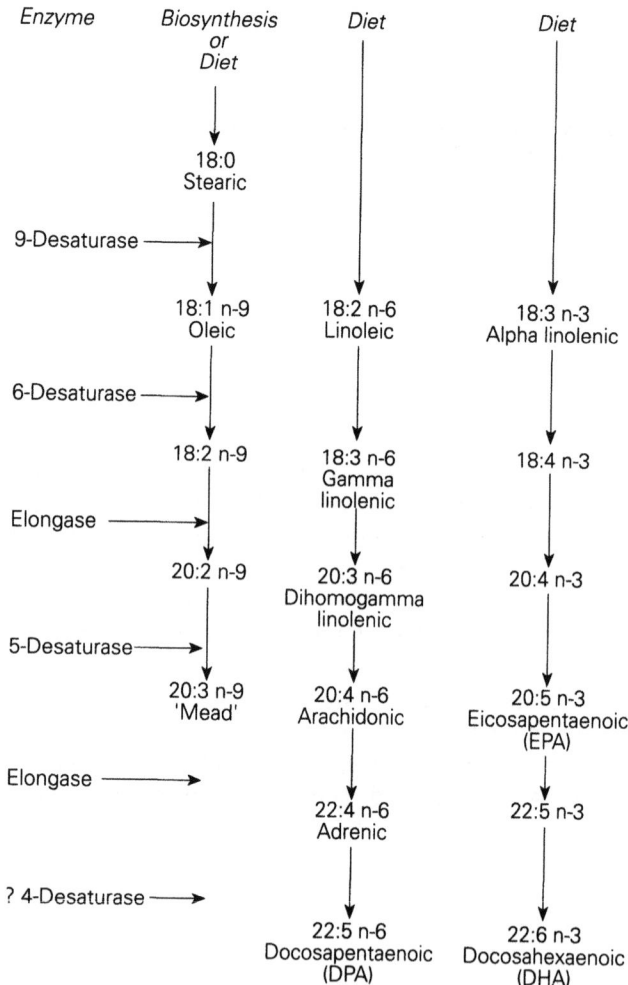

Figure 7.2 Metabolic transformations of the three major unsaturated fatty acid families by desaturation and elongation.

7.1.2 Monounsaturation and the 9-desaturase enzyme

The introduction of a single double bond into a saturated fatty acid chain is catalysed by a desaturase enzyme. Since the position in the molecule at which the double bond is introduced is usually between carbon atoms 9 and 10, the enzyme responsible is called 9-desaturase. It is present universally in both plants and animals. (N.B. This enzyme is often referred to as delta-9-desaturase, the delta indicating numbering from the carboxyl carbon.)

The 9-desaturase enzyme present in the microsomes of rat liver has been purified. The cloning of its gene and expression in *E. coli* (Thiede *et al.*, 1986; Strittmatter *et al.*, 1988) represents a major breakthrough in lipid biochemistry. It should soon be possible to analyse the molecular genetics of this enzyme and, subsequently, the molecular genetics of other fatty acid desaturases. The enzyme is a single polypeptide chain of 53,000 Daltons; it contains one atom per molecule of non-haem iron and is a hydrophobic protein which is deeply embedded in microsomal membranes. Fatty acid desaturation requires the presence of cytochrome b_5 and the flavoprotein NADH-cytochrome b_5 oxidoreductase as well as the desaturase enzyme itself.

This desaturase enzyme complex is widespread in animal tissues and rates of desaturation are normally quite rapid. The reaction involves the removal of two adjacent hydrogen atoms from methylene groups at positions 9 and 10 of the saturated fatty acid. It requires the presence of molecular oxygen and the end-products are water and the corresponding monounsaturated fatty acid. The reaction has many of the characteristics of a group of enzymes, known as mixed-function oxygenases, that are involved in the detoxication of many substances in the body, mainly in the liver. However, the precise mechanism for the removal of a pair of hydrogen atoms and the involvement of the associated electron-transport chain is not known. A major technical problem is that the 9-desaturase enzyme is membrane-bound.

The activity of 9-desaturase is markedly decreased by fasting (protein deprivation) and diabetes (glucose deprivation), whereas protein consumption and the administration of insulin restore its activity (Brenner, 1989). These effects are due to changes in the amount of the enzyme and take place via repression or induction of enzyme synthesis.

Consumption of diets rich in cholesterol increases 9-desaturase activity and results in increased ratios of monounsaturated/saturated fatty acids (eg 16:1 to 16:0 and 18:1 to 18:0). This acts

by a series of enzymes that collectively form the multi-enzyme complex, fatty acid synthetase. The end-product is usually the 16-carbon saturated fatty acid palmitic acid (16:0), or sometimes the 18-carbon saturated fatty acid stearic acid (18:0).

The enzymes involved in fatty acid biosynthesis are active when high carbohydrate, low fat diets are eaten, but suppressed when high fat diets are eaten. Polyunsaturated fatty acids (PUFA) are particularly active in suppressing fatty acid biosynthesis *de novo* and recent evidence suggests that the effects are determined at the level of gene expression.

There is little need for synthesis of saturated fatty acids in man because the diet supplies adequate amounts. Cell membranes do, however, need unsaturated fatty acids to maintain a degree of fluidity and could not function if saturated fatty acids alone were available. A mechanism for the introduction of double bonds exists and this is referred to as desaturation (Figure 7.2).

to 'fluidise' membrane phospholipids, thereby off-setting the 'hardening' effects of their enhanced cholesterol content. Thus the 9-desaturase could participate in the regulation of cell-membrane fluidity when monounsaturated fatty acids are not readily available from the diet.

The substrate for the desaturase enzyme is the coenzyme A thioester of a saturated fatty acid. Thus, stearoyl-CoA is desaturated to oleoyl-CoA. In a similar way, palmitoyl-CoA is desaturated to palmitoleoyl-CoA.

The product of the reaction, the monounsaturated fatty acyl-coenzyme A, can readily transfer its fatty acyl group to a lipid, as described in Section 7.3.

7.1.3 Polyunsaturation

7.1.3.1 Polyunsaturation in plants and marine algae

Plants, unlike animals, can insert additional double bonds into oleic acid (18:1 n-9) between the existing double bond at the 9 position and the methyl terminus of the molecule. An oxygen-dependent 12-desaturase enzyme acts on oleic acid to form linoleic acid (18:2 n-6) which can undergo further desaturation at the n-3 position by an oxygen-dependent 15-desaturase enzyme to yield alpha linolenic acid (18:3 n-3).

Many marine plants, especially the unicellular algae that are present in phytoplankton, can also effect chain elongation and further desaturation of 18:3 n-3 by inserting more double bonds between the existing double bond at the 9 position and the terminal carboxyl group to yield n-3 PUFA with 20 and 22 carbon chains (C20 and C22) and 5 and 6 double bonds respectively. It is the formation of these long-chain n-3 PUFA by marine algae, and their efficient transfer through the food chain to fish, that accounts for the abundance of C20 and C22 n-3 PUFA in marine fish oils.

7.1.3.2 Limitations of desaturation and the need for chain elongation in animals

Monounsaturated fatty acyl-CoA esters can become substrates for the 6-desaturase enzyme which is present in a wide variety of animal cell membranes (Figure 7.2). Thus, oleic acid (18:1 n-9) can be converted into the di-unsaturated fatty acid (18:2 n-9). However, the 6-desaturase operates more commonly on 18:2 n-6 and 18:3 n-3 supplied in the diet from plant sources to form the tri-unsaturated fatty acid, gamma linolenic aid (GLA) (18:3 n-6) and the tetra-unsaturated fatty acid (18:4 n-3) respectively.

Unsaturated fatty acids with more than three or four double bonds are produced by the elongation of the 18 carbon chain by two carbon atoms, and the insertion of another double bond between the carboxyl group and the first double bond. Thus, after elongation of the C18 unsaturated fatty acids by two carbon atoms, the enzyme 5-desaturase can insert a further double bond between carbon atoms 5 and 6 to yield arachidonic acid (20:4 n-6) and eicosapentaenoic acid (EPA) (20:5 n-3) respectively. After yet another elongation, the enzyme 4-desaturase can insert another double bond between carbon atoms 4 and 5 to yield docosapentaenoic acid (DPA) (22:5 n-6) and docosahexaenoic acid (DHA) (22:6 n-3) respectively. This sequence of desaturations and elongations produces the variety of long-chain polyunsaturated fatty acids necessary for membrane structure and eicosanoid production (Chapter 8).

7.1.4 Substrate competition for 6-desaturase and its consequences

The 6-desaturase enzyme is responsible for the introduction of a double bond at position 6 in each of the first members of the three fatty acid families (the n-9, n-6 and n-3 families) and this has important consequences. All these fatty acids can compete for the same 6-desaturase enzyme and can therefore influence the metabolism of fatty acids of other families. The reaction rates are different, however, for the different substrates. Alpha linolenic acid (18:3 n-3) is preferred to linoleic acid (18:2 n-6) which is, in turn, preferred to oleic acid (18:1 n-9).

Quantitatively, the most important pathway is that of the n-6 series in which linoleic acid (18:2 n-6) is converted into arachidonic acid (20:4 n-6). Normally, the diet contains sufficient linoleic acid, so that, given its high affinity for the 6-desaturase enzyme, this metabolic pathway is continuously able to supply the arachidonic acid needed by body tissues. If the absolute amount of linoleic acid in the diet is low, other C18 unsaturated fatty acids compete more successfully, resulting in the pre-dominance of pathways other than the n-6 pathway. Since the most abundant unsaturated fatty acid is usually oleic acid (18:1 n-9), its desaturation and elongation are enhanced, resulting in an accumulation of the end product of the n-9 family, all-cis-5,8,11-eicosatrienoic acid, the so-called 'Mead' acid (20:3 n-9), which is not normally found in tissues in other than trace amounts. The significance of this alternative pathway is further discussed in the context of essential fatty acid (EFA) deficiency in Section 7.2.

7.1.5 Control of desaturation and elongation

Whereas much is known about the physiological control of fatty acid synthetase, the multi-enzyme complex responsible for saturated fatty acid biosynthesis (see Section 7.1.1), the same cannot be said of the various desaturases. This reflects the great difficulty in isolating the membrane-bound enzymes in a pure form.

7.1.5.1 The 6-desaturase enzyme

The 6-desaturase from rat liver microsomes has been purified to homogeneity. The purified enzyme is a single polypeptide chain of 66,000 Daltons which contains one atom of non-haem iron per molecule. It is considerably less hydrophobic than 9-desaturase and consequently it penetrates less deeply into the membrane. When reconstituted, with cytochrome b_5 and NADH-cytochrome b_5 oxidoreductase, the purified 6-desaturase catalyses the oxygen-dependent desaturation of the thiol ester of linoleic acid (18:2 n-6) to the thiol ester of GLA (18:3 n-6). The enzyme has been purified but its gene has yet to be cloned.

The activity of 6-desaturase varies between species and between individual tissues of the same species. Thus, rat and mouse liver have high 6-desaturase activity, whereas rabbit, guinea pig and human livers have low activities.

As already mentioned, the activity of 6-desaturase depends on the number of double bonds in its substrate. It is also subject to end-product inhibition. GLA (18:3 n-6) is the primary product of 6-desaturation of linoleic acid (18:2 n-6) and also a potent inhibitor of 6-desaturase. Similarly, arachidonic acid (20:4 n-6), EPA (20:5 n-3) and DHA (22:6 n-3) also inhibit 6-desaturation of linoleic acid (18:2 n-6).

Complex interactions involving the 6-desaturase enzyme are possible in tissues metabolising linoleic acid and alpha linolenic acid simultaneously, but they are not understood. The determination of an optimal dietary ratio of n-6/n-3 PUFA will have greater scientific basis when all the fatty acid desaturases have been purified and studied in detail.

The activity of 6-desaturase in animals is decreased by fasting and restored when glucose is given. It is enhanced in rats fed on high-protein diets and decreases with age. Feeding a high-cholesterol diet to rats reduces the 6-desaturase activity. This effect, which is the opposite to that obtained with 9-desaturase, is reversed by feeding the animals a low-cholesterol diet.

The activities of desaturases are under endocrine control (Brenner, 1989) (see Figure 7.3). 6-desaturase activity is reduced in insulin-deficient

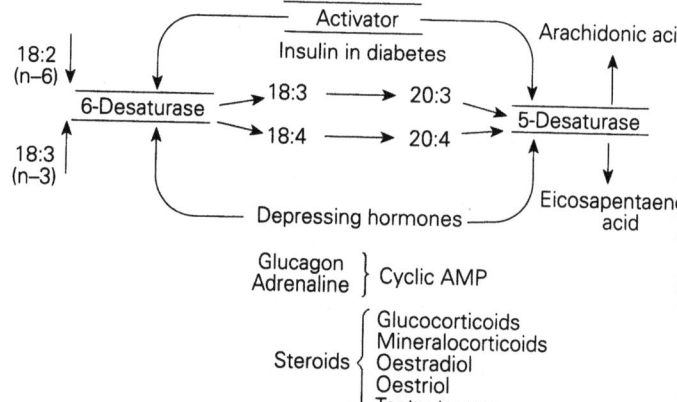

Figure 7.3 Effect of different hormones on 6- and 5-desaturases. (From Brenner, 1990)

rats and is increased by insulin administration. Glucagon and adrenalin, which are antagonistic to insulin, depress 6-desaturase activity, as do glucocorticoids such as dexamethasone. 6-desaturase activity is sensitive to thyroid hormones with both hyperthyroidism and hypothyroidism causing decreases in activity. Ethanol also depresses 6-desaturase activity, and alcoholism in man is accompanied by decreased conversion of linoleic acid (18:2 n-6) to DGLA (20:3 n-6). The involvement of the latter as a precursor of 1-series prostaglandins has led to the suggestion that impairment of prostaglandin production may underlie some of the symptoms of chronic alcoholism (Horrobin, 1981).

7.1.5.2 The 5-desaturase enzyme

The species and tissue distribution of 5-desaturase, which has not been purified to homogeneity, generally parallels that of 6-desaturase. There is some evidence that Inuits have elevated plasma levels of DGLA (20:3 n-6) and it has been suggested that this is due to an impairment of 5-desaturase activity. This raises the possibility that genetic differences may occur in the metabolic pathways producing PUFA.

5-desaturase is much more active on fatty-acyl CoAs than on fatty-acyl groups in phospholipids. The enzyme converts DGLA (20:3 n-6) to arachidonic acid (20:4 n-6), eicosatetraenoic acid (20:4 n-3) to EPA (20:5 n-3) and 20:2 n-9 to 20:3 n-9. 5-desaturase can also be active on certain *trans* isomers of 18:1 and 20:1 although other *trans* isomers of 18:1 can partially inhibit the enzyme. Likewise, partially hydrogenated fish oil, which is rich in various *trans* isomers of 20:1 and 22:1, inhibits both 6-desaturase and 5-desaturase when fed to rats. The same is true for partially hydrogenated soyabean oil which is rich in *trans* isomers of 18:1, although the inhibition is less than with fish oil.

5-desaturase responds to dietary changes and hormones in an essentially similar manner to 6-desaturase (Figure 7.3) (Brenner, 1989). Thus the enzyme is depressed in rats with insulin-dependent diabetes and stimulated by insulin administration. It is depressed by glucagon, cAMP and glucocorticoids such as dexamethasone. ACTH decreases both 5- and 6-desaturases in both the adrenal gland and liver of rats. Oestradiol inhibits 5-desaturase activity in rat hepatocytes. The activity of 5-desaturase in rat liver is decreased when the animals are given a fat-free diet and induced by subsequently giving linoleic acid or GLA. The induction effect is blocked by cycloheximide, indicating synthesis de novo of the enzyme. The decrease of 5-desaturase activity in EFA deficiency in the rat is paralleled by a decrease in n-6 PUFA formation, whereas the 6-desaturase activity is generally increased. This indicates that the enzymes are controlled by different mechanisms.

The consumption of cholesterol-rich diets by rats decreases 5-desaturase as well as 6-desaturase, reinforcing the decreased formation of arachidonic acid (20:4 n-6) and the increased level of linoleic acid (18:2 n-6).

7.1.5.3 The 4-desaturase enzyme

Little is known about 4-desaturase. Microsomal preparations and cell culture systems in vitro generally show little 4-desaturase activity towards appropriate fatty acid substrates. There is, therefore, no real evidence that the double bond at position 4 in docosapentaenoic acid (DPA) (22:5 n-6) and in DHA (22:6 n-3) is introduced in the same way as the other double bonds.

It has recently been suggested that production of DHA (22:6 n-3) and DPA (22:5 n-6) from their respective precursors, 22:5 and 22:4, may occur via elongation to 24:5 and 24:4 respectively, followed by operation of a 6-desaturase enzyme on both these compounds to produce 24:6 and 24:5 respectively. These may then be retroconverted to 22:6 and 22:5 by the peroxisomal beta oxidation pathway (see Section 7.5.1.2). Thus the putative 4-desaturase enzyme may in reality be a 6-desaturase. These possible conversions have great significance for control of PUFA biosynthesis since they involve both chain elongation and chain shortening as well as desaturation, and may involve peroxisomes as well as enzymes of the endoplasmic reticulum.

Moreover, the competitive interactions that exist between 18:3 n-3, 18:2 n-6 and 18:1 n-9 for the enzyme 6-desaturase will be greatly complicated by adding competitions of both 24:5 n-3 and 24:4 n-6 for the same enzymes. This means that conversion of 18:2 n-6 to 20:4 n-6 will compete not only with conversion of 18:3 n-3 to 20:5 n-3 but

also with conversion of 20:4 n-6 to 22:5 n-6 and with the conversion of 20:5 n-3 to 22:6 n-3. The latter two conversions will also be competitive. The implications of this situation for whether the precursor or the product of a given PUFA series should be presented in nutritional supplementation are great (see Chapter 9). The issue is one of major importance for future studies.

The conversion of docosapentaenoic acid (22:5 n-3) to DHA (22:6 n-3) decreases with age. Retinal lipids normally have an abundance of DHA (22:6 n-3) and so diminished vision in old age might be related to reduced production of DHA (22:6 n-3).

7.1.5.4 Elongases

Relatively little is known of the chain-elongating enzymes (elongases) involved in converting PUFA with 18 carbon chains to their C20 and C22 derivatives, although studies with rat liver microsomes have established that the different PUFA series can compete for chain elongation. For example, the chain elongation of linoleic acid (18:2 n-6) to 20:2 n-6 is inhibited by other C18 fatty acids, with 18:3 n-3 being the most potent inhibitor. It is also inhibited by arachidonic acid (20:4 n-6), EPA (20:5 n-3) and DHA (22:6 n-3). The chain elongation of GLA (18:3 n-6) to DGLA (20:3 n-6) is partly inhibited by alpha linolenic acid (18:3 n-3) and DHA (22:6 n-3). It is not inhibited by linoleic acid (18:2 n-6), arachidonic acid (20:4 n-6) or EPA (20:5 n-3). There is evidence also from cultured cells that added monounsaturated fatty acids decrease cellular levels of PUFA by inhibiting chain elongation reactions.

7.1.5.5 Rate limitation

It has been clearly established in cell-free systems that the activities of the elongases are greater than the activities of the desaturases (Sprecher, 1989). Alpha linolenic acid (18:3 n-3) is converted at a higher rate to its C20 and C22 products than linoleic acid (18:2 n-6) and the activity of 6-desaturase is probably the rate-limiting step. However, it is not known with confidence in what proportions C20 and C22 products will be produced from given mixtures of linoleic acid and alpha linolenic acid fed to any animal.

7.2 ESSENTIAL FATTY ACIDS (EFA)

7.2.1 Discovery of EFA and description of EFA deficiency

Burr and Burr (1930) described how acute deficiency states could be induced in rats by giving them fat-free diets and how these deficiencies could be eliminated by adding specific fatty acids to the diet. It was originally thought that only fatty acids related to linoleic acid were responsible for this effect, and they were referred to as 'essential fatty acids' (EFA). EFA deficiency can be produced in a variety of animals including man but the condition is best documented in the laboratory rat. The disease is characterised by skin signs such as dermatosis, and the skin becomes more permeable to water. Growth is retarded, reproduction is impaired and there is degeneration or impairment of function in many organs of the body, especially the kidney. Biochemically, EFA deficiency is characterised by changes in the fatty acid compositions of many cell membranes whose functions are impaired, including the production of metabolic energy associated with beta-oxidation of fatty acids in the mitochondria.

Well-documented EFA deficiency in man is rare, but was first seen in children given fat-free diets who developed a skin condition similar to that produced in rats. The skin abnormalities and other signs of EFA deficiency disappeared when linoleic acid was added to the diet. More recently it has become apparent that EFA deficiency is secondary to a wide variety of disorders, including protein-energy malnutrition and fat malabsorption (e.g. as a result of major bowel surgery, cystic fibrosis or Crohn's disease). Failure to include appropriate amounts of EFA during total parenteral nutrition can also lead to signs of EFA deficiency.

7.2.2 Structural features of EFA

Extensive feeding experiments in which EFA-deficient animals were given diets containing fatty acids of defined structure revealed that other members of the n-6 family had EFA activity in relation to eliminating the gross symptoms of deficiency, whereas those of the n-9 families had none. The double bonds need to be of the *cis* configuration but the presence of a *trans* double bond is not incompatible with EFA activity as long as the basic n-6 pattern of *cis* double bonds is also present. Thus, columbinic acid (*trans*-5, *cis*-9, *cis*-12–18:3 n-6) has EFA activity. The EFA potencies of members of the n-6 family differ considerably depending on the biological test used for assessing EFA activity and the structure of the fatty acid. Thus arachidonic acid and GLA, which are further along the desaturation-elongation sequence than linoleic acid, have considerably more potency than linoleic acid itself.

Alpha linolenic acid (18:3 n-3) has less potency as an EFA than linoleic acid, but several of its longer chain more highly unsaturated derivatives

have greater activity. Certain tissues, e.g. brain, neural tissue and the retina of the eye, are particularly rich in n-3 fatty acids and it has been traditionally assumed that they perform a specific function in these tissues (see Chapter 9).

7.2.3 Definition of EFA

The essential fatty acids, or their metabolic derivatives, are fatty acids which are required for normal growth and physiological integrity and cannot be synthesised in adequate amounts by the body. Thus linoleic acid (18:2 n-6) and alpha linolenic acid (18:3 n-3) are now known to be true essential fatty acids. Derivatives such as arachidonic acid (20:4 n-6), docosahexaenoic acid (DHA) (22:6 n-3) and eicosapentaenoic acid (EPA) (20:5 n-3) might be called 'conditionally essential' in circumstances such as those which exist in the premature infant where they cannot be synthesised by the body in adequate amounts.

7.2.4 Assessment of EFA status

Many of the signs of EFA deficiency are difficult to quantify and, therefore, do not lend themselves to providing sound measurements of EFA status. Changes in water permeability can be measured, but although fairly sensitive, they are too laborious for general use. Biochemical tests, based on changes in membrane fatty acid composition, have therefore gained wide acceptance and the concept of the triene/tetraene ratio was introduced (Mohrhauer and Holman, 1963a, b; Holman, 1970). The ratio is based on the 'switch' in metabolism from the n-6 to the n-9 pathway during EFA deficiency resulting in the accumulation of the 20-carbon tri-unsaturated acid (Mead acid), which only occurs in trace amounts in healthy tissue. Holman (1970) originally suggested an arbitrary value of the triene/tetraene ratio of 0.4 to indicate the onset of EFA deficiency. More recently, he has used the technique of serum PUFA 'profiling' as a more sensitive indicator of different types of EFA deficiency states (Holman and Johnson, 1981; Holman, 1986a). In this technique, a large number of fatty acid constituents of the serum phospholipid fraction are analysed by capillary gas-liquid chromatography and the ratio of the concentration of individual isomers to their concentration in a normal tissue is compared.

7.2.5 Cell- and tissue-specific functions

Cell- and tissue-specific functions that depend

critically on PUFA are now well known. A striking example occurs in rat skin in which the epidermal lipids are rich in very long chain acyl glucosyl ceramides (Hansen, 1989) (see Chapter 18). The long chain acyl substituent in these lipids is linoleic acid (18:2 n-6) which is linked via its carboxyl group to the methyl group of 34:1 n-9 to generate an extremely long chain C52 fatty ester. These lipids form an intercellular matrix that apparently forms a water permeability barrier by which the animal avoids excessive water loss. Alpha linolenic acid (18:3 n-3) cannot substitute for linoleic acid (18:2 n-6) in the matrix and this important role of 18:2 n-6 obviously accounts for several of the major EFA deficiency symptoms of rats.

Likewise, the importance of n-6 PUFA in reproduction in the rat and the failure of alpha linolenic acid (18:3 n-3) to sustain reproduction in this species reflects not only the importance of prostaglandins in parturition, but also the abundance of DPA (22:5 n-6) in rat testis and sperm, rather than DHA (22:6 n-3) which is the case for other species, including man (Leat, 1989). Although species-specific differences occur in the relative abundance of 22:6 n-3 and 22:5 n-6 in brain and retina, it remains the case that animal brain and retina contain large amounts of n-3 PUFA, especially DHA (22:6 n-3) (Bazan, 1989).

At the biochemical level, cell- and consequently tissue-specific concentrations (and presumably functions) of both n-6 and n-3 PUFA are of major importance in determining the EFA requirements of animals and such functions are often species-specific. These specific functions co-exist with the more generalised functions specifically of n-6 PUFA in animal cells which relate to their involvement in the production of eicosanoids that regulate cell metabolism (see Chapter 8). However, because eicosanoids are involved in inter-cellular signalling, their physiological functions are expressed both within and between tissues, i.e. they are also tissue-specific. Such a generalised requirement for n-6 PUFA leaves open the question of whether n-3 PUFA are required for modulating eicosanoid production from n-6 PUFA (see Chapter 8). Tissue-specific 'structural' functions of PUFA, as in the epidermis, retina or brain, may be fulfilled without the requirement of PUFA for eicosanoid production since it is known that columbinic acid (trans-5,cis-9,cis-12–18:3 n-6), which cannot be converted into prostaglandins, can restore some, but not all, of the EFA deficiency symptoms in rats.

7.3 ESTERIFICATION AND ACYLGLYCEROL BIOSYNTHESIS

7.3.1 Importance of esterification

Fatty acids are rarely found in living cells in the form of the free carboxylic acids. This is because their detergent-like properties render them toxic and disruptive to cells and tissues, and the small amounts that exist free are invariably bound to proteins, e.g. plasma albumin (see Chapter 6). Most of the fatty acids in the body are present in the form of esters, usually with glycerol, as triacylglycerols or phosphoglycerides. The conversion of the fatty acid to its thiolester with coenzyme A (a derivative of the B group vitamin, pantothenic acid) is a prerequisite for its formation into an ester.

7.3.2 Positional specificity (see Figure 7.4)

Fatty acids (as CoA esters) are first attached to glycerol-3-phosphate, a product of glucose metabolism. Two separate enzymes catalyse the esterification of fatty acids in positions 1 and 2. These enzymes have preference for particular types of fatty acid (see Chapter 1). Saturated acids and trans monounsaturated acids are preferentially esterified at position 1 and cis unsaturated acids, especially PUFA, are preferentially incorporated at position 2. Thus, diacylglycerol-3-phosphate or phosphatidic acid is formed. The next reaction in the sequence is the loss of the phosphate group from position 3 catalysed by phosphatidic acid phosphatase. The final step is the esterification of the third fatty acid at position 3, which normally contains a mixed population of fatty acids, to form a triacylglycerol. In the formation of phospholipids, the phosphate is not lost but is linked to a moiety such as choline to form phosphatidyl choline (PtdCho), through the reaction of phosphatidic acid and activated choline. Likewise, phosphatidyl ethanolamine (PtdEtn) can be formed through the reaction of phosphatidic acid and activated ethanolamine, and phosphatidyl serine (PtdSer) through the reaction of phosphatidic acid and activated serine.

7.3.3 Acyl exchange

The positional specificity of fatty acids in lipids is determined during their biosynthesis. However, the fatty acid composition of intact lipids can be modified. Fatty acids can be released from each position on the glycerol backbone by specific lipases or phospholipases so that other esterifying

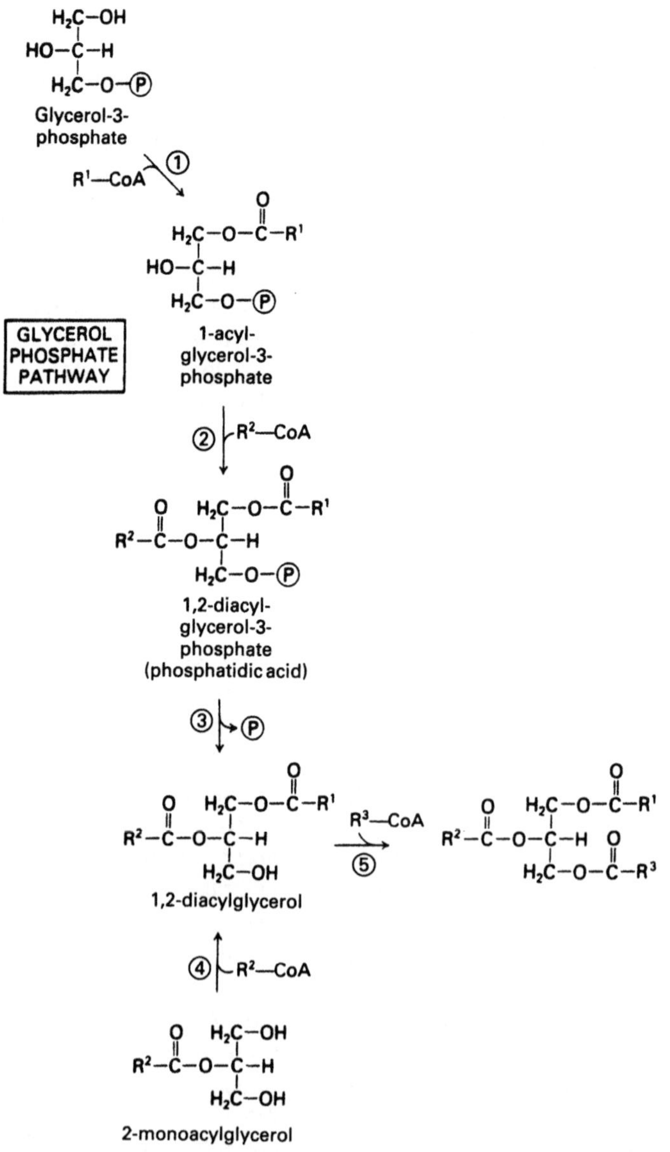

Figure 7.4 The biosynthesis of triacylglycerols. (From Gurr and Harwood, 1991)

enzymes can replace the original fatty acid with a different one. In addition, acyl exchange enzymes can exchange existing fatty acyl groups in phospholipids with free fatty acyl CoA derivatives. In this way, a great variety of molecular species of lipids can be elaborated to enable the cell to adapt to different environmental conditions.

It is the specificity of both the acyl synthetases involved in biosynthesis *de novo* and the acyl exchange enzymes that determines the fatty acid compositions of phospholipids in different cells. These can differ widely: e.g. the fatty acid composition of brain, retina and liver are very different. In no case has the relative roles of the two mechanisms (acyl synthetases and acyl exchange enzymes) been defined.

7.3.4 Control of esterification

The complex metabolic controls involved in esterification pathways are not well understood, but it is likely that phosphatidic acid phosphatase and the activity of the enzymes transferring the phosphocholine or phosphoethanolamine group to the phosphatidic acid play major roles in determining relative rates of triacylglycerol and phosphoglyceride biosynthesis. Thus a coordinated regulation of triacylglycerol and phosphoglyceride synthesis probably exists.

An example of this involves certain fatty acid analogues that induce proliferation of peroxisomes and cause hypolipidaemia. These analogues enhance phospholipid biosynthesis and simultaneously depress triacylglycerol biosynthesis: i.e. less triacylglycerol is formed in and exported from liver, while the induced liver peroxisomes initiate the oxidation of free fatty acids through chain shortening processes (see Section 7.5.1.2). The availability of free fatty acids for oxidation is reinforced by the enzyme fatty acyl CoA hydrolase migrating to the microsomes with the enzymes involved in esterifying fatty acyl CoA into glycerides. Thus any fatty acyl CoA formed in the system is not available for esterification in the microsomes but is instead hydrolysed to free fatty acids. These controls operate to enhance fatty acid oxidation and to depress triacylglycerol formation in liver, their overall effect being to reduce serum lipid without causing build up of triacylglycerol in the liver (Skorve *et al.*, 1990).

7.4 ACYLGLYCEROL BREAKDOWN (LIPOLYSIS)

The breakdown of triacylglycerols and phosphoglycerides is known as lipolysis and is tightly controlled by lipases.

7.4.1 Triacylglycerol lipases

Adipose tissue contains the hormone-sensitive lipases, triacylglycerol lipase and monoacylglycerol lipase. Triacylglycerol lipase can cause lipolysis of both triacylglycerols and diacylglycerols; monoacylglycerol lipase can effect lipolysis of monoacylglycerols.

In the fasting state, demanding the consumption of fuel reserves, the low concentrations of insulin turn off the biosynthetic pathways and release the inhibition of triacylglycerol lipase within the cell. The activity of this enzyme is controlled by a complex 'cascade' mechanism illustrated in Figure 7.5. The lipase must first be converted into an active form. This conversion is regulated by cyclic AMP (cAMP) produced from ATP by the enzyme adenylate cyclase. The activity of the cyclase is under the control of catecholamine hormones, such as noradrenaline, which stimulate its activity, or prostaglandins (formed from dietary PUFA) which inhibit its activity. Another feature of the control is the destruction of cAMP by a specific phosphodiesterase which is inhibited by xanthines, such as caffeine.

The demand for the consumption of fuel reserves can be regarded as a form of metabolic stress. This is characterised by a low activity of insulin relative to 'stress' hormones: catecholamines, corticotropin, glucocorticoids and glucagon. Such a hormonal spectrum can occur in starvation, diabetes, trauma

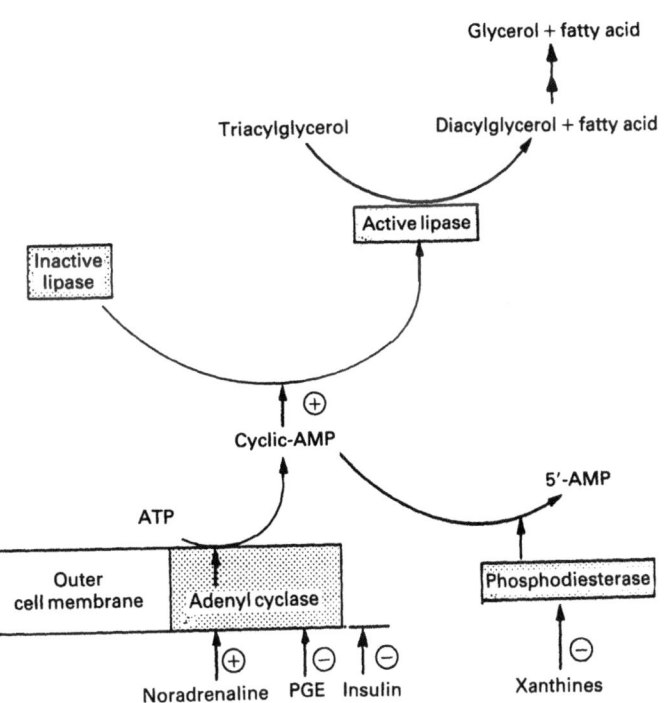

Figure 7.5 The control of lipolysis. (From Gurr and Harwood, 1991)

and under the influence of certain toxins. Insulin secretion is primarily a response to the ingestion of nutrients.

Fatty acids released by lipolysis are transported out of the cell, bound to plasma albumin and transported to tissues, such as muscles including cardiac muscle, that utilise fatty acids as major sources of energy. Whether fatty acids are directed into beta-oxidation or triacylglycerol synthesis in the recipient tissues may be determined by competition for available acyl-CoA molecules by the acyltransferases involved in the esterification of acylglycerols and the carnitine palmitoyl transferase of the mitochondrial membrane. The activity of the latter enzyme is increased during starvation when the animal most needs to oxidise fatty acids as a reserve source of energy.

7.4.2 Phospholipases

The phospholipids present in cellular membranes are not hydrolysed to yield free fatty acids for energy production other than in extreme circumstances, and apart from acyl exchange reactions are generally metabolically stable. However, important phospholipases, phospholipase A_2 and C, are involved in cell signalling processes involving the eicosanoids and the 'inositol cycle'. These phospholipases are considered in Chapter 8.

7.5 OXIDATION OF FATTY ACIDS

Fatty acids are major sources of energy for all aerobic cells. When the energy content of animal diets is low, or when there is a high demand for energy such as during growth, reproduction or sustained mechanical work, fatty acids are mobilised from the triacylglycerols in adipose tissue by the action of hormone-sensitive lipase (see Section 7.4.1). The released fatty acids are transported in the plasma bound to albumin (see Chapter 6) and delivered to cells. Here they are sequestered by 'fatty-acid binding proteins' before being converted in the cytosol to CoA esters by fatty-acid CoA ligase.

7.5.1 Beta-oxidation

Beta-oxidation (Figure 7.6) is a metabolic pathway whereby a fatty acid is broken down, step by step, two carbon atoms at a time. The chemical energy locked up in the carbon compounds is released in a controlled way and used to generate ATP needed for biochemical synthesis. On average, fatty acids provide 37 kJ/g (9 kcals/g).

All fatty acids are potentially substrates for beta-oxidation and can therefore yield energy. However, they are not all oxidised at equal rates and, under certain circumstances, EFA may be spared from oxidation to perform more important physiological roles.

7.5.1.1 Mitochondrial pathway

Beta-oxidation takes place in the mitochondria of cells. Mitochondria are elaborate systems of membranes containing the enzymes involved in beta-oxidation and many other metabolic processes.

The overall stoichiometry of fatty acid oxidation is:

$$C_{16}H_{32}O_2 + 23O_2 \rightarrow 16CO_2 + 16H_2O + 137 \text{ ATP}$$

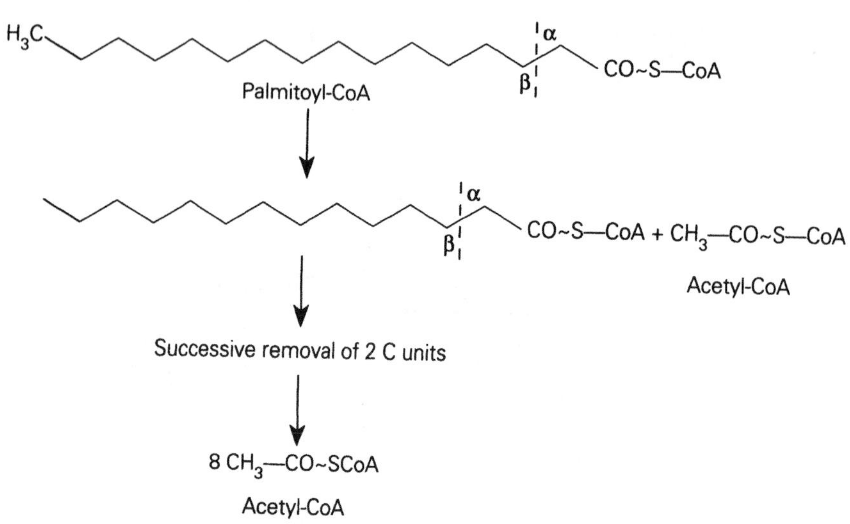

Figure 7.6 Overview of beta-oxidation of fatty acids. (From Murray *et al.*, 1988)

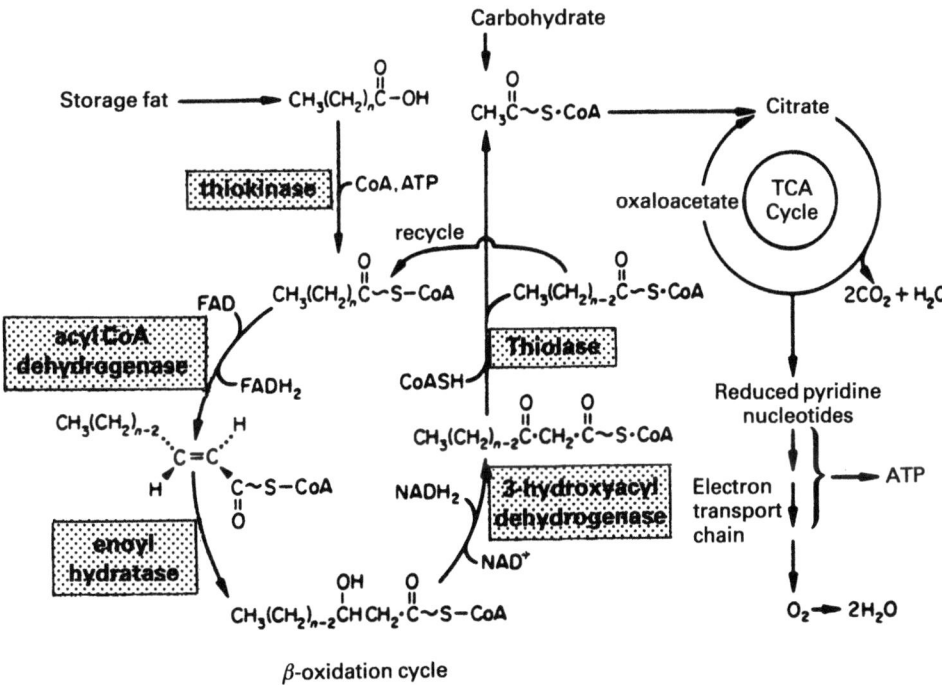

Figure 7.7   Beta-oxidation and energy metabolism in mitochondria. (From Gurr and Harwood, 1991)

Fatty acyl CoAs cannot cross the mitochondrial membrane; they first have to be converted into their carnitine derivatives.

After entry of the fatty-acyl carnitines into the mitochondrion, the fatty acyl CoAs are re-constituted and are catabolised by the beta-oxidation pathway. The first step in beta-oxidation (Figure 7.7) is a flavoprotein-dependent (FAD) dehydrogenation of a fatty acyl CoA to form a *trans* 2,3 enoyl CoA which is hydrated to a L-3-hydroxyacyl CoA. The latter is converted by an NAD-dependent dehydrogenase to a 3-ketoacyl CoA. This is then cleaved by the enzyme thiolase in the presence of CoA to yield acetyl CoA and a fatty-acyl CoA chain which is two carbon atoms shorter than the starting substrate. The cycle is repeated to completion, the acetyl CoA molecules going into the tricarboxylic acid cycle and the mitochondrial electron-transport chain to be oxidised to yield the end products CO_2, H_2O and ATP.

Trans monounsaturated fatty acids can be oxidised directly by the above pathway providing the *trans* double bond is on an even-numbered carbon atom, since such isomers will be converted during beta-oxidation to a *trans* 2,3 enoyl CoA. If the *trans* bond in a monounsaturated acid starts at an odd-numbered carbon atom (from the carboxyl end of the molecule), or if the fatty acid is *cis*-monounsaturated, isomerase enzymes can generate the required intermediates.

Until recently, the above reactions were thought to account for the oxidation of all fatty acids including PUFA. However, it is now known that this

is not the case (Osmunden and Hovik, 1988). Certain PUFA, e.g. DHA (22:6 n-3), arachidonic acid (20:4 n-6) and GLA (18:3 n-6), are metabolised by different pathways, principally peroxisomal, to yield the *trans* 2,3-*cis* 4,5 dienoyl CoAs. A key enzyme involved in converting the latter derivative into a form that can be beta-oxidised is NADPH-2,4 dienoyl CoA oxidoreductase. The requirement for NADPH in a system geared to generate NADH means that the metabolic control of these oxidations is complex. Apart from this, the oxidation of PUFA is the same as for other fatty acids.

7.5.1.2 Peroxisomal pathway and retroconversion

Peroxisomes are cellular organelles which are particularly prevalent in the liver and kidney. They also have an active beta-oxidation system, but this differs from the mitochondrial system in several important respects:

(i)   The peroxisomal system is inducible by various agents including high fat diets, especially those rich in long chain monounsaturated fatty acids such as erucic acid (22:1 n-9), i.e. peroxisomes proliferate to deal with the excess substrate to be oxidized.

(ii)   Fatty acyl CoAs can enter directly into the peroxisomal beta-oxidation system without the involvement of carnitine.

(iii)   The first enzyme in the beta-oxidation scheme in peroxisomes is a fatty-acyl oxidase rather than a flavoprotein dehydrogenase. The remaining enzymes of beta-oxidation in peroxisomes are functionally identical to the corresponding mitochondrial enzymes.

(iv) The peroxisomal beta-oxidation system generally stops after very few 'rounds' of oxidation, i.e. it is primarily concerned with producing chain-shortened fatty acids which can be transferred to the mitochondria for complete oxidation.

Peroxisomes have a major role in shortening long chain fatty acids such as erucic acid (22:1 n-9), DHA (22:6 n-3) and DPA (22:5 n-6) (Norum et al., 1989). Retroconversion is the term given to partial breakdown of fatty acids by a beta-oxidation type of process. Diets rich in long chain monounsaturated fatty acids, e.g. high-erucic acid rape seed oil or partially hydrogenated fish oil, are in fact efficient inducers of the peroxisomal beta oxidation in various tissues, notably liver. Peroxisomes have a high affinity for PUFA and are currently considered to be the principal site of chain shortening of these fatty acids in cells. They are therefore of significance in the retroconversions of PUFA, such as those converting DPA (22:5 n-6) to arachidonic acid (20:4 n-6) and those converting DHA (22:6 n-3) to EPA (20:5 n-3).

Retroconversion is employed, not for the generation of energy, but for shortening the carbon chain of fatty acids. This enables the cell to take a 'short cut' to the production of specific fatty acids if longer chain precursors are available.

There has been uncertainty about the extent of retroconversion of C22 to C20 PUFA. However, recent studies with cultured human cells have clearly established that levels of C22 and C20 PUFA can be regulated by rates of elongation and retroconversion. It appears that elongation to DHA (22:6 n-3) is favoured for EPA (20:5 n-3), whereas retroconversion to arachidonic acid (20:4 n-6) is favoured for DPA (22:5 n-6) (Rosenthal et al., 1991).

7.5.2 Lipid peroxidation

7.5.2.1 Basic mechanisms

Peroxidation involves direct attack on the unsaturated fatty acids in lipid molecules by oxygen. The reaction may be chemical, catalysed by metal ions (autoxidation), or enzymic, catalysed by the enzyme lipoxygenase. Lipid peroxidation is initiated when a hydrogen atom is removed from a methylene (-CH$_2$-) group in the hydrocarbon chain of a lipid molecule (Stage 1 in Figure 7.8). The molecular species that is formed is chemically reactive and is known as a free radical (see British Nutrition Foundation, 1991). There are many initiators of lipid peroxidation. One is 'singlet' oxygen, which itself may be formed from the normal form (ground state) of oxygen by 'sensitisers' such as chlorophylls, bilirubin, porphyrins and haem compounds, in the presence of light. Another common initiating species is the hydroxyl radical, generated from superoxide anions (O^{2-}) by ferric ion catalysis. Sensitisers and ferric ions are common in foods and biological tissues.

After initiation, the process of propagation gives rise to chemical rearrangements of the lipid molecules and further reaction with oxygen to form fatty acid peroxy- and hydroperoxy-radicals (Stage 2 in Figure 7.8). Once the process has started, it continues by a chain reaction of propagation steps which can be terminated (Stage 3 in Figure 7.8) in a number of ways:

(i) two lipid radicals can combine to form polymeric products; or

(ii) fatty acid peroxyradicals can undergo cyclisation and the resulting cyclic endoperoxide breaks down into a range of end-products characteristic of lipid peroxidation (including malondialdehyde, other medium chain aldehydes, oxyacids and hydrocarbons).

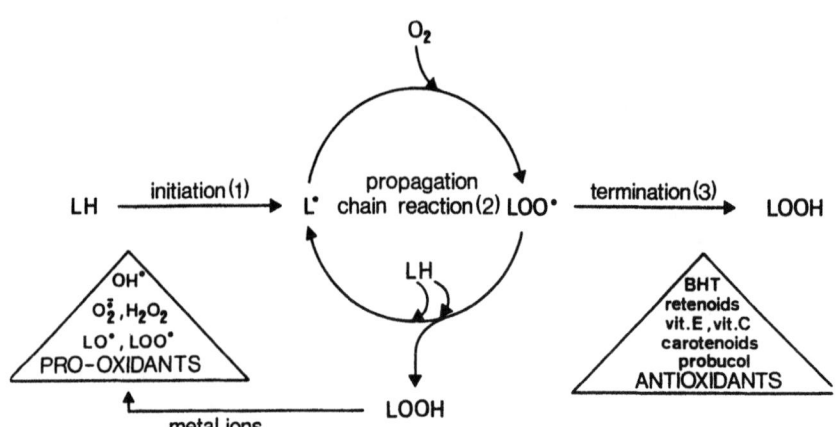

Figure 7.8 Lipid peroxidation.

7.5.2.2 Controlled lipid peroxidation

Peroxidation occurs naturally in living tissues, the different steps being catalysed by specific enzymes. Controlled peroxidation of PUFA can lead to the formation of eicosanoids (see Chapter 8) that are involved in the regulation of muscular contraction, inflammatory responses and platelet aggregation (see Chapters 13 and 19). Free radical peroxidation mechanisms are also present in some types of white blood cells as one of the mechanisms for destroying bacteria.

7.5.2.3 Unwanted lipid peroxidation of food lipids

When lipids in food are exposed to oxygen, they can undergo peroxidation. This can occur at room temperature over a long period of time. Peroxidation is a particular hazard if the fats in food are in contact with traces of metal ions (e.g. iron, copper) or haem compounds. Peroxidation proceeds much faster if the temperature is raised, as in cooking. Peroxidation in food lipids can also be catalysed by tissue peroxidases, especially the lipoxygenases in plants, and this activity can be destroyed by boiling briefly. Cholesterol, often regarded as an inert substance, is also susceptible to peroxidation.

Because the most susceptible lipids are found in oils which are rich in EFA, one nutritional effect of peroxidation could be to reduce the EFA content of edible fats. The overall nutritional significance is generally minimal, however, since losses are usually small in relation to the total content of EFA supplied by these oils. The losses of vitamin E are also small and the vitamin E/linoleic acid ratio of most oils after deep frying is acceptable. The effect of repeated deep frying is not known.

Reactive species like free radicals and peroxides are potentially damaging to cells. Evidence from animal studies using radioactively labelled lipid peroxides indicate, however, that very few are absorbed intact and therefore cannot enter tissues and cause damage, except perhaps to the gut itself. There is some evidence that smaller molecular weight degradation products of lipid peroxides, such as hydroperoxyalkenals, are more readily absorbed and thus become potentially toxic. Nevertheless, the first effect of oxidative changes on edible fats is the deterioration in flavour and appearance. In most cases, oxidised fats are rendered unpalatable long before the changes have reduced nutritional value or created toxicity.

7.5.2.4 Unwanted peroxidation in vivo

The control of peroxidation in vivo is sometimes inadequate and damage can be caused by peroxidative reactions and the accumulation of the end-products of lipid peroxidation in tissues. The insoluble auto-fluorescing material composed of oxidized lipid and protein (ceroid) found in a variety of circumstances in mammalian tissues is particularly abundant in animals deficient in vitamin E. An important factor determining the extent of peroxidation in vivo is the availability and location of metal ion catalysts of hydroperoxyl radical formation (Figure 7.8). For example, the disorganisation of cellular structures containing iron or copper can accelerate radical formation. This process can be self-catalytic since superoxide anions can release iron from ferritin thus providing the source of more initiation steps.

7.5.2.5 Protection against peroxidation

One important factor in ensuring protection against uncontrolled lipid peroxidation is a well-organised membrane structure, especially in organelles involved in oxidative metabolism such as the mitochondria. Any tendency for membrane structure to break down will increase the chances of peroxidative damage. A second important protective factor is the presence of adequate antioxidant capacity. Antioxidants can be either preventative antioxidants or chain-breaking antioxidants.

Preventative antioxidants reduce the rate of chain-reaction initiation by converting peroxide radicals into inactive non-radical species: the enzyme catalase (which degrades hydrogen peroxide) and glutathione peroxidase are examples. The latter enzyme requires selenium for its activity and this probably explains the dietary requirements for this element and its apparent role in protection against peroxidation.

Chain-breaking antioxidants interfere with chain propagation by trapping chain propagation radicals, for example LOO· or L· (Figure 7.8). Examples are the enzyme superoxide dismutase and vitamin E. Vitamin E is the major lipid soluble, chain-breaking antioxidant in human tissues. Vitamin C and the carotenoids are involved in the reconstitution of reduced vitamin E.

8

FUNCTIONS OF UNSATURATED FATTY ACIDS

The major roles of lipids can be described conveniently as:

- energy storage;
- energy-providing;
- structural;
- metabolic.

Individual lipids may have different roles to play at different times or sometimes they play different roles at the same time. This chapter will focus mainly on the structural roles of lipids since unsaturated fatty acids have unique roles to play here.

8.1 ENERGY STORAGE ROLE

8.1.1 Adipose tissue

8.1.1.1 Fatty acid composition of storage lipids

Adipose tissue represents the largest reservoir of storage lipids. Storage lipids (triacylglycerols) tend to contain more saturated and monounsaturated fatty acids than structural lipids (phospholipids), although the composition of human storage triacylglycerols can be influenced by the composition of the diet. Fatty acids are mobilised from adipose tissue (lipolysis) to meet energy demands at times when dietary energy is limiting. The release of stored fatty acids is regulated by the amounts and types of different dietary components and by hormones, whose secretion may also be regulated in part by the diet (see Chapter 7).

Storage lipids may be derived directly from the fat in the diet or they may be synthesised in the liver, mammary gland or adipose tissue from dietary carbohydrates (see Chapter 7). The capacity of these tissues to synthesise fatty acids is geared to the needs of the body and is under dietary and hormonal control. The range of fatty acids that can be made is limited, usually to palmitic, stearic and oleic acids. Human adipose tissue has the enzymic machinery for fatty acid synthesis, but activities are very low. When there is little fat in the diet, the fatty acid pattern of adipose tissue is mainly dependent on the liver's biosynthetic activity and is characteristic of the species. The introduction of fat into the diet can suppress the synthetic activity of the tissues and mechanisms operate to transport dietary fatty acids into the storage lipids so that the fat composition is more characteristic of the diet.

Because of the relatively high proportion of energy derived from fat in most industrialised countries, the diet is much the most important determinant of the composition of the fatty acids in adipose tissues.

8.1.1.2 Effects of dietary unsaturated fatty acids on storage lipids

Giving vegetable oils rich in linoleic acid (18:2 n-6) to animals and man causes an enrichment of this fatty acid in plasma and storage triacylglycerols (Field and Clandinin, 1984; Becker, 1989). The low levels of arachidonic acid (20:4 n-6) in these lipids may also be elevated, though not markedly, after giving linoleic acid due to biosynthesis in the liver. Alpha linolenic acid (18:3 n-3) is markedly elevated in the plasma triacylglycerols of rats given linseed oil in the diet and there is some deposition in adipose tissue. Likewise, EPA and DHA are enriched in the storage lipids of rats given certain fish oils containing these acids, but less so than 18:2 n-6 in rats fed vegetable oils (Jandacek et al., 1991). Very recently, accumulation of EPA and DHA has been measured in the adipose tissue triacylglycerols of human beings consuming fish oils in the diet (Lin and Conner, 1990).

If rats are given cod liver oil, which is rich in the long chain monounsaturated fatty acids gadoleic (20:1 n-9) and cetoleic (22:1 n-11), their cardiac triacylglycerols (but not phosphoglycerides) are enriched in these fatty acids. Giving diets containing a high proportion of the monounsaturated acid erucic acid (22:1 n-9) (present in older varieties of rapeseed oil) can also result in the transitory accumulation of triacylglycerols rich in erucic acid in the cardiac muscle of a variety of experimental animals (Norum *et al.*, 1989).

There are numerous examples of dietary hydrogenated oils, which contain different amounts of fatty acids (mainly monounsaturates) with *trans*-unsaturation, causing marked elevations of *trans* fatty acids in triacylglycerols of plasma, adipose tissue and milk (British Nutrition Foundation, 1987). These substitutions occur mainly in positions 1 and 3 of triacylglycerols. High dietary inclusions of C18 *trans*-unsaturated fatty acids may also cause increases in the *trans*-monounsaturated fatty acid content of tissue phosphoglycerides. This occurs mainly at position 1 (British Nutrition Foundation, 1987).

Starvation of rats whose diets originally contained vegetable oils rich in linoleic acid results in preferential mobilisation of 16:0 and 18:1 and a selective retention of linoleic acid and arachidonic acid. Starvation of rats whose diets originally contained fish oils rich in n-3 fatty acids results in the selective retention of DHA in triacylglycerols but a selective removal of EPA (Cunnane, 1989).

8.1.1.3 Analysis of adipose tissue as an indicator of dietary fatty acids

The lipids in adipose tissue are slowly, but continuously, being replaced. The continual breakdown and resynthesis of lipids is called lipid turnover. Lipid replacement may occur

(i) by complete synthesis of the lipid from its simplest precursors;
(ii) by replacement of parts of the molecules;
(iii) by replacement of whole lipid molecules that have been transported from another site and that have gone through the lipolysis and re-esterification cycle.

Turnover allows a finer degree of metabolic control in a dynamic system than would be possible in a more static system. Although in some tissues lipids turnover very rapidly, in adipose tissue turnover is so slow that the fatty acid composition of adipose tissue gives a reasonable reflection of the habitual fat consumption over a long period of time.

The biopsy technique can be used to obtain small pieces of adipose tissue for fatty acid analysis. It does not give a direct measure of the dietary content of all types of fatty acid because the fatty acids are not necessarily deposited in direct proportion to their content in the diet. It is, however, most useful for the essential fatty acid, linoleic acid. Different amounts of linoleic acid can be fed in diets and corresponding amounts can be found in

adipose tissue.

The technique has also been useful for monitoring the dietary intake of *trans* fatty acids. It is less useful for some of the longer chain polyunsaturated fatty acids, e.g. arachidonic acid, which tend to accumulate in adipose tissue to a lesser extent than would be predicted from their content in the diet. This is probably because they are preferentially esterified into the structural phospholipids, leaving less to be stored in the fat tissue.

8.1.2 Storage lipids in other tissues

The triacylglycerols in the fat globules of milk can be regarded as an energy store for the newborn baby.

Other tissues, such as the liver and heart of mammals, can accumulate fat in the form of small globules but only for short-term use. The extensive accumulation of fat in mammalian liver and heart is a pathological condition. However, some species of fish, particularly gadoids, normally store fat in the liver rather than subcutaneously or around the internal organs.

8.2 ENERGY-PROVIDING ROLE

Fatty acids are mobilised from adipose tissue to meet the demands for energy at times when dietary energy is limiting, for example in starvation or in strenuous exercise. The release of stored energy is regulated by the amounts and types of different dietary components and by hormones, whose secretion may also be regulated in part by diet. Thus, the post-prandial elevation of the hormone insulin in the blood is one of the factors that ensures that fatty acids from circulating lipoproteins are shunted into adipose tissue via the action of lipoprotein lipase, whilst the hormone-sensitive lipase that breaks down the stored fat is inhibited. During fasting, the relative deficiency of insulin and sufficiency of glucagon allows the hormone-sensitive lipase to release fatty acids from the stored fat. They are then mobilised into the bloodstream and carried in a complex with albumin to other tissues where they can be broken down by the process of beta-oxidation (described in Chapter 7) to yield energy that drives metabolic reactions.

Different types of fatty acids may be oxidised to different extents. There is some evidence that the essential fatty acids may be conserved for their more vital membrane functions at the expense of the non-essential fatty acids.

8.3 STRUCTURAL ROLES IN MEMBRANES

8.3.1 Membrane structure

Current theories of membrane structure envisage that most of the phospholipid is present as a bimolecular sheet with the fatty acid chains in the interior of the bilayer (the 'fluid mosaic' model of Singer and Nicholson, 1972). Membrane proteins are located at the internal or external face of the membrane, or projecting through from one side to the other (Figure 8.1).

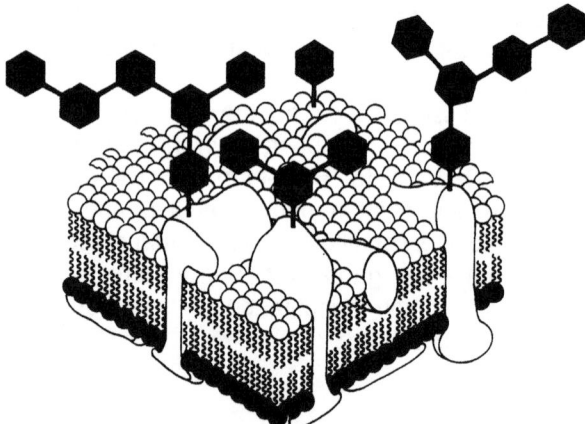

Figure 8.1 Structure of cellular membrane: the Singer–Nicholson fluid-mosaic proposal for membrane structure. Carbohydrate groups on lipids and proteins face the extracellular space.

There may be polar interactions between the phospholipid headgroups and ionic groups on the proteins as well as hydrophobic interactions between the fatty acid chains and hydrophobic amino acid sequences. Lipid molecules are quite mobile along the plane of the membrane but there is limited movement across the membrane. The patterns of lipid molecules on each side of some membranes which exhibit 'membrane asymmetry' are quite different.

The lipid provides a flexible structure in which are located the proteins that control many metabolic activities within the membrane. The proteins in question may be enzymes, transporters or receptors for substances such as hormones, antigens or cell-growth factors. The physical properties of the membrane are strongly influenced by the lipid composition. They are important because they are regulated in the face of environmental changes (diet, temperature, etc) by subtle differences in the proportions of amphiphilic lipids and sterols and changes in the fatty acid composition of the lipids.

An important feature of the physical properties of membranes is the degree of freedom for molecules to move about in the membrane, generally described

as membrane fluidity. Fluidity is, in part, related to the packing of lipids in the bilayer. Among the most important factors that affect this packing are:

(i) the nature of the fatty acid chains;
(ii) the amount of cholesterol (in animal membranes);
(iii) interactions, both polar and non-polar, between lipids and proteins.

8.3.2 Role of unsaturated fatty acids in membrane structure

Unsaturated fatty acids are found in membranes mainly esterified in phosphoglycerides.

The presence of unsaturation in the chains of fatty acids affects their shape and their ability to pack together. Saturated chains can be constrained to pack together in crystalline arrays that give low fluidity. With the introduction of one double bond, a bend in the molecule is formed and the space occupied by the fatty chain is much increased. The chains, therefore pack less well together and the fluidity increases. There is not a linear relationship, however, between the degree of unsaturation and the fluidity. After the introduction of two or three double bonds, the fluidity begins to level off and even decrease again. This is because the highly polyunsaturated fatty acids begin to adopt a helical configuration that allows closer packing again (Stubbs and Smith, 1990).

Computer modelling studies have confirmed that the minimal energy conformation of glyceride-linked DHA can either be an 'angle-iron' shape or, more likely, a helix (Applegate and Glomset, 1986).

Consequences of helicity in PUFA molecules are:

(i) The molecules are shortened: e.g. DHA (22:6 n-3) is significantly shorter than arachidonic acid (20:4 n-6) which is shorter than stearic acid (18:0).
(ii) The molecules pack together better than kinked chains so that they form more compact structures.

Substitution of *trans*-monounsaturated for *cis*-monounsaturated fatty acids may lead to a decrease in the fluidity of the lipid bilayer because the *trans*-unsaturated acids can pack together more like saturated acids. However, when dietary *trans*-monounsaturated fatty acids are incorporated into membranes, there is little influence on overall fluidity because the *trans*-unsaturated acids replace saturated fatty acids at position 1 of phosphoglycerides rather than replacing *cis*-unsaturated fatty acids at position 2.

The stability of mammalian membranes is crucially dependent upon the types of unsaturated fatty acids that are incorporated into the phosphoglycerides of the bilayer. Stability is highly dependent on the presence of the essential fatty acids and the longer chain, more highly polyunsaturated fatty acids derived from them. Thus, in the condition of essential fatty acid deficiency (see Chapter 7) essential fatty acids of the n-6 family are replaced by non-essential fatty acids of the n-9 family. The resulting membrane structure is more permeable to water and the efficiency of diverse metabolic processes that occur in the membrane matrix is much reduced.

The contribution of lipids containing specific combinations of fatty acids to the structure of particular membranes is discussed in Section 8.3.6.

8.3.3 Membrane traffic

In general, different cellular membranes have different phosphoglyceride compositions and, in a given membrane such as the plasma membrane which surrounds the cell, phosphoglycerides are sited asymmetrically in the two bilayers. The phosphoglycerides, PtdIns and PtdSer are located predominantly on the cytoplasmic face, whereas PtdCho is found predominantly on the serosal face. Precisely how this is achieved is not known in detail, although two mechanisms may be involved (Van Meer, 1989):

(i) Transfer of the phosphoglycerides from the endoplasmic reticulum where they are synthesised by means of phospholipid binding proteins is especially well characterised for the PtdCho transfer proteins. These are involved particularly in the transfer of PtdCho from the endoplasmic reticulum to the outer mitochondrial membrane.
(ii) Transfer of phospholipids to their final membrane destinations can also take place as vesicles, often generated on intracellular membranes known as the Golgi apparatus. The preferential location of both sphingomyelin and cholesterol in the plasma membrane rather than in internal cell membranes is due to concentration and selection of these lipids in Golgi vesicles.

The molecular selection mechanisms underlying the sorting of phospholipids for their final destinations in the cell are currently unknown. In membrane trafficking, lipids may well be selected in association with specific membrane proteins (Lisanti and Rodriguez-Boudan, 1990).

8.3.4 Establishment of the fatty acid composition of membrane lipids

8.3.4.1 Biosynthetic mechanisms

Individual phosphoglycerides in membranes are characterised by specific combinations of fatty acids in positions 1 and 2 of the molecules. It is not known to what extent these characteristic fatty acid compositions are determined by acylation reactions within the endoplasmic reticulum, or by subsequent selection of particular phosphoglyceride molecular species in the various transport processes described in Section 8.3.3. Selectivity can, in principle, occur at any one of a number of stages in lipid biosynthesis (see Chapter 7) (Gurr and Harwood, 1991):

(i) cellular uptake of free fatty acids (involving fatty acid binding proteins);
(ii) activation of free fatty acids to CoA derivatives;
(iii) chain elongation and/or shortening and further desaturation reactions;
(iv) acylation reactions during biosynthesis of phosphatidic acid followed by selection of diacylphosphoglyceride species;
(v) acyl transfer reactions with preformed phosphoglycerides;
(vi) membrane sorting and trafficking processes (Section 8.3.3).

Many of these processes exhibit species and tissue specificity.

8.3.4.2 Influence of dietary fat on membrane lipid composition

While the composition of membrane lipids is less susceptible to dietary influence than that of storage lipids (Section 8.1.1), modification of the fatty acid composition by dietary fatty acids is well established and, in recent years, has been particularly well studied for PUFA of the n-3 and n-6 series.

The observed changes are not always readily interpretable in terms of tissue lipid biochemistry. For example, when rats are given a diet in which linoleic acid predominates, arachidonic acid (20:4 n-6) is the major PUFA in the liver phosphoglycerides PtdCho and PtdEtn, while 22-carbon PUFA of the n-6 family are present in very low concentrations. In rats given diets containing mainly alpha linolenic acid (18:3 n-3), the major fatty acids in these liver phosphoglycerides are EPA (20:5 n-3) and DHA (22:6 n-3). When the diet contains appreciable amounts of both linoleic acid and alpha linolenic acid, the major PUFA of liver phosphoglycerides are arachidonic acid and DHA. These results do not directly reflect the activities of the fatty acid desaturases and elongases in rat liver microsomes (Sprecher, 1989, 1991) (see Chapter 7).

Nevertheless, if the conversions of linoleic acid and alpha linolenic acid into their higher PUFA metabolites are studied *in vivo* in the livers of rats that are essential fatty acid-deficient (Chapter 7) with dietary levels of linoleic acid and alpha linolenic acid that are below minimum dietary requirements, and taking into account the inhibitory effects of the n-3 series upon the n-6 series, then interpretable dose–response relationships exist (Lands, 1991). This underlines the importance of working within the responsive range of essential fatty acids in nutritional metabolic research and offers hope that situations involving more complex dietary mixtures of fatty acids, and tissues other than liver, may ultimately be understood in quantitative terms.

Similar effects in enriching membrane phospholipids with the long chain metabolites of linoleic acid and alpha linolenic acid administered in the diet occur in tissues other than the liver, e.g. platelets and neutrophils. However, the proportion of C20 and C22 PUFA ultimately present in tissue membrane lipids can vary widely: e.g. the major phosphoglycerides of both platelets and neutrophils contain much more C20 than C22 PUFA from either C18 dietary precursor (Sprecher, 1989). Since these tissues, as well as heart muscle cells, are not active in PUFA biosynthesis, the observed effects are a combination of liver biosynthetic activities and selective uptake and esterification reactions in the tissues in question.

Extreme examples of tissues with highly selective uptake and esterification mechanisms are brain and retina (see Chapter 9). Postnatal photoreceptor cells, and probably brain cells, appear to derive most of their high concentrations of DHA from the liver or the diet rather than by their own synthetic activity. However, the retina has both extracellular and intracellular fatty acid binding proteins with a high affinity for DHA. Moreover, DHA is activated to its CoA ester by a high affinity acyl-CoA synthetase in retina (Bazan, 1989). Brain also has a PtdEtn fatty acyltransferase selective for DHA (Masuzawa *et al.*, 1989). Processes such as these are fundamental in concentrating DHA in the brain, retina and testis.

Increases in the n-3 PUFA content of the phospholipids PtdCho, PtdEtn and PtdSer have been consistently observed when animals and human beings are given diets supplemented with fish oils rich in EPA and DHA. The composition of PtdIns, which is dominated by a molecular species containing stearic acid (18:0) at position 1 and arachidonic acid (20:4 n-6) at position 2, remains relatively unchanged by such diets. The physiologi-

cal consequences of such changes in composition, especially in platelets and neutrophils, will be described in later sections.

In experiments where n-3 PUFA supplements are provided as triacylglycerols, ethyl esters or as the free fatty acids, the synthesis of fatty acids by the liver *de novo* is largely suppressed. The liver may, however, be involved in the modification of the dietary fatty acids by chain elongation in the case of EPA or retroconversion in the case of DHA. Therefore the spectrum of fatty acids delivered to cells is not necessarily the same as that administered in the diet. Additional factors in determining the ultimate fatty acid composition of the cellular phosphoglycerides of the target cells include:

(i)   the capacity to oxidise the fatty acids, whether completely by beta-oxidation or partially by retroconversion;
(ii) the potential for selective introduction of fatty acids at any of the stages in phosphoglyceride biosynthesis (Section 8.3.4.1);
(iii) the dietary dose and its rate of absorption;
(iv) the developmental and physiological state of the animal.

It is not, at present, possible to predict outcomes with confidence, but dietary fish oils can be said to raise levels of long chain n-3 PUFA substantially in the major phosphoglycerides in most body tissues.

8.3.5 Specific fatty acid combinations in membrane lipids

8.3.5.1 Polyunsaturated fatty acids

The impressive advances in molecular biology in the last few decades have inevitably focused much more attention on the protein than the lipid components of membranes. Membrane proteins can now be purified with relative ease; many of their genes, especially those for receptor proteins and associated signal transduction processes, have been cloned. Lipids, with one or two notable exceptions, have been generally relegated to a rather non-specific, somewhat inconsequential background role in membrane structure and function. However, even very slight changes in phosphoglyceride fatty acids can cause marked changes in membrane properties. This is especially true for the PUFA in the phosphoglycerides. For example, the substitution of the Mead fatty acid (20:3 n-9) for arachidonic acid (20:4 n-6) generates a membrane that is less compact and more permeable to ions (Evans and Tinoco, 1978). Similarly, the substitution of arachidonic acid (20:4 n-6) by EPA (20:5 n-

3) has marked effects on eicosanoid production in the membranes (see Section 8.5).

This has implications for tissues rich in n-3 PUFA including retina (see Section 8.3.6.5), brain and testis. Mammalian brain, particularly the 'cephalin' fraction (i.e. mainly PtdEtn), is also a rich source of DHA. Moreover the PtdEtn fraction of brain is rich in 'plasmalogens', i.e. phosphoglycerides containing a vinyl ether-linked chain at position 1 instead of an ester-linked chain. Ethanolamine plasmalogens from mammalian, including human, brain have 16:0, 18:0 and 18:1 substituents in approximately equal abundance at position 1 with DHA, EPA and arachidonic acid in approximately equal abundance at position 2. Diacyl PtdEtns are much richer in 18:0 as well as in DHA suggesting that 18:0/22:6 n-3 is a major molecular species in this phosphoglyceride. The content of PUFA, especially DHA, decreases in both lipids with increasing age.

The preceding sections have established that fatty acids are incorporated into individual membrane lipid classes with a high degree of specificity and into specific positions in the individual phosphoglycerides, consistent with their possessing highly specific functions. The highly characteristic distribution of 18:0 and 20:4 n-6 on positions 1 and 2 of PtdIns has been mentioned. To understand fully the functions of fatty acids in membranes requires a consideration of specific molecular species of not only the more common phospholipids, PtdCho, PtdSer, PtdEtn and Ptd Ins, but also the generally less studied classes such as cardiolipin, sphingomyelin, cerebrosides, gangliosides and sulphatides.

8.3.5.2 Monounsaturated fatty acids

This marked molecular specificity in membrane lipids is seen with fatty acids other than PUFA. Very long chain monounsaturated fatty acids such as erucic acid (22:1 n-9) and nervonic acid (24:1 n-9) are not readily incorporated into phosphoglycerides and indeed are relatively rare in natural phosphoglycerides. However, in rats fed diets rich in erucic acid (22:1 n-9), this fatty acid is preferentially incorporated into cardiolipin which itself is preferentially located in mitochondria (Blomstrand and Svensson, 1974). The same occurs with partially hydrogenated fish oil, though only the *cis* and not the *trans* isomers are incorporated into cardiolipin from the oil (Blomstrand and Svensson, 1983). Nervonic acid (24:1 n-9) is so named because of its relative abundance in brain associated traditionally with sphingomyelin. In the human brain, it accounts for some 20–25% of the total fatty acids.

8.3.6 Role of specific lipids in bilayer structures

Two general structural roles for lipids can be considered in biomembranes:

(i) that which relates to the bulk phase of the membrane or the bilayer *per se*;
(ii) that which relates to the direct interaction of lipids with proteins (see Section 8.3.7).

It is known that different and specific lipid compositions are used to produce membranes with different bulk structural properties. Some examples follow.

8.3.6.1 The role of membrane lipids in maintaining fluidity

The retailoring of individual phospholipids by acyl exchange reactions is of major significance in maintaining the fluidity of the membrane (Lynch and Thompson, 1984). For example, acyl exchange between 18:0/18:0 and 18:1/18:1 to generate 18:0/18:1 is a very efficient means of 'fluidising' membranes. Other processes, including altering phospholipid head groups, altering the overall fatty acid composition of the membrane, and altering the cholesterol content, are also important.

Of special significance is the relatively recent acceptance that the abundance of C20 and C22 PUFA, especially n-3 PUFA, in membranes can no longer be accounted for simply in terms of membrane fluidity (see Section 8.3.2). This is because increased unsaturation above a certain number of double bonds, exceeded in C20 and C22 PUFA, does not necessarily translate into increased 'fluidity', either in the sense of the potential of the molecules to undergo segmental, rotational or lateral motions, or in the sense of their ability to form ordered lipid domains as reflected by their phase transition temperatures (Dratz and Deese, 1986).

8.3.6.2 The apical membrane of epithelial cells

It is important that the apical membrane of epithelial cells has a low permeability to ions and water. Hydrogen bond interactions between the carbohydrate groups of glycolipids located on the outer (serosal) side of the bilayer can generate a high viscosity membrane with the required characteristics (Van Meer, 1988).

8.3.6.3 The erythrocyte membrane

The red blood cell is a dramatic example where specific fatty acids can alter cell shape. Specific phosphatidyl choline (PtdCho) molecules can be transferred experimentally into the erythrocyte membrane using a phospholipid exchange protein (Jos *et al.*, 1985). The incorporation of 16:0/18:1 PtdCho into the membrane generates 'normal' (biconcave) cells, 16:0/16:0 PtdCho generates convex cells and the incorporation of 18:2/18:2 PtdCho generates concave cells.

8.3.6.4 Myelin

With its highly specific lipid composition, myelin is a classical example of an ion-impermeable, insulating cell membrane whose unique structure and properties are probably determined largely by specific interactions between membrane proteins and phospholipids and the cytoskeleton of the cell.

8.3.6.5 Rods

The bilayer regions of the outer segments of rods have a very high content of di22:6 n-3 (DHA) phosphoglycerides (Tinoco, 1982). This clearly represents a structural specialisation to generate a highly structured but fluid membrane, possibly with an enhanced elasticity. The membrane may then be able to accommodate the high frequency of conformational changes undergone by the high density of rhodopsin molecules contained within it.

A recent striking discovery is the presence in retinal lipids of very long chain PUFA (VLCPUFA) with even-numbered chain lengths from 22 to 36 carbon atoms (Aveldano, 1987). Phosphoglycerides containing VLCPUFA in rod outer segments seem to be closely associated with the dominant membrane protein, rhodopsin.

8.3.7 Lipid – protein interactions

The second general role of lipids in biomembranes relates to their direct molecular interactions with proteins at the lipid–protein interface. The fact that highly specific detergents are required to crystallise a membrane protein such as bacteriorhodopsin (Roth *et al.*, 1989) implies that equally specific interactions between phospholipids and the protein occur in the native membrane. Phosphoglycerides containing long chain PUFA may fulfil this role in mammalian rod outer segments. Interactions of this type, though currently undefined, could be of obvious importance not only in stabilising the hydrophobic helix within the membrane but possibly also in facilitating the folding of the protein into its active conformation within the membrane and in its subsequent function (Dratz and Deese, 1986; Neuringer and Conner, 1989).

Such interactive roles of specific lipids with proteins in facilitating and participating in conformational changes necessary for membrane protein function are well reflected in the many examples of the activities of purified and partially purified membrane proteins being influenced by their lipid environment, other than simply in terms of bulk membrane 'fluidity'. Thus the molecular specificity of lipids in biomembranes is intimately related to both the structural and functional specificities of biomembranes. This applies both to roles of specific lipids in determining particular bulk structural properties of membranes, and to roles of specific lipids in interacting with proteins to determine particular metabolic functions of membranes. Thus the importance of interactions between proteins and other molecules in their immediate environment applies as much to the organic (membrane) phase of the cell as it does to the aqueous (cytosoluble) phase. However, the organic phase is particularly challenging chemically and this has undoubtedly delayed full appreciation of the specific interactions that can occur within cell membranes. Some examples of specific lipid-protein interactions follow.

8.3.7.1 Transporter proteins

The catalytic parameters of the erythrocyte glucose transporter (V_{max}, K_m, turnover number) are markedly influenced by the nature of the lipids associated with it, such as head group composition, and the degree of unsaturation and chain length of constituent fatty acids (Carruthers and Melchier, 1986).

8.3.7.2 Enzymes

The activities of Ca/Mg ATPase from sarcoplasmic reticulum (fundamental in muscle contraction), adenyl cyclase (fundamental in agonist-induced formation of cAMP) and 5-nucleotidase (fundamental in generating the extracellular agonist adenosine) are all markedly influenced by the levels of n-6 and n-3 PUFA in the membrane lipids. Thus in animals fed diets enhanced in fish oils rich in EPA (20:5 n-3) and DHA (22:6 n-3), the activity of Ca/Mg ATPase is decreased, adenyl cyclase is increased and 5-nucleotidase is increased (Kinsella, 1990). These diets result in enhanced levels of 22:6 n-3 in phosphoglycerides in all of the cell membranes in question.

8.3.7.3 'Policeman' role

Phosphatidyl inositol (PtdIns) has a well-defined role in locating, and probably directing, various enzymes to the external surface of the lipid bilayer, where they act extracellularly (Cow *et al.*, 1986). Such enzymes include acetylcholinesterase and 5-nucleotidase, and in all cases the enzymes are anchored in the bilayer externally to the cell

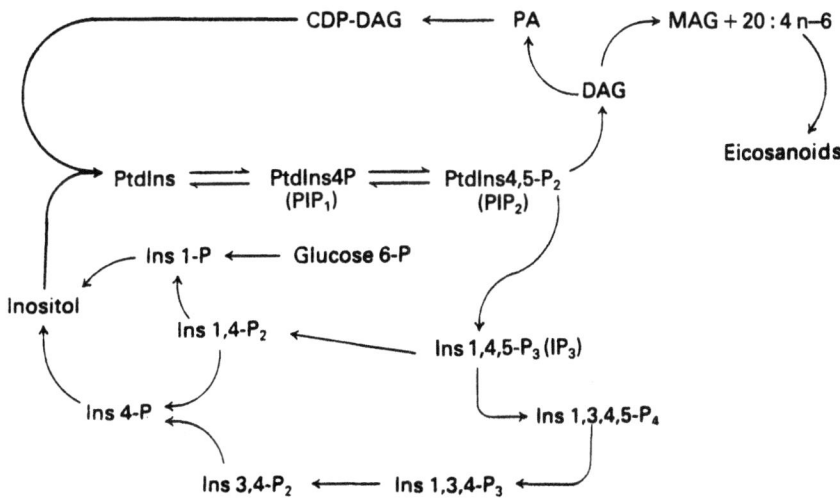

Figure 8.2   The inositol–lipid cycle.   CDP-DAG = CDP-diacylglycerol; PA = phosphatidic acid; MAG = monoacylglycerol; PIP_1 = phosphatidyl inositol 4 phosphate; PIP_2 = phosphatidyl inositol 4,5 biphosphate; IP_3 = inositol 1,4,5 triphosphate; PtdIns = phosphatidyl inositol; Ins = inositol; DAG = diacylglycerol.

through being linked to PtdIns via glycosyl substituents on the proteins.

8.4   ROLE OF UNSATURATED FATTY ACIDS IN MEMBRANE METABOLIC CONTROL VIA THE INOSITOL LIPID CYCLE

8.4.1   Mechanism

A major example of membrane lipid metabolism of central importance in cellular physiology is the inositol lipid cycle. This cycle is now known to underlie the responses of very many cells to a range of hormones, neurotransmitters and cell growth factors (see Figure 8.2).

To initiate the cycle, an extracellular agonist (e.g. acetyl choline or an alpha adrenergic agonist) interacts with a specific cell receptor. The receptor/agonist complex then interacts with a specific G-protein system containing alpha, beta and gamma subunits to generate GTPase activity. This results in activation of a specific phospholipase C (PLC). PLC then cleaves PtdIns 4,5 bisphosphate (PIP_2) to generate two important, short lived second messengers: inositol 1,4,5 trisphosphate (IP_3) and diacylglycerol (DAG). IP_3 is converted through the sequential action of a series of phospho-monoesterases to yield inositol. DAG is converted sequentially to diacylglycerophosphate and then to cytidyl-diacylglycerophosphate (CDAGP). CDAGP is coupled with inositol to yield cytidyl-monophosphate (CMP) and PtdIns. PtdIns is then phosphorylated sequentially by two ATP kinases to yield, first, PtdIns 4 phosphate (PIP_1) and then, finally, the

starting molecule PIP_2 is regenerated. In this way, a whole series of hormones and neurotransmitters can generate two discrete intracellular messengers, IP_3 and DAG.

A different mechanism for activating the cycle appears to hold for cell growth factors since these extracellular agonists interact with receptors that frequently possess tyrosine kinase activity (Boyer et al., 1989). At least one substrate of the latter is a PLC isoenzyme that is activated by phosphorylation to cleave PIP_2. The net effect is the same, i.e. the production of IP_3 and DAG. At least five immunologically distinct PLCs are now known from mammalian tissues and at least four of these are apparently represented by separate gene products.

8.4.2   Role of the second messengers (IP_3 and DAG)

The inositol lipid cycle generates the two important second messengers, IP_3 and DAG.

IP_3 activates the release of intracellular stores of calcium ions from the endoplasmic reticulum and probably also facilitates the entry of calcium ions into the cell (see British Nutrition Foundation, 1989).

The role of DAG as a second messenger is less well defined. Together with calcium ions and PtdSer, it activates a protein kinase C that phosphorylates numerous intracellular proteins with, so far, largely undefined functions. There is evidence, however, that one of the systems activated through phosphorylation is a membrane transport system that can cause alkalinisation of the cell interior.

The elevation of both intracellular calcium and

alkalinisation are essential for cell division. These metabolic controls are therefore very important in processes involving the rapid division of cells in the immune response and in tumour growth.

8.4.3 Importance of the inositol lipid cycle in nutrition

The inositol lipid cycle, which is central to the control of cell metabolism throughout the body and whose regulation underlies very many fundamental physiological control processes, not least those relating to cell division and growth, is intimately involved in lipid nutrition and metabolism in at least three ways:

(i)   The specific PLC that hydrolyses PIP_2 is a plasma membrane bound enzyme. As with other membrane enzyme systems, including the G proteins and adenyl cyclases, it can in principle be influenced by the nature of the lipid phase in which it operates.

(ii) The dominant species of PtdIns in the plasma membrane has the fatty acid composition 18:0/20:4 n-6. Many isoenzymes of PLC exist, often with high tissue specificity (e.g. in the retina). There is also evidence that the molecular species composition of PtdIns may be tissue-specific within a given species and also species-specific for given tissues. The significance of this, particularly how it relates to the EFA intake of the species and how it may influence the action of PLC on PtdIns, especially via isoenzymes, is currently unknown.

(iii) The possibility of variations in the fatty acid composition and molecular speciation of PtdIns raises some questions. Are there consequences for the activation of protein kinase C by DAG since the latter presumably reflects the molecular species composition of its precursor PtdIns? In particular, can the content of C20 and C22 n-3 and n-6 PUFA be altered in PtdIns and its derived DAG by dietary means? Similar considerations apply to the fatty acid and molecular species composition of the PtdSer necessary for the co-activation of protein kinase C with DAG and calcium ions, since the activation process appears to occur in association with the cell membranes. This complex area is not understood but the potential influence of dietary EFA on the biochemistry of the phosphatidyl inositol cycle clearly represents an area of high priority for future studies.

It has been reported that eicosapentaenoic acid (EPA), added as the free fatty acid to the medium used to culture bovine aorta endothelial cells, inhibits the production of platelet-derived growth factor (PDGF) by the cells (Fox and Dicorleto, 1988). Several lines of evidence suggest that oxidative damage to the EPA is necessary for the observed inhibitory effect. Whether this is principally a physiological or a pathological effect is not known but this finding alone justifies a very high priority for this topic in future work on EFA nutrition. Some of these considerations will be amplified in following chapters dealing with thrombogenesis (Chapter 13), tumour formation (Chapter 16) and the immune response (Chapter 19).

8.5 ROLE OF UNSATURATED FATTY ACIDS IN MEMBRANE METABOLIC CONTROL VIA EICOSANOID FORMATION

The second major area where unsaturated fatty acids can exert membrane metabolic control is via their critical role as precursors of the eicosanoids (Johnson et al., 1983). These are a complex group of highly biologically active, often short-lived compounds with 20 carbon atoms produced by cells to act in their immediate environment.

8.5.1 Discovery of eicosanoids

While research was being undertaken to determine which fatty acids have EFA activity and how this was related to chemical structure, it was discovered that the human uterus, on contact with fresh human semen, was provoked into either strong contraction or relaxation (Kurzrok and Lieb, 1930). Von Euler (1934) then showed that a fatty acid fraction in lipid extracts from sheep seminal plasma caused marked stimulation of smooth muscle. The active factor was named prostaglandin and was shown to exhibit a variety of physiological and pharmacological properties at extremely low concentrations. The chemical structures of the prostaglandins suggested that the substances might originate from arachidonic acid and this was subsequently proved in a series of very elegant biochemical experiments using radioactively labelled arachidonic acid.

8.5.2 Types of eicosanoids (see Figure 8.3)

Eicosanoids are hydroxylated derivatives of C20 PUFA that can be categorised into four groups:

(i)   cyclic products, formed by an initial peroxidation reaction, that are known collectively as the prostanoids and include prostaglandins, prostacyclins and thromboxanes (see Section 8.5.3);

(ii) linear products, initiated by the 5-hydro-peroxidation of PUFA by a 5-lipoxygenase, known collectively as leukotrienes (LT) (see Section 8.5.4);

(iii) other lipoxygenase products of PUFA such as those formed by the action of 12-lipoxygenase on arachidonic acid or EPA;

(iv) mono-oxygenase products of PUFA formed by the action of cytochrome P_{450}: e.g. the hydroxylation of arachidonic acid at the methyl carbon atom or the adjacent carbon atom.

8.5.3 Prostanoids

Prostanoids are a group of metabolically related compounds of great physiological importance which are formed in every tissue. Because of their extreme potency and their short circulation time, they are considered local hormones. They all contain 20 carbon atoms, arranged in a (bi)cyclic structure, carrying two side chains, one of which is terminated by a carboxyl group (COOH). The other side chain contains a hydroxyl group (OH) at carbon number 15 and is terminated by a methyl group (CH_3). Consequently prostanoids can be considered cyclic hydroxy fatty acids. Depending on the configuration of the cyclic part of their molecule, prostanoids are divided into prostaglandins (PGs, containing a ring of five carbon atoms), thromboxanes (TXs, containing a ring of six carbon atoms) or prostacyclins (containing a ring of five carbon atoms).

8.5.3.1 Functions

The C20 polyunsaturated fatty acids can generate a whole range of related compounds but with subtle differences in structure, which exert a range of profound physiological activities at very low concentrations. The prostaglandins have the ability to contract smooth muscle, to inhibit or stimulate the adhesion of blood platelets and to cause constriction or dilation of blood vessels with related influence on blood pressure (see Chapter 14).

In addition to the prostaglandins themselves, the prostanoids include the metabolites of prostaglandin H. These include the prostacyclins and thromboxanes. These have essentially opposing physiological effects. Prostacyclins, formed in the wall of blood vessels, are among the most powerful known inhibitors of platelet aggregation. They relax the arterial walls and promote a lowering of blood pressure. Thromboxanes, found in platelets, stimulate the platelets to aggregate (an important mechanism in wound healing), contract the arterial wall and promote an increase in blood pressure. The balance between these activities is important in maintaining normal vascular function (see Chapter 13).

All prostanoids are produced locally near to their sites of action, are released in minute quantities, act rapidly, and are quickly destroyed by a battery of degradative enzymes. The breakdown products are excreted in urine and their measurement has been employed to estimate their daily production by the body.

8.5.3.2 Biosynthesis of prostanoids (Figure 8.4)

The first step in the biosynthesis of prostanoids is the release of the essential fatty acid (EFA) as the free fatty acid from the membrane phospholipid in which it is stored. This release is catalysed by phospholipase A_2. The EFA substrate then binds to the enzyme cyclo-oxygenase. This multi-functional enzyme is a single polypeptide with a mass of 70,000 Daltons containing two haem groups per molecule. It has two activities:

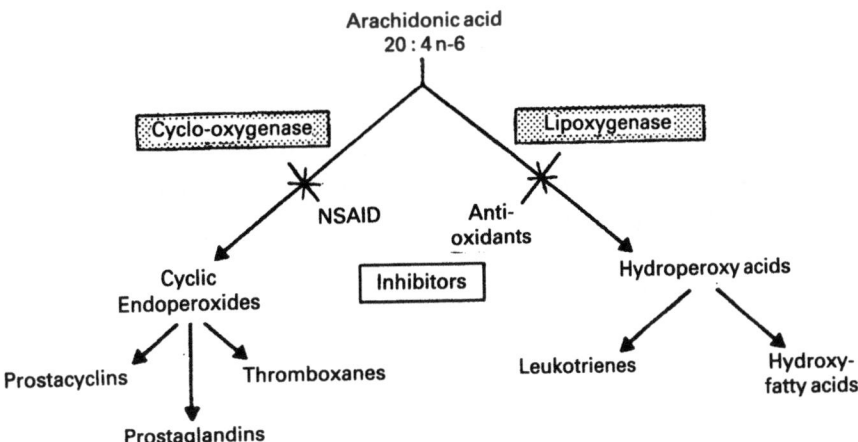

Figure 8.3 Types of eicosanoid.

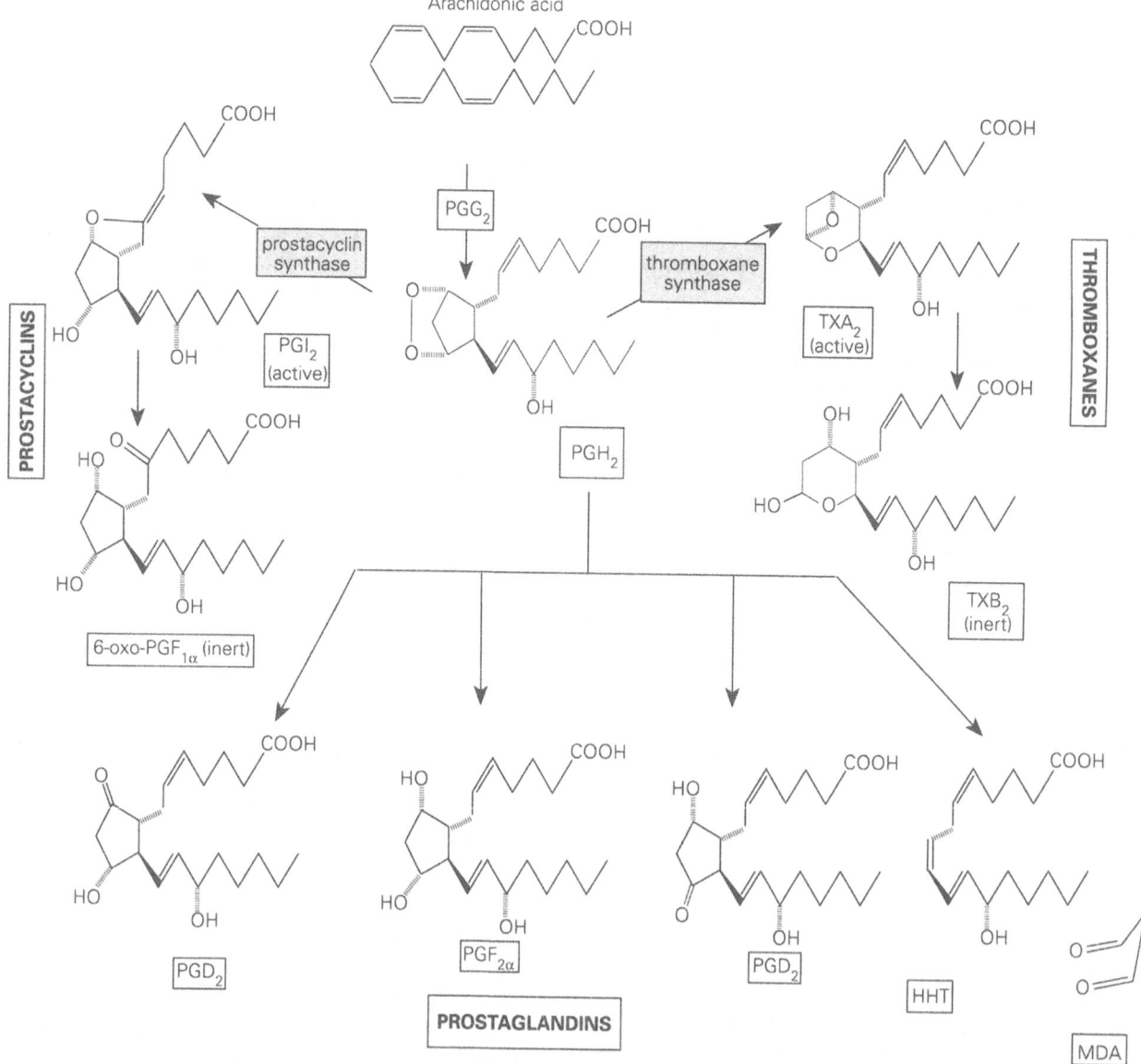

Figure 8.4   Prostanoids formed from arachidonic acid. (From Johnson *et al.*, 1983)

(i)   oxygen atoms are inserted in the folded fatty acid chain to yield a cyclic endoperoxide. This intermediate is prostaglandin G (PGG) (Figure 8.5).

(ii)   catalytic activity reduces the peroxide function of PGG to an hydroxy group and thus forms prostaglandin H (PGH).

Prostaglandin H (PGH) is the key intermediate for conversion into a wide range of other prostaglandins and eicosanoids. Figure 8.4 illustrates the range that can be produced if arachidonic acid is the substrate for cyclo-oxygenase. The balance of activities of the enzymes that catalyse these reactions determines the pattern of eicosanoids formed in any given tissue. Diet is one of the factors affecting these patterns as discussed in Section 8.5.5.

PGG and PGH are unstable precursors of all other members of the prostanoid family. Thus platelets and many other tissues contain thromboxane synthetase that catalyses conversion of the endoperoxide intermediates to thromboxane A_2 (TXA$_2$) which is inherently unstable and decays to the inert TXB$_2$. In other tissues, prostacyclin synthetase forms PGI$_2$ from the unstable endoperoxides. PGI$_2$ is also unstable, though less so than TXA$_2$, and decays to an inert form. Other tissues may contain enzymes that isomerise the endoperoxides to PGE, PGF and PGD, although these reactions may also occur by non-enzymic rearrangements.

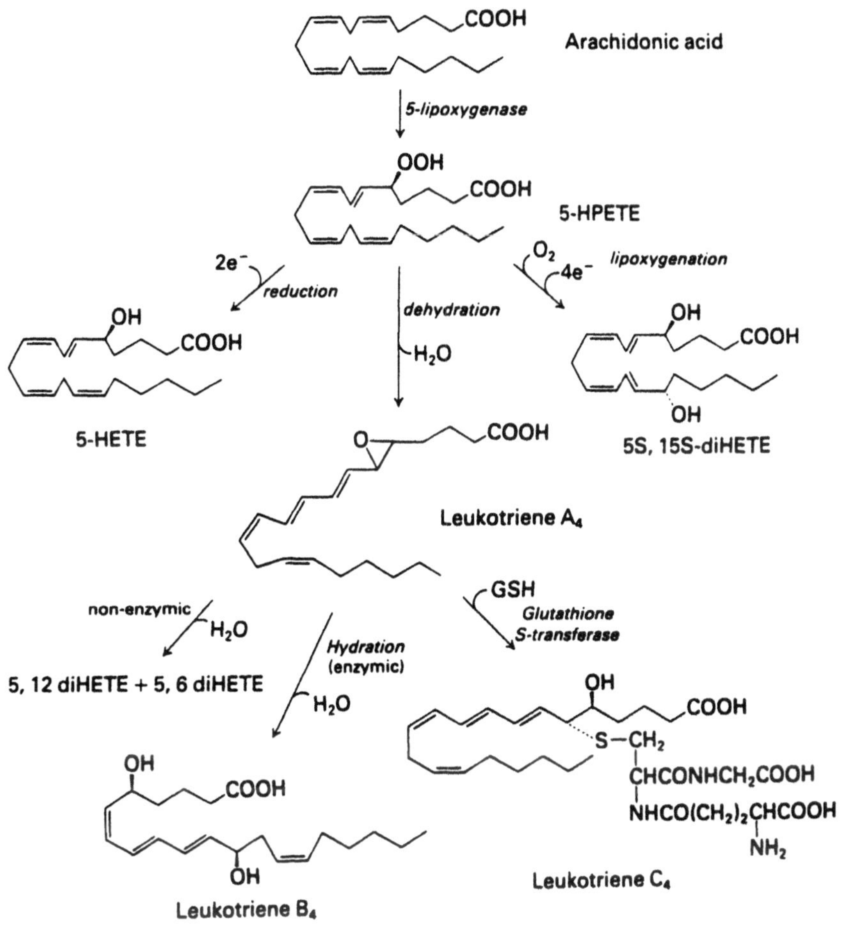

Figure 8.5 Biosynthesis of leukotrienes. (From Gurr and Harwood, 1991)

8.5.4   Leukotrienes

8.5.4.1   Functions of leukotrienes

An alternative route for the conversion of C20 unsaturated fatty acids into biologically active oxygenated products is by 5-lipoxygenase peroxidation. The end products of peroxidation are called leukotrienes, the name derived from the cells (leukocytes) in which they were originally recognised.

Leukotrienes have a range of potent biological activities including the contraction of respiratory, vascular and intestinal smooth muscles. Some are chemotactic agents for cells of the immune system (see Chapter 19).

8.5.4.2   Biosynthesis of leukotrienes

In the linear pathway (Figure 8.5), the enzyme 5-lipoxygenase catalyses the oxygen-dependent hydroperoxidation of arachidonic acid (20:4 n-6) to yield 5-hydroperoxyeicosatetraenoic acid (5-HPETE) or the unstable intermediate, 5-epoxy derivative leukotriene A4 (LTA4).

LTA4 can be further metabolised in two ways (Figure 8.5):

(i)   LTA4 can be converted by an epoxide hydrolase to yield LTB4 which is a powerful chemo-attractant produced by leukocytes to attract additional white cells to sites of tissue damage.

(ii) LTA4 can react with glutathione S-transferase to form the sulphidopeptidyl leukotrienes known as LTC4, LTD4 and LTE4. These are known collectively as the 'slow reacting substances A' (SRS-A).

The human 5-lipoxygenase has been purified and its gene has been cloned. The enzyme is unstable, has a mass of 78,000 Daltons and requires ATP and calcium ions, as well as other undefined protein and membrane fractions, for optimal activity. The 5-lipoxygenase activating protein (FLAP) has been identified in polymorpho-nuclear leukocytes, and has a mass of 18,000 Daltons. Its function appears to be to facilitate the transfer of lipoxygenase from the cytosol to the plasma membrane. Inhibitors of FLAP block the biosynthesis of 5-lipoxygenase.

8.5.5 Effect of dietary unsaturated fatty acids on eicosanoid formation

8.5.5.1 Prostanoids

Arachidonic acid gives rise to the 2-series prostaglandins as shown in Figure 8.4, but essential fatty acids of both n-6 and n-3 families can give rise to similar products differing in numbers and patterns of double bonds (see Figure 8.6). Dihomogamma linolenic acid (DGLA) (20:3 n-6) generates the 1-series PG, and eicosapentaenoic acid (EPA) (20:5 n-3) generates the 3-series PG (see Figure 8.7). Competitive effects occur between these precursor fatty acids in the cyclo-oxygenase reaction. Thus, when membrane phospholipids have elevated levels of n-3 PUFA after dietary administration of fish oils, formation of 2-series PG from arachidonic acid is depressed and formation of the 3-series derivatives is elevated. In general, the 3-series derivatives are less pharmacologically active than the 2-series.

It was once thought that the main criterion of essentiality in a fatty acid was its ability to be converted into a physiologically active eicosanoid.

There are now known to be exceptions to this rule. Although it is difficult entirely to separate the membrane and eicosanoid functions of EFA, certain EFA may contribute to the integrity of membranes in ways that have little to do with their conversion into eicosanoids (see Sections 8.3.6 and 8.3.7).

In addition to the prostaglandins themselves, there are also prostacyclins and thromboxanes which are metabolites of prostaglandin G. Again, although arachidonic acid generates the 2-series prostacyclins and thromboxanes, DGLA and EPA generate the 1- and 3-series of these compounds.

8.5.5.2 Leukotrienes

Arachidonic acid (20:4 n-6) is converted into the 4-series LT. Similarly dihomogamma linolenic acid (20:3 n-6) is converted into the 3-series LT and eicosapentanoic acid (20:5 n-3) is converted into the 5-series LT. As in the case of cyclo-oxygenase, competititve effects occur between these precursor fatty acids in the 5-lipoxygenase reaction and, in particular, EPA (20:5 n-3) and DHA (22:6 n-3) competitively inhibit the formation of the 4-series

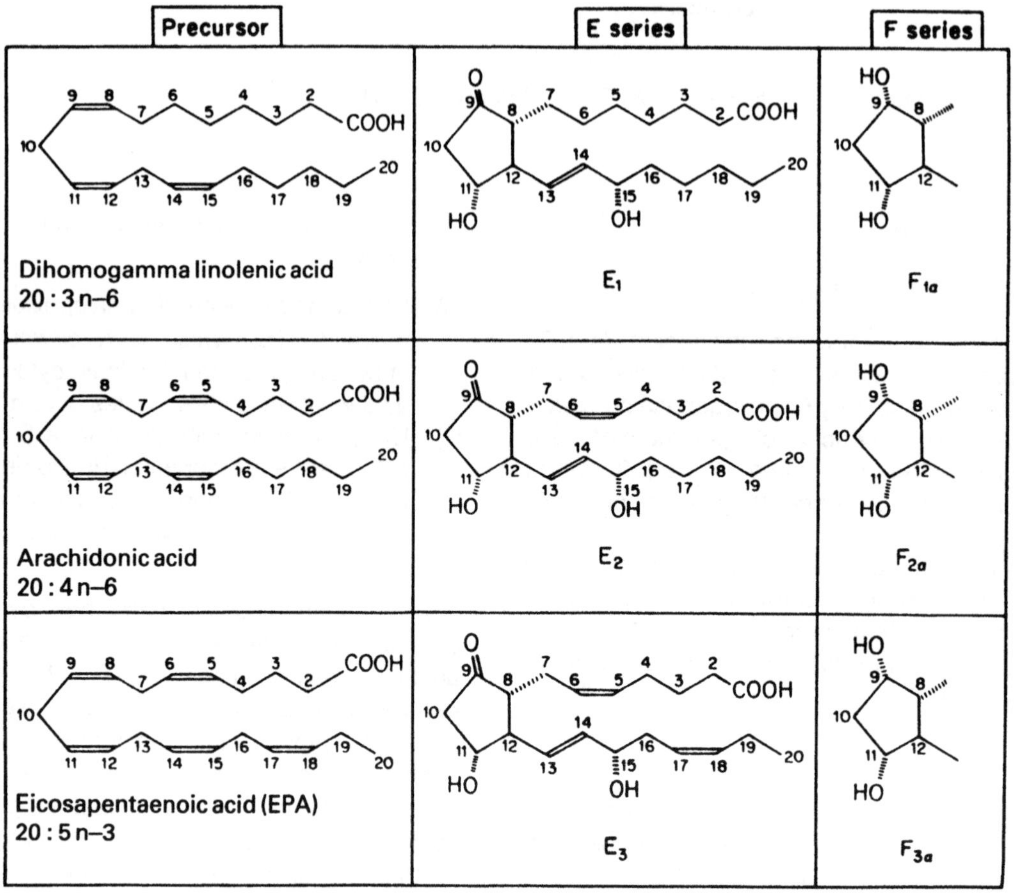

Figure 8.6 Different unsaturated fatty acids precursors produce different prostaglandins. (From Gurr and Harwood, 1991)

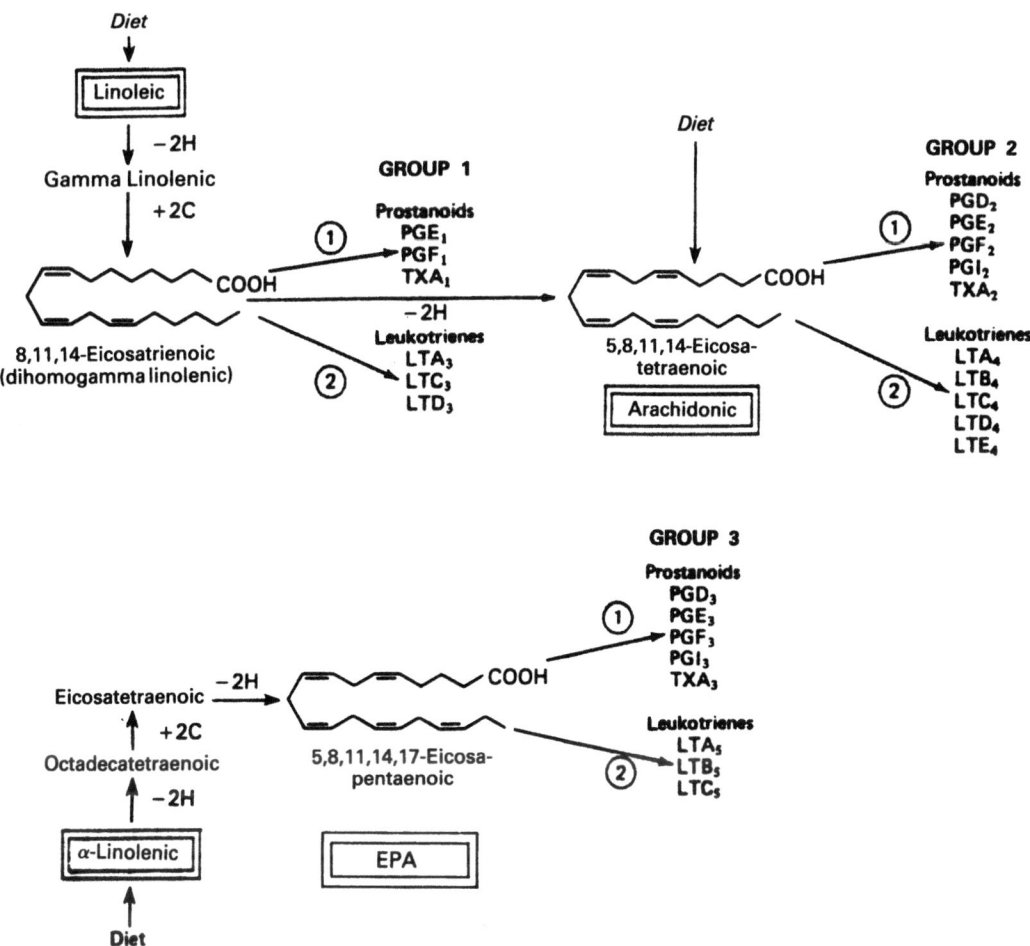

Figure 8.7   The three groups of eicosanoids and their biosynthetic origin. (From Murray *et al.*, 1988)

LT from arachidonic acid (20:4 n-6). Thus, elevation of membrane phospholipid levels of n-3 PUFA by feeding fish oils to animals depresses the formation of 4-series LT and increases the formation of 5-series products. The latter are generally less active pharmacologically than the 4-series.

8.5.5.3   Overall effect and dependence of eicosanoid formation on fatty acid substrate

Figure 8.7 is a summary of the biosynthetic origins of the eicosanoids. It shows how the range of eicosanoids that can be produced depends on which fatty acid provides the substrate for either cyclo-oxygenase (to produce prostaglandins, prostacyclins and thromboxanes) or lipoxygenase (to produce the leukotrienes).

8.5.6   Control of eicosanoid formation

8.5.6.1 Control via phospholipase

Formation of eicosanoids requires the hydrolysis of plasma membrane phospholipids by phospholipase A_2 to generate the free PUFA that are the substrates for the cyclo-oxygenase and the lipoxygenases. The phospholipases in question are activated by a range of natural tissue-specific agonists interacting with specific cell membrane receptors: e.g. the thrombin-induced formation of prostaglandins by platelets (see Chapter 13). The mechanism of activation of the phospholipase A_2 involves specific G protein systems similar to those already described for activation of phospholipase C in the inositol lipid cycle. Many compounds which inhibit phospholipase A_2 block the production of eicosanoids in isolated cells and tissues, although these compounds tend to be non-selective in their actions.

Glucocorticoids appear to inhibit eicosanoid production by controlling phospholipase A_2 activity and this effect is caused by the generation of protein(s) collectively termed lipocortins. Some of these proteins have been cloned and at least one has eicosanoid inhibitory and anti-inflammatory activity.

8.5.6.2   Control via intracellular calcium

Elevation of intracellular calcium also induces

prostaglandin production, probably through activation of phospholipase A_2. However, controversy still exists over the precise substrate for the phospholipases which produce PUFA for eicosanoid production. PtdCho, PtdEtn, PtdIns and phosphatidic acid have all been implicated as substrates in various tissues. Hydrolase activity may also liberate PUFA from diacylglycerols.

8.5.6.3 Interactions with inositol lipid cycle

Complex interactions between the inositol lipid cycle and the eicosanoids can occur under physiological conditions. Eicosanoids are among the natural activators of the cycle. Such complexity is well illustrated by considering the mode of action of the leukotriene LTD_4 in mammalian cell systems (Crook et al., 1989).

LTD_4 interacts with a specific cell receptor which, through a G protein system, activates a phospholipase C to hydrolyse PIP_2 to IP_3 and DAG. DAG, together with the calcium ions released through the action of IP_3, activates a protein kinase C. This in turn activates a DNA isomerase by phosphorylation. The active gene produces mRNA which is translated into a specific protein that activates a phospholipase A. This then acts on plasma membrane phospholipids to release arachidonic acid. Cyclo-oxygenase and 5-lipoxygenase can then act in the usual way to generate more prostaglandins and leukotrienes. Thus, the original LTD_4 signal is successfully amplified.

8.5.6.4 Implications for dietary manipulation

These highly efficient cascade reactions illustrate the great complexity of the processes which can amplify the original signal. The processes are highly cell-specific and typical of the biochemical controls underlying phenomena such as blood clotting and tissue inflammation that will be dealt with in Chapters 13 and 19. Their complexity should not obscure the basic fact that both the production and mode of action of the eicosanoids are fundamental membrane phenomena. They can, in principle, be influenced by perturbating membrane structure more or less specifically. However, because of the very direct link between eicosanoids and the essential fatty acids, eicosanoid metabolism is particularly amenable to dietary manipulation.

9

UNSATURATED FATTY ACIDS AND EARLY DEVELOPMENT

9.1 DEVELOPMENT OF THE FETAL BRAIN

9.1.1 Background

During the last trimester of pregnancy, the human fetal brain experiences a rapid growth spurt, increasing in weight four or five fold during this time (Clandinin et al., 1980; Martinez, 1989) and utilising more than 50% of the energy supply. The most active period of brain cell division is during fetal development. After birth, the emphasis switches to myelination, growth in cell volume and synaptic formation. The most abundant fatty acids in brain cellular membranes are arachidonic acid (20:4 n-6), adrenic acid (22:4 n-6) and docosahexaenoic acid (DHA) (22:6 n-3). Particularly high concentrations occur in the membranes of neuronal synapses and the retina. The photoreceptor contains high concentrations of di 22:6 n-3 phosphoglycerides.

The proportion of DHA (22:6 n-3) in the ethanolamine phosphoglycerides increases steadily in the fetal brain and in the retina during the last trimester of pregnancy. These cell membrane phosphoglycerides contain over 50% of the polyunsaturated fatty acids (PUFA). Evidence from animal studies suggests that retinal function and learning ability are permanently impaired if there is a failure in the accumulation of sufficient DHA during development (Galli et al., 1977; Sinclair and Crawford, 1973; Yamamoto et al., 1987; Neuringer et al., 1988). A similar effect was suggested in a case study of a human infant (Holman et al., 1982).

9.1.2 The delivery of long chain PUFA to the brain

The key issue regarding brain development is that the brain does not use the C18 fatty acids and only uses the C20–C22 fatty acids in any appreciable quantity (Crawford and Sinclair, 1972). Consequently, the mechanism and efficiency of conversion of C18 fatty acids to longer chain fatty acids

becomes a crucial issue (see Chapter 7).

The human fetus obtains long chain PUFA by placental transfer. The concentration of PUFA desaturation and elongation products is progressively increased from the maternal circulation to the placenta then to the fetal liver and finally the fetal brain.

The concentrations of arachidonic acid (20:4 n-6) and DHA (22:5 n-3) are substantially greater in the blood of the umbilical cord than in the maternal circulation. The proportion of DHA rises from 6% of total fatty acids in maternal plasma to 18% in the developing fetal brain. Conversely, the concentration of linoleic acid is reduced in the umbilical cord both at term (Olegard and Svennerholm, 1970) and mid-term (Crawford et al., 1976). This could suggest that the placenta is able to desaturate and elongate linoleic acid (18:2 n-6) to form arachidonic acid. However, the increase in concentration of DHA is unlikely to arise from elongation and desaturation of alpha linolenic acid (18:3 n-3) or EPA (20:5 n-3) because there is relatively little of either present. If placental desaturation occurs in the human, it is a slow process. During rapid growth, the demand is more likely to be met by selective incorporation of particular fatty acids into phosphoglycerides (see Section 9.1.3).

Post-natally, human milk provides a source of long-chain PUFA (see Table 9.1). Some biosynthesis does occur but there is evidence that this is not very efficient in the neonate (Putnam et al., 1982).

9.1.3 Selective incorporation of PUFA into phosphoglycerides

Fatty acids are distributed into various lipid classes, i.e. phosphoglycerides, triacylglycerols and free fatty acids. Studies in rats using universally labelled linoleic acid (18:2 n-6) and alpha linolenic acid (18:3 n-3) have shown that 50–60% of both dietary fatty acids can be recovered from expired CO_2 within 24 hours. Approximately 60% of the remain-

Table 9.1   The fatty acid composition of human milk from various countries (as % fatty acids)

Country	Linoleic acid (18:2 n-6)	Alpha linolenic acid (18:3 n-3)	Arachidonic acid (20:4 n-6)	DHA (22:6 n-3)	References	Group
USA	14.5	–	–	–	Mellies et al. (1979)	
USA	14.7	1.4	0.4	–	Clarke et al. (1982)	
USA	15.8	0.8	0.4	0.1	Putnam et al. (1982)	
USA	15.6	1.0	0.6	0.23	Bitman et al. (1983)	
USA	17.2	1.3	0.1	–	Borschel et al. (1986)	
USA	15.3	0.8	0.4	0.1	Harris et al. (1984)	
USA	14.7	1.5	0.28	0.06	Finley et al. (1985)	omnivores
USA	18.4	1.6	0.3	0.04	Finley et al. (1985)	vegetarians
USA	17.6	0.4	–	–	Thompson et al. (1985)	
USA	14.5	0.7	0.68	0.29	Specker et al. (1987)	omnivores
USA	28.8	2.8	0.67	0.27	Specker et al. (1987)	vegans
UK	7.2	0.8	–	–	DHSS, 1977	
UK	6.9	0.8	0.54	0.59	Sanders et al. (1978)	
UK	31.7	1.5	0.72	0.23	Sanders et al. (1978)	vegans
UK	9.7	0.7	0.19	0.29	Hall, 1979	
UK	23.8	1.36	0.32	0.14	Sanders and Reddy (1992)	vegans
UK	19.5	1.25	0.38	0.30	Sanders and Reddy (1992)	vegetarians
UK	10.9	0.49	0.38	0.37	Sanders and Reddy (1992)	omnivores
Australia	10.8	0.6	0.4	0.32	Gibson et al. (1981)	
Canada	12.0	0.9	0.5	0.36	Clandinin et al. (1987)	
Canada	12.7	0.6	0.7	0.4	Innis et al. (1988)	
Canada	11.5	0.5	0.6	1.4	Innis et al. (1988)	arctic Inuit
Greenland	8.6	–	0.6	3.3	Bang et al. (1985)	
Sweden	12.9	0.7	0.4	0.3	Janson et al. (1981)	
Finland	11.4	–	–	–	Vuori et al. (1982)	
W Germany	10.0	0.8	0.39	0.16	Harzer et al. (1983)	
W Germany	10.8	0.8	0.36	0.22	Koletzko et al. (1985)	

der is found in triacylglycerols and 40% in phosphoglycerides with less than 1% of the labelled acid available as free fatty acid.

In contrast, only 15% of arachidonic acid (20:4 n-6) and DHA (22:6 n-3) is oxidised. For the most part, they are preferentially incorporated into phosphoglycerides and they may not normally be available to be used as substrates for fatty acid oxidation or desaturation.

This process of selective incorporation into phosphoglycerides could explain the increased availability of arachidonic acid and DHA to the fetal brain (Sinclair, 1975; Leyton et al., 1987).

9.1.4   Effect on birthweight and head circumference

Hansen et al. (1969) reported that maternal serum linoleic acid and arachidonic acid correlated with the levels of these PUFA in blood from the umbilical cord. Analysis of the diet during pregnancy of 93 women in London, showed significant associations between dietary intakes of arachidonic

acid and DHA and birthweight, head circumference and placental weight (Doyle et al., 1982; Crawford et al., 1986; Koletzko and Braun, 1991). Intakes of these fatty acids were positively correlated with the arachidonic acid content of phosphoglycerides in maternal and cord blood. The correlations between fatty acid intake and birthweight and head circumference were stronger for arachidonic acid than for DHA (Crawford et al., 1989). This is consistent with the observation that the selective incorporation by the human placenta is greater for arachidonic acid than it is for DHA (Olegard and Svennerholm, 1970; Crawford et al., 1976). Raised triene/tetraene ratios in the endothelium of the umbilical arteries have been observed in low birthweight babies (Ongari et al., 1984; Crawford et al., 1990).

PUFA do not occur in isolation from other nutrients and low dietary intakes of several vitamins and minerals are also strongly correlated with low birthweight and with head circumference. In general, the n-6 PUFA occur in foods with B vitamins, tocopherol and certain minerals, whereas the n-3 PUFA tend to occur in foods with tocopherol, ascorbic acid and beta-carotene.

9.2 ESSENTIAL FATTY ACID (EFA) REQUIREMENTS IN EARLY DEVELOPMENT

9.2.1 Pre-conceptual requirements

There is evidence that nutritional deficiencies (for example folic acid), prior to and during early pregnancy, adversely affect fetal growth and development in many, if not all, mammals (Hurley, 1979; Sinclair and Crawford, 1973; Giroud, 1970; Wynn and Wynn, 1981; Smithells, 1981; MRC Vitamin Study Research Group, 1991). Prior to conception, there appears to be a requirement for fat deposition for reproduction (Frisch, 1977).

Retrospective studies (Wynn and Wynn, 1981; Villar and Rivers, 1988) and prospective studies (Doyle et al., 1990) also imply that maternal nutrition prior to or at the time of conception is significant for the growth of the fetus, particularly the head. Thus the evidence is consistent with embryonic and fetal cell division being most sensitive to poor nutrition during the first few weeks of pregnancy (Giroud, 1970). Furthermore, it has been established for the neural system that retardation or abnormalities of cell division cannot be made good later (Dobbing, 1972).

The application of this principle to the subsequent functional efficiency of other organs has yet to be tested but there is evidence that congenital heart defects (Wynn and Wynn, 1981), the immune system (Chandra, 1976), development of the vascular system (Barker and Osmond, 1987) and the control of blood pressure (Barker et al., 1990) are influenced by dietary factors during very early development.

9.2.2 EFA requirements in pregnancy

During pregnancy, there are nutritional requirements for maternal increase in blood volume (1,200 g), placenta (650 g), uterus (900 g), mammary gland (405 g), liquor aminii (800 ml), fat deposition (4,000–4,500 g), in preparation for the fetal growth spurt and lactation and for the growth of the fetus itself (3,500 g) (Hytten and Leitch, 1971).

The evidence for EFA requirements during pregnancy has been reviewed previously (FAO/WHO, 1978; Galli and Socini, 1983) and the specific questions relating to EFA function have been discussed by Uauy et al. (1989) and Bazan (1989).

The FAO/WHO Consultation (1978) calculated the incremental requirements for the different components of pregnancy. They estimated that 1–1.5% of total energy should be provided as EFA as an additional requirement for pregnancy based on the quantitative measurements of physical growth (Hytten and Leitch, 1971). Clandinin et al. (1981) estimated the accretion of long chain EFA during the last trimester for fetal liver, brain and adipose tissue to be at a rate of 4.13 g of EFA per week: i.e. 0.59 g/day.

Although the efficiency of EFA utilisation may increase during pregnancy, some EFA may also be oxidised for energy. Until there is good evidence to the contrary, it is wise to assume that the quantitative demands for mother and fetus need to be met. A study comparing the EFA intakes of prospective mothers found that mothers who produced low birthweight babies had EFA intakes below the FAO/WHO recommendations whereas the intakes of mothers who produced normal weight babies were at, or above, the FAO/WHO absolute amounts (Crawford et al., 1989).

9.2.3 EFA requirements in lactation

The amount of linoleic acid in breast milk varies over a wide range according to diet (see Tables 9.1 and 9.2). The amount of arachidonic acid, DHA and other long chain fatty acids is relatively constant in breast milk from women eating mixed diets despite wide cultural differences in food selection practices. Inuit women, however, who eat diets rich in long chain n-3 PUFA produce characteristically high levels of DHA in breast milk (see Table 9.2).

Table 9.2 The fatty acid composition of human milk by geographical region or group (as % fatty acids)

	Linoleic acid (18:2 n-6)	Alpha linolenic acid (18:3 n-3)	Arachidonic acid (20:4 n-6)	DHA (22:6 n-3)
North America	14.2	0.9	0.5	0.2
Australia	10.8	0.6	0.4	0.3
Sweden	12.9	0.7	0.4	0.3
UK	8.7	0.8	0.4	0.4
W Germany	10.4	0.8	0.4	0.2
Vegans and vegetarians	24.4	1.7	0.5	0.2
Inuit	10.1	0.5	0.6	2.4

The difficulty in arriving at a single figure to describe EFA requirement in lactation was recognised by the FAO/WHO committee (1978) which calculated an additional EFA requirement of 2–4% of total energy based on the amount needed to replace the EFA lost in breast milk for mothers who are breast feeding.

It may be better, however, to describe requirements not as a proportion of the energy, but as absolute amounts. A mother whose diet contains insufficient energy may appear to have an adequate intake of EFA when expressed as a proportion of her energy intake, but this could be inadequate for fetal membrane growth. The EFA available for this purpose could be further reduced as part may be oxidised to help meet the energy deficit.

Using a figure of 10% of the fatty acids as linoleic acid and alpha linolenic acid (see Table 9.2), the amount of EFA secreted in breast milk is about 5 g/day. The energy stored during pregnancy can provide one third of the energy cost of lactation over the first 3 months assuming there is a proportional reduction of the fat store and that the fat store is fully utilised. This means that two thirds of the EFA secreted has to be provided from the diet. Mothers who breast-feed their babies for longer than 3 months will need to replace the whole amount.

9.2.4 Neonatal EFA requirements

The balance of protein, minerals and lipids in milk is related to the post-natal requirements of the neonate. The milk of fast growing herbivores is characterised by a high protein and mineral content consistent with body and skeletal growth (e.g. the cow, rhinoceros and horse) whereas the milks of those species exhibiting slow body growth (e.g. the great apes) contain the lowest concentrations of protein and minerals. Contrasts in PUFA content of milk which relate to post-natal brain development are also apparent. On the whole, the brains of rats and cats develop post-natally whereas the bulk of brain development occurs pre-natally in the guinea pig (Widdowson, 1970). It is not surprising then that rat and cat milks contain higher proportions of long chain PUFA than are found in the milk of guinea pigs.

Newborn human babies exhibit slow body growth coupled with rapid neural and vascular development. It is appropriate, then, that human milk is relatively low in protein and minerals but rich in the types of fatty acids used in neural and vascular development.

There are as yet no adequate quantitative data on which to set neonatal requirements. The early data published by Hansen et al. (1963) indicated that provision of about 1% of energy as linoleic acid corrected skin lesions in babies who had been fed test milks with no added EFA (Soderhjelm et al., 1970). It is now recognised that EFA are relevant to the wellbeing of tissues other than skin and for these the requirements for linoleic acid, alpha linolenic acid and the long chain derivatives are almost certainly different.

Since human milk provides preformed long chain EFA derivatives (arachidonic acid and DHA), there is a good case to be made that these should also be present in any infant milk formula which provides the main nutrient intake of the infant.

There is evidence from neonatal rat studies that linoleic acid and alpha linolenic acid are not as potent as arachidonic acid or DHA. Sinclair's (1975) isotope incorporation studies suggested an efficiency ratio of linoleic acid to arachidonic acid or alpha linolenic acid to DHA of 1 to 4 for liver and more than 1 to 10 for the brain membranes during growth. Frankel (1980) suggested a smaller ratio of 1 to 3 to correct deficiency symptoms in the skin and a 1 to 5 ratio was suggested from the comparison of linoleic acid and arachidonic acid with the triene/tetraene ratio. Increasing the amount of linoleic acid, however, displaces arachidonic acid in cell membranes (Galli, 1983). It may not be possible, therefore, to compensate for a lack of arachidonic acid and DHA in infant formulae by providing an equivalent amount of linoleic and alpha linolenic acid instead.

9.2.5 The premature and low birthweight infant

The premature infant faces two disadvantages: first, a possible legacy of a nutritional deficit; and second, a lack of placental nourishment. These babies have a higher risk of neuro-developmental disorders, the babies with low birthweights having the greatest risks (Davies and Stewart, 1975; Dunn, 1986; Marlow and Chiswick, 1985).

Measurements in maternal and cord blood showed that low birthweight was associated with a low level of arachidonic acid in cord blood (Crawford et al., 1989). This is consistent with reduced prostacyclin synthesis in the umbilical arteries (Ongari et al., 1984) and the raised triene/tetraene ratio in the endothelial phosphoglycerides of low birthweight infants (Ongari et al., 1984; Crawford et al., 1990). It is therefore likely that a premature baby who has also been growth retarded may well have experienced a period of nutritional deficit during fetal growth which involved a deficit of EFA.

A prospective study looking at the effects of post-natal feeding in pre-term infants on subsequent IQ found that those infants who received breast milk for one month, post-partum had significantly higher IQs at 8 years than those who did not (Lucas *et al.*, 1992). This finding was significant after adjusting for confounding variables such as mother's education and social class. The authors speculated that this difference may be due to the presence of long chain polyunsaturated fatty acids such as DHA which were not present in the alternatives.

The premature infant is exposed to a much reduced intake of arachidonic acid and DHA compared to what it would have obtained via placental transfer had it proceeded to term. Carlson *et al.* (1986) and Uauy *et al.* (1989) showed that the level of arachidonic acid and DHA fell and the levels of linoleic acid rose in plasma phosphoglycerides of premature babies during postnatal development. This suggests that, whatever the desaturation ability of the premature infant, it cannot convert linoleic acid to arachidonic acid at a rate which approaches the supply it would have obtained from the placenta. Friedman *et al.* (1979) similarly described a reduction of both long chain derivatives and of prostaglandin synthesis in the presence of what was thought to be an adequate supply of linoleic acid.

A similar situation has been noted in the full-term infant (Crawford *et al.*, 1976; Putnam *et al.*, 1982) and in infants fed intravenously on 'Intralipid' which contained no long chain EFA. However, in the full-term infant, brain cell myelination is more important than cell division and the consequences of EFA deficiency might not be so serious.

Although survival rates of low birthweight infants have improved, there has been an increase in severe neuro-developmental handicap amongst low birthweight infants (Hagberg *et al.*, 1989; Pharoah *et al.*, 1990). Retinopathy of prematurity, which was once considered to be due to excess oxygen and inadequate vitamin E, is still a serious problem. The possibility that DHA deficits in retinal membranes may be relevant to inadequate retinal development is currently a matter of research (Uauy *et al.*, 1990) along with a more general concern about nutrition and neural development (Lucas *et al.*, 1990).

Although questions regarding neural development in the premature and low birthweight infant are of obvious interest in view of the role of arachidonic acid and DHA in neural development, it is also true that arachidonic acid and DHA are required for vascular development and the regulation of blood flow. Barker and Osmond (1986, 1987, 1989) have presented epidemiological evidence suggesting that poor maternal and childhood nutrition are related to subsequent increased adult risk of cardiovascular disease and hypertension. Indeed, the development of any tissues where membrane growth and organisation are specialist features (e.g. the lungs, endocrine and immune systems) could be jeopardised by an inadequate supply of the EFA during gestation and early infancy.

10

UNSATURATED FATTY ACIDS AND CORONARY HEART DISEASE

10.1 INTRODUCTION

Over the last 30 years, research has contributed enormously to our knowledge of risk factors for coronary heart disease (CHD). However, there are still considerable gaps in our understanding of its aetiology and this disease remains the major cause of death in middle and old age within the UK. However, in England and Wales, between 1980 and 1988, CHD mortality rates have decreased by 17–35% among middle-aged men (Table 10.1). The 55–64 year age group shows the smallest percentage decrease (17%), but of course this reflects the greatest decrease in absolute number of deaths. Data on the relationships between the consumption of unsaturated fatty acids and CHD are available from several sources: cross-cultural comparisons, secular trends, case-control comparisons, cohort studies and intervention trials. Each of these sources of evidence will be discussed in the following sections.

10.2 MONOUNSATURATED FATTY ACIDS

Interest in monounsaturated fatty acids (MUFA) arose from two main sources:

(i) the recognition that people in Mediterranean countries have a high intake of MUFA and an associated low prevalence of CHD;
(ii) the reported cholesterol-lowering effect of MUFA.

Nevertheless, few studies have been made to investigate the relationship between MUFA and CHD. Intakes of these fatty acids have been determined in several cohort (prospective) studies. These are longitudinal surveys of a defined population to investigate associations between various factors and incidence (new occurrence) of a disease. However, data on intakes of MUFA provide no consistent evidence of an association with risk of CHD (Figure 10.1). Mean MUFA intakes

Table 10.1 Changes in male death rates from coronary heart disease in England and Wales (per million population)

	35–44 years	45–54 years	55–64 years
1980	543	2625	7123
1981	508	2495	6807
1982	459	2366	6754
1983	443	2330	6824
1984	402	2176	6777
1985	414	2150	6700
1986	406	2093	6397
1987	398	1938	6192
1988	352	1816	5899
Reduction 1980–1988	35%	31%	17%

(Source: Office of Population, Censuses and Surveys, 1982–1990)

(expressed as a percentage of energy intake) were slightly higher in CHD cases than in non-cases in four studies, lower in those who developed CHD in one study and identical in both groups in one study.

Within the UK, data on trends in MUFA intakes are available from the National Food Survey (Ministry of Agriculture, Fisheries and Food) (see Chapter 3). Over the last decade, these data indicate little change in the percentage of food energy derived from MUFA (e.g. 16.0% in 1980, 15.3% in 1984 and 15.2% in 1988).

Thus there is very little evidence to support the notion that intakes of MUFA have been an overriding factor in the decreased mortality from CHD during the 1980s.

10.3 n-6 POLYUNSATURATED FATTY ACIDS

10.3.1 Cross-cultural comparisons

A cross-sectional survey of random population samples of men aged 40–49 years from four

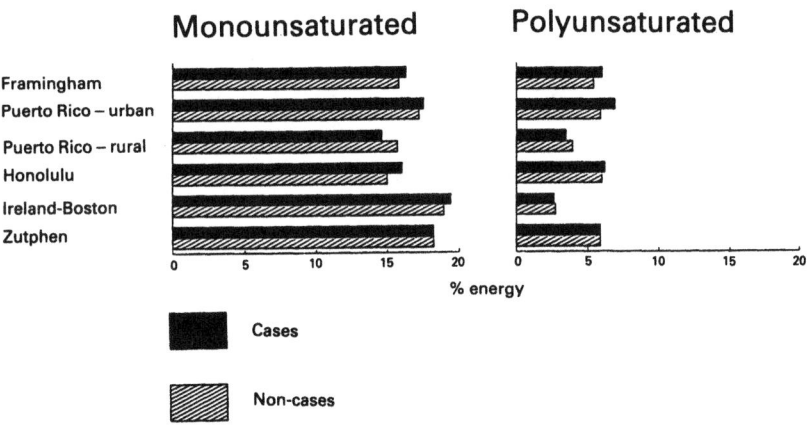

Figure 10.1 Energy intake from unsaturated fatty acids in relation to likelihood of CHD. Sources of data: Framingham, Gordon *et al.* (1981); Puerto Rico, Garcia-Palmieri *et al.* (1980); Honolulu, McGee *et al.* (1984); Ireland–Boston, Kushi *et al.* (1985); Zutphen, Kromhout and Coulander (1984).

European regions showed that the proportions of linoleic acid (18:2 n-6), dihomogamma linolenic acid (DGLA) (20:3 n-6) and arachidonic acid (20:4 n-6) in adipose tissue was lowest in North Karelia in Finland which had the highest CHD mortality rate, highest in Italy which had the lowest CHD mortality rate, and intermediate in Scotland and south west Finland (Riemersma *et al.*, 1986).

10.3.2 Time trends

National Food Survey data for 1980–88 show a steady increase in the percentage of dietary energy derived from polyunsaturated fatty acids (PUFA) (mainly linoleic acid) from 4.6% to 6.4% and an increase in the PUFA:saturated fatty acid ratio (P/S) from 0.24 to 0.40 (see Chapter 3). This is, however, a relatively small dietary change and is unlikely to account for the reduction in CHD mortality rates within the UK over the same period.

10.3.3 Case-control studies

In some case-control studies, lower levels of linoleic acid in the diet, plasma or adipose tissue were found in men with evidence of CHD than in those without evidence of CHD (Simpson *et al.*, 1982; Wood *et al.*, 1984), while others have found no difference (Thomas *et al.*, 1987; Fehily *et al.*, 1987). The main disadvantage of this type of study is that measurements are usually made after the onset of the disease, so that it cannot be established whether any differences in intake are a cause or an effect of the disease. However, in the Edinburgh (Wood *et al.*, 1984) and Caerphilly (Thomas *et al.*, 1987; Fehily *et al.*, 1987) studies, an attempt was made to overcome this problem by

selecting subjects who were unlikely to have altered their diet – those with ECG evidence of CHD but no symptoms, or new cases of CHD.

10.3.4 Cohort (prospective) studies

10.3.4.1 General population studies

The Seven Countries Study (Keys, 1980) was a longitudinal, prospective study of 16 cohorts of middle-aged men. A strong positive association was found between the mean percentage of dietary energy obtained from saturated fatty acids for each cohort and CHD mortality over the next 10 years ($r = 0.84$). However, there was no association between intakes of PUFA and CHD mortality.

Data from six within-population cohort studies, relating to the percentage of dietary energy obtained from PUFA (Figure 10.1), indicate that differences in intakes between those who developed CHD and those who did not were very small and inconsistent. Intakes tended to be higher in people with evidence of CHD in three cohorts, lower in those people in two cohorts and identical in both groups in one study. Mean intakes among those who developed CHD and those who did not were not reported for the Western Electric Study (Shekelle *et al.*, 1981), but risk of death from CHD was found to be inversely related to the percentage of energy from dietary PUFA. There were 13.5%, 10.4% and 10.1% coronary deaths in the low, middle and top third respectively when the participants were ranked according to their PUFA intake. A Finnish prospective survey also showed an inverse relationship between baseline n-6 PUFA (linoleic acid in serum phospholipids) and 5-year CHD incidence (Miettinen *et al.*, 1982).

The ratio of total PUFA/total saturated fatty acids (P/S ratio) in the diet was lower in people who

Table 10.2 Cohort studies: mean dietary P/S ratio and CHD incidence

Study	Reference	CHD cases	Non-cases
Framingham	Gordon et al. (1981)	0.43	0.39
Puerto Rico–Urban	Garcia-Palmieri et al. (1980)	0.58	0.50
Puerto Rico–Rural		0.26	0.33
Honolulu	McGee et al. (1984)	0.47	0.48
UK	Morris et al. (1977)	0.15	0.16
Ireland–Boston	Kushi et al. (1985)	0.15	0.16
Zutphen	Kromhout et al. (1985)	0.34	0.34

developed CHD compared with those who did not in four cohorts (Table 10.2). The differences in the ratio ranged from 2% in the Honolulu study to 21% in the Puerto Rico rural cohort. In the Framingham Study and the Puerto Rico urban cohort, people who developed CHD had mean P/S ratios 10% and 16% higher respectively than those who did not. Mean P/S ratio was identical in both groups in the Zutphen study. It is noteworthy that the differences between the studies in mean dietary P/S ratio are considerably greater than the differences within studies. Table 10.2 shows a three- to four-fold range in dietary P/S ratio between the studies. This is partly due to variation in intakes between these countries, but also partly due to variation in the timing of the dietary surveys.

10.3.4.2 Vegetarians

That vegetarians have a lower risk of CHD than non-vegetarians is well known. However, the reason for their lower CHD mortality is not clear. Vegetarians differ from non-vegetarians in many ways and not just by the avoidance of meat – they tend to be leaner, smoke less and they consume fewer alcoholic beverages. Seventh-Day Adventists are an ideal population group among which to investigate the effects of vegetarianism on health: alcohol and tobacco are prohibited and therefore will not confound the relationship between vegetarianism and disease risk; and most meats, fish and eggs are discouraged but not prohibited, resulting in a wide range of consumption of these items. A 20 year follow-up of 27,529 California Seventh-Day Adventists showed a positive association between meat consumption and CHD mortality (Snowdon, 1988).

In another attempt to correct for lifestyle differences, subjects were recruited via UK health food stores and health food magazines so as to obtain a group of people with a general interest in health, but containing both vegetarians and non-vegetarians. This 12 year follow-up study of 10,896 people also showed that CHD mortality was lower among vegetarians than among non-vegetarians (Burr and Butland, 1988). In both the above

studies, vegetarians were leaner than non-vegetarians and had lower plasma cholesterol concentrations, probably due to a lower intake of saturated fatty acids, although the diets of vegetarians may differ from those of non-vegetarians in other important respects: e.g. higher intakes of nuts/seeds, pulses and vitamin C (Burr et al., 1981).

10.3.4.3 General considerations

Cohort studies overcome the main problem of case-control studies since measurements are made prior to the event, i.e. the occurrence of CHD. However, there are several disadvantages with this type of study. Data must be collected from large numbers of people, all of whom must then be followed up for a lengthy period to obtain a sufficient number of CHD events to ensure a high probability of detecting differences in the initial measurements between those who subsequently develop the disease and those who do not. Among the studies mentioned in this section, there is considerable variation in the number of subjects studied and in the duration of follow-up, the number of CHD cases on which results are based varying from 30 (Zutphen) to 309 (Honolulu).

The ability of a study to determine the different risks associated with a high or low consumption of a particular dietary component is influenced not only by the number of CHD events but is also dependent upon the accuracy of the measurement of intake of that dietary component and on its distribution within the survey population. If the dietary data are of poor quality and/or the range of intake narrow, this will tend to reduce the probability of demonstrating an association between intake and disease incidence, if such a relationship exists. In almost all the CHD cohort studies, for example, nutrient intakes were estimated from either 24 hour recall or dietary history/questionnaire. Although these methods may provide adequate estimates of group mean intakes, they may not yield sufficiently good estimates of individual intakes to establish reliable quantitative associations. The range of intakes between populations will probably be much greater than those within a population.

There will always be many differences between populations other than diet and it is difficult to ascertain which of the many differences are causal factors. Even in within-population studies, dietary variables tend to be related to each other as well as to factors such as smoking habit and physical activity. Hence people who choose to consume a particular type of diet will probably be different in many ways from those who do not consume this type of diet. Interpreting data from observational studies such as case-control and cohort studies is therefore very difficult.

10.3.5 Intervention studies

Direct evidence of a relationship between diet and CHD can only be obtained from intervention studies. In these studies, one group of subjects is given dietary advice and then both those given advice (intervention group) and those not given advice (control group) are followed up to ascertain whether the intervention influences subsequent CHD risk and survival. It is important that subjects are randomised to the intervention and control groups as this is the best way of ensuring that the two groups are similar at baseline. Intervention trials are, of course, costly and require that a large number of people be involved for many years.

A number of primary prevention and secondary prevention trials have been conducted to investigate whether intervention reduces subsequent risk of CHD. Trials are described as 'primary prevention' if the subjects have no evidence of CHD at recruitment, and as 'secondary prevention' if the subjects have evidence of CHD (e.g. angina) on entry to the trial.

10.3.5.1 Secondary prevention trials

Five randomised controlled trials, in which the effect of a reduction in the intake of saturated fatty acids and an increased intake of PUFA were investigated in the secondary prevention of CHD, are summarised in Table 10.3. Dietary P/S ratios in the intervention group ranged from 1.0 to 2.6. Most of these studies, however, had a very low probability of detecting a beneficial effect of the intervention as the sample sizes were inadequate. In order to detect a 25% reduction in CHD mortality at least 1600 subjects would be needed. It is hardly surprising therefore that these trials did not show any benefit on CHD incidence or survival. However, a trial with over 2000 subjects (Burr et al., 1989) also showed no effect on CHD incidence or on total mortality over a two year period. Nevertheless, it could be argued that two years is insufficient for a benefit to occur, or that the achieved difference in dietary P/S ratio between those given advice and those not given advice was inadequate.

10.3.5.2 Primary prevention trials

Primary prevention trials are considerably more costly than secondary prevention trials since they require many thousands of subjects in order to have a high probability of detecting a benefit of the intervention. This is because the mortality rate among healthy men is much lower than among those who already have CHD. The primary prevention trials which have been conducted have generally been multifactorial, the intervention 'package' including advice about other CHD risk factors in addition to diet. It is therefore very difficult to ascertain to what extent any benefit of the intervention package is due to the dietary component. The results of randomised primary prevention trials are summarised in Table 10.4.

Table 10.3 Summary of secondary prevention randomised controlled dietary trials

Trial	Number of men	Duration (years)	Results
n-6 PUFA			
Rose et al. (1965)	80	2	No difference in morbidity or mortality
Leren (1970)	412	11	Reduction in CHD mortality No effect on total mortality
MRC Research Committee (1968)	393	2–6	No effect on CHD mortality or total mortality
Woodhill et al. (1978)	458	2–7	No effect on survival
Burr et al. (1989)	2033	2	No effect on CHD mortality or total mortality
n-3 PUFA			
Burr et al. (1989)	2033	2	Reduction in CHD mortality Reduction in total mortality

Table 10.4 Summary of primary prevention randomised controlled dietary trials

Trial	Reference	Number of men	Duration (years)	Results
Los Angeles Veterans Administration	Dayton et al. (1969)	846	<8.3	No effect on CHD mortality No effect on total mortality
Oslo	Hjerman et al. (1986)	1,232	8	Reduction in incidence of MI and sudden death Reduction in total mortality (ns)
MR FIT	Multiple Risk Factor Intervention Trial Research Group (1982)	12,866	7	No effect on CHD mortality No effect on total mortality
Gothenburg	Wilhelmsen et al. (1986)	30,000	10	No effect on CHD mortality No effect on total mortality
WHO	WHO European Collaboration Group (1986)	60,881	6	Reduction in CHD events Small effect on total mortality (ns)

ns = not statistically significant

The Los Angeles Veterans Administration Study (Dayton et al., 1969) was a mixed primary and secondary prevention study involving 846 men aged 54–88 years (mean 65 years) living in an institution. Men were randomised to the intervention or control group and the two groups ate in different canteens. The intervention group was given a diet low in saturated fatty acids and rich in PUFA and the control group received a typical American diet. There were fewer CHD events in the intervention group, but the difference was not statistically significant. There was also no significant difference in total mortality.

In the Oslo study (Hjermann et al., 1986), 1232 men aged 40–49 years at high risk of CHD were randomised to dietary and smoking habit intervention or to a control group and then followed up for eight years. Initial serum cholesterol levels were high (7.5–9.8 mmol/l) and were reduced by 13% in the intervention group compared with 3% in the control group (net difference of 10%). Mean fasting serum triacylglycerols were 20% lower in the intervention group compared with the control group. Eighty percent of men in both groups smoked tobacco daily at the start of the study. Only 24% of smokers stopped smoking in the intervention group compared with 17% of the control group. Eight years after the start of the trial, the total number of coronary events (fatal plus non-fatal) was significantly (47%) lower in the intervention group than in the control group. Total mortality was 39% lower in the intervention group but this was not statistically significant, presumably because of the small sample size.

In the Multiple Risk Factor Intervention Trial (MR FIT) (Multiple Risk Factor Intervention Trial Research Group, 1982), 12,866 high-risk men aged 35–57 years were randomly assigned either to a special intervention programme consisting of treatment for hypertension, counselling for cigarette smoking and dietary advice for lowering blood cholesterol levels, or assigned to their usual sources of health care in the community. Over an average follow-up period of seven years, risk factor levels declined in both groups, but to a greater degree in the intervention group. The difference in mean serum cholesterol level between the intervention and control group was only 2%, however. There was no difference in CHD mortality or total mortality.

The Gothenberg Study (Wilhelmsen et al., 1986) was a randomised controlled trial among 30,000 men aged 47–55 years. Men were randomly allocated to the intervention group or one of two control groups. Intervention was multifactorial and dietary advice aimed primarily at those with high serum cholesterol levels. There appeared to be little change in cholesterol level in any of the groups; mean serum cholesterol values after four years of intervention were only 1% lower than those for the controls. After 10 years of follow-up, there was no significant effect on CHD mortality or on total mortality.

The WHO European Collaborative Study was a randomised controlled trial of multifactorial prevention of CHD among 60,881 men aged 40–59 years, employed in 80 factories in Belgium, Italy, Poland and the UK (WHO European Collaborative Group, 1986). Factories were assigned on a random basis to the intervention or control group rather than the individual men. The men in the intervention group received advice on a cholesterol-lowering diet, control of smoking, overweight and blood pressure, and regular exercise.

The control group received no advice. Overall, the trial showed that the intervention was associated with reductions of 10% in CHD events (fatal plus non-fatal) and 5% in total mortality (not statistically significant). There were differences between the centres in the effects of the intervention on both risk factors and mortality. The greatest effects were reported from the Belgian centre and this may have been due to a more aggressive intervention policy.

For all intervention trials, interpretation of the results is complicated because not every subject in the intervention group complies with the advice given and some in the control group may alter their diet of their own volition, in the direction of the intervention. The probability of detecting a beneficial effect of the intervention, if it occurs, is thus largely dependent on the actual difference in intake between the intervention and control groups. In most of the randomised controlled trials described, the effect of the dietary advice on intakes was not estimated. If the effect on plasma cholesterol concentration is regarded as an objective measure of compliance, then there are undoubtedly considerable differences between the studies. Intervention studies have usually been conducted with middle-aged subjects to obtain a sufficient number of outcome events. However, this poses an additional difficulty in interpreting the results of these trials: if the mechanism for an effect of partial replacement of saturated fatty acids by n-6 PUFA is via atherosclerosis, then the effect may be less easy to demonstrate in middle-aged subjects than it would be in younger subjects.

Overall, the results of randomised controlled trials suggest that advice to partly replace saturated fatty acids by n-6 PUFA may reduce the risk of a CHD event but has not been shown to reduce risk of death from all causes. How one interprets these findings, in terms of the likely benefits of dietary changes for the population, depends on whether one views a reduction in CHD incidence (with a probable improvement in quality of life) or a lack of effect on risk of overall death as the most important outcome.

10.4 n-3 POLYUNSATURATED FATTY ACIDS

10.4.1 Cross-cultural comparisons

The observation that CHD is rare in Greenland Inuit is well-known. Their low risk has been attributed to a high intake of n-3 PUFA, particularly eicosapentaenoic acid (EPA, 20:5 n-3) and docosahexaenoic acid (DHA, 22:6 n-3) obtained from the consumption of marine animals and fish. The low CHD mortality rate in Japan is also well known and it has been suggested that it is related to their high intake of fish. A recent review of the evidence for an

association between CHD mortality and fish consumption (Burr, 1989) indicated some inconsistencies and concluded that differences in CHD mortality between countries cannot be explained primarily in terms of fish intake.

10.4.2 Time trends

Since the beginning of this century, there has been a considerable fall in fish consumption by people in the UK. However, there have also been considerable changes in other CHD risk factors such as smoking habit and physical activity, so that changes in CHD mortality cannot be attributed solely to changes in fish intake. Moreover, there has been a reduction in CHD mortality in recent years (Table 10.1) but very little change in fish consumption during the same years (Ministry of Agriculture, Fisheries and Food, 1980–88).

10.4.3 Cohort studies

The relationship between fish, or long chain n-3 PUFA consumption, and the incidence of CHD has been investigated in several cohort studies. Of these, five demonstrated an inverse association (Miettinen et al., 1982; Kromhout et al., 1985; Shekelle et al., 1985; Norell et al., 1986; Dolecek et al., 1989) and three reported no association (Curb and Reed, 1985; Vollset et al., 1985; Lapidus et al., 1986). A major difference between these studies appears to be in the amounts of fish consumed by the subjects (Figure 10.2). For example, in the Zutphen and Western Electric studies in which a protective effect of fish was found, habitual fish consumption was low with few subjects eating fish twice a week or more. By contrast, in the Honolulu and Norway studies in which no protective effect of fish was found, habitual fish intake was high and few subjects rarely or never ate fish. It is therefore possible that consumption of a moderate amount of fish is protective against CHD and that continued consumption of large amounts may not provide additional protection.

In the two Swedish studies, the amounts of fish consumed were not stated. Norell et al. (1986) reported a protective effect of fish, the risk of CHD being 15% lower in those classified as having a high fish consumption than in those eating little or no fish. Lapidus et al. (1986) found no association between fish intake and CHD, but the number of CHD events was small and fish intake was calculated from a single 24 hour recall for each subject.

In the Finnish prospective study (Miettinen et al., 1982), baseline levels of EPA in the serum

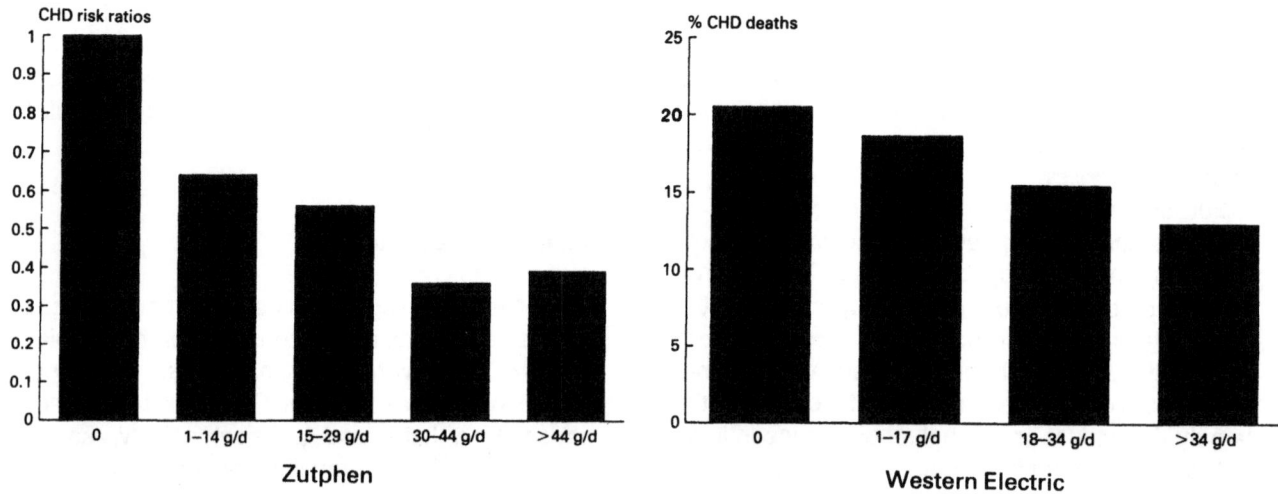

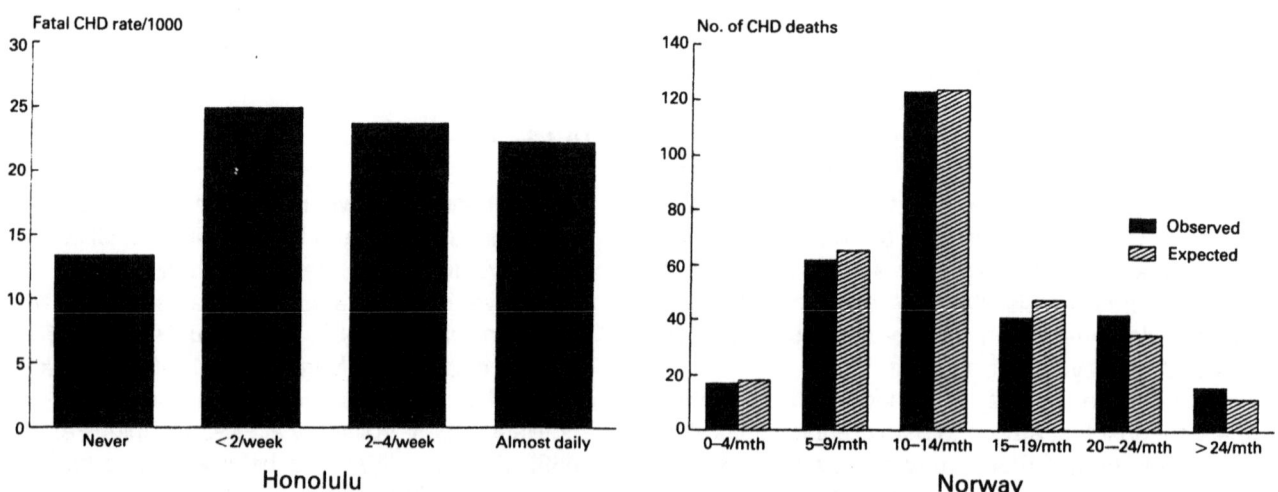

Figure 10.2 Cohort studies: fish consumption and CHD. Fish consumption expressed as g/day in Zutphen and Western Electric studies and in fish meals/week or month in Honolulu and Norway studies. Sources of data: Zutphen, Kromhout *et al.* (1985); Western Electric, Shekelle *et al.* (1985); Honolulu, Curb and Reed (1985); Norway, Vollset *et al.* (1985).

phospholipids were lower in men who subsequently developed CHD than in those who did not. A preliminary communication from the MR FIT Research Group (Dolecek *et al.*, 1989) reported an inverse association between the intake of n-3 PUFA and death from CHD and death from all causes over a period of 10 years.

None of these studies distinguished clearly between lean and fatty fish so that the effect of the long chain n-3 PUFA might be deduced, but the results suggest that a fairly small intake of fish is protective against CHD death.

10.4.4 Intervention studies

To date, only one randomised controlled trial of the effect of fish consumption on mortality has been conducted (Burr *et al.*, 1989). In this study, 2033 men who had suffered a recent myocardial infarc-

tion (MI) were randomly allocated to a 'fish advice' or 'no fish advice' group. Those allocated to the fish group were advised to consume at least two portions per week of fatty fish. Men who could not eat this amount were supplied with fish oil capsules as a partial or total substitute so that their intake of long chain n-3 fatty acids was increased. At the end of two years, there were 29% fewer deaths among those advised to eat fish than those not given this advice (Table 10.3). This difference was statistically significant and was attributable to a reduction in CHD deaths. Surprisingly there were more non-fatal heart attacks in the 'fish advice' group than in the 'no fish advice' group, although the number of non-fatal events was small and the differences not statistically significant.

These results suggest that moderate amounts of fish, or fish oil, reduce the risk of death after an acute myocardial infarction. The mean intake of EPA in the 'fish advice' group was 2.3 g/week, corresponding to about 290 g fatty fish. Addition of fatty fish to the diet led to a small reduction in intake of other foods of animal origin (meat, meat products, cheese and eggs), but this had a negligible effect on the intake of saturated fatty acids. Thus the effect on mortality is unlikely to be due to dietary changes other than the increase in fish consumption. The mechanism for the effect may be fairly quick-acting, since the effect on mortality appeared within the first 6 months of the trial.

In conclusion, there is now considerable evidence that fish consumption is protective against CHD mortality. Nevertheless, only one randomised controlled trial has so far been conducted and further trials are needed to investigate the effect of dietary fish in the primary prevention of CHD.

11

UNSATURATED FATTY ACIDS AND PLASMA LIPIDS

11.1 POPULATION STUDIES

Before reviewing the experimental evidence relating dietary lipids to plasma lipid concentrations, a relevant question is: how are levels of plasma lipids influenced by normal habitual diets? It is also important to know to what extent diet is important within the context of other lifestyle factors and how genetic background can affect plasma lipids (see Section 11.4).

11.1.1 Between-population studies relating dietary intake of unsaturated fatty acids to plasma lipids

Between-population studies, both prospective and retrospective, have generally demonstrated an association between the percentage of energy derived from saturated fatty acids and total plasma cholesterol concentrations. Indeed they have frequently been used to support recommendations for the prevention of coronary heart disease (WHO, 1990; DHSS, 1984; Department of Health, 1991). The Seven Countries Study (Keys, 1980) has been the most influential. It was a prospective study of diet and risk factors for cardiovascular disease carried out in localities in Holland, Japan, Greece, USA, Yugoslavia, Finland and Italy. Two-thirds of the variation in serum cholesterol concentration between populations could be predicted from differences in the percentage of energy derived from saturated fatty acids and the intake of cholesterol using equations derived from experiments on healthy volunteers. Polyunsaturated fatty acid (PUFA) intakes were similar in all countries and did not predict plasma cholesterol concentrations. Neither was total fat intake associated with the serum cholesterol concentration.

11.1.2 Within-population studies

It has been harder to demonstrate differences in serum cholesterol concentration within populations and to relate these to fatty acid intakes (Shekelle et al., 1981; Fehily et al., 1988; Berns et al., 1990; Gregory et al., 1990).

In the USA, the correlation coefficient between the P/S ratio and plasma lipid concentration was small (Shekelle et al., 1981). The Tecumseh study could not demonstrate a significant relationship between fat, cholesterol or other macronutrients in the diet and plasma lipids (Nichols et al., 1976).

The 'Caerphilly Study' (Fehily et al., 1988) provided evidence for a positive association between saturated fatty acid intake and low density lipoprotein (LDL) cholesterol in 653 men aged 45–59 years. This evidence is more reliable than most, being based on a weighed inventory method of diet assessment, with a short interval between the diet survey and the cholesterol measurement. Nevertheless, diet as a whole accounted for only 1–7% of the variance in plasma lipoprotein concentrations and only 1.9% of the variance in plasma total cholesterol.

It is probably more difficult to demonstrate an effect of fatty acids on plasma lipids within populations because there is often little variance in fatty acid intakes. Thus most of the variance in plasma cholesterol concentration with the population reflects genetic variance. However, if groups with widely varying fatty acid intakes are compared then a significant relationship between serum cholesterol and saturated fatty acid intake as a percentage of the food energy can be demonstrated. For example, comparisons between strict vegetarians and meat-eaters show marked differences in plasma cholesterol concentrations that can be explained almost entirely by differences in saturated fatty acid intakes (Sanders and Roshanai, 1984). Similar comparisons have been made between French farmers from different parts of France who have different fat intakes (Renaud, 1986).

11.1.3 Population studies relating unsaturated fatty acids in adipose tissue to plasma lipids

A different experimental approach which provides evidence that there is only a weak relationship between the type of dietary fat habitually consumed and plasma lipids comes from the study of Berry et al. (1986). Adipose tissue fatty acid composition was measured as an indicator of habitual diet and, by implication, the type of dietary fat eaten. This method provides a much more reliable measure of the intake of polyunsaturated fatty acids over a long period of time than short- or long-term recall studies or even seven-day weighed inventories (see Chapter 8). In this study, adipose tissue composition explained only a small proportion (1–19%) of the variance in plasma lipids in subjects with plasma lipid concentrations in the normal range. The size of the population studied (413) and the very wide variations in the P/S ratio found in the adipose tissue indicated that there were wide differences in the composition of fat consumed (although this method is unable to provide information on the amount of fat consumed).

11.1.4 Intervention studies

There is ample evidence to confirm that plasma lipoprotein concentrations can be raised or lowered by dietary fat modification under strictly controlled experimental conditions such as those found in a metabolic ward of a research unit (see Section 11.2). What is the evidence that such changes can be achieved in everyday life?

A study in London set out to compare subjects on a typical UK diet with a practical fat-modified diet achieved by trimming the fat from meat. Watts et al. (1988) demonstrated an 8.6% reduction in plasma cholesterol concentration when subjects were given a diet containing 35% energy as total fat, 14% as saturated fatty acids and a P/S ratio of 0.5 (resembling the recommendations of the 1984 COMA Panel (DHSS, 1984)) compared with a diet contributing 42% energy as fat, 21% as saturated fatty acids and a P/S ratio of 0.27 (similar to the 'average UK diet' at that time). A reduction of 18.5% of total cholesterol was obtained with a more extreme diet contributing 27% energy as total fat with 8% as saturated fatty acids and a P/S ratio of 1.0. Regarding the relevance of this study to practical measures for the general population, the limitations were that there were only 15 subjects whose initial mean plasma cholesterol concentration was fairly high (8.1 mmol/l) and that they were studied for only four weeks under fairly strict supervision.

A study conducted in Helsinki (Frick et al., 1987) examined the changes in two groups, one treated with a drug to lower cholesterol, the other a placebo. Both groups were given intensive dietary advice again similar to that recommended by the 1984 COMA Panel. The study lasted five years and at the start of the trial, the average plasma cholesterol was 7.5 mmol/l compared with 6.3 mmol/l for the population as a whole. Plasma cholesterol concentrations in the group treated with the drug and diet decreased by 8% but the total plasma cholesterol of the group given dietary advice alone actually rose by 3%.

The Oslo study is much quoted as an example of the ease with which plasma cholesterol can be lowered by giving relatively simple dietary advice to 'free-living' subjects (Hjermann et al., 1981). It cannot be concluded, however, that the results of this study are entirely relevant to practical measures for the general population for the following reasons. The subjects (1232 healthy 40–49 year old men) were selected on the basis that their plasma cholesterol concentrations fell in the range of 7.5–9.8 mmol/l which is at the top of the distribution curve. Men who had high plasma lipids but who were already on a diet which would have been expected to reduce plasma cholesterol were eliminated from the study. Therefore, the study consisted of a group of men selected as being more likely than others to respond to diet. It is unlikely that a normal unselected population could achieve this degree of lipid lowering under normal living conditions.

Ramsey et al. (1991) argued that many long-term studies in the general populations fail to lead to a significant reduction in plasma cholesterol concentration. The failure of some studies to show that dietary advice can achieve lasting reductions in plasma cholesterol concentration could be the failure of the subjects to comply with the diet: a problem well known in the management of obesity. Some well controlled studies which have achieved good dietary compliance have shown that long-term dietary fat modification does lead to significant reduction in plasma cholesterol (Kromhout et al., 1987; Wood et al., 1991; Renaud et al., 1986). However, no more than a 20% reduction (typically less, in the order of 10–15%) in plasma cholesterol can be achieved by dietary fat modification alone (Grundy, 1991).

11.2 EXPERIMENTAL STUDIES

11.2.1 The effect of saturated fatty acids on plasma lipids

Early studies suggested that fatty acids with a chain length of less than 12 carbon atoms failed to raise plasma total cholesterol levels (Keys, 1970). It has also been known for some time that saturated fatty acids with a chain length of 12–16 carbon atoms increase plasma total and LDL cholesterol concentrations.

There is, however, now some doubt about palmitic acid (16:0) which Hayes et al. (1991) have shown to have minimal effect on LDL cholesterol compared with lauric (12:0) and myristic (14:0) acids in experiments with monkeys. In contrast, some studies in middle-aged men (Mattson and Grundy, 1986) have shown that palm oil which is rich in palmitic acid (16:0) and oleic acid (18:1) is hypercholesterolaemic. Not all human studies have shown this effect (see Cottrell, 1991 for review).

11.2.2 Comparison of modified fat diets with low fat/high carbohydrate diets

Modified fat diets are defined here as those which contain less than 10% energy from saturated fatty acids and between 30 and 35% energy from total fat. The energy from saturated fatty acids is replaced with either linoleic acid or oleic acid. Modified fat diets lower LDL cholesterol compared to diets high in saturated fatty acids.

Low fat diets (and therefore high carbohydrate diets), with only 25% energy from fat, also lower LDL cholesterol (Mensink et al., 1989a; Grundy, 1989; Ginsberg et al., 1990) but they have a number of disadvantages compared with modified fat diets:

(i)   they are less palatable;
(ii) they lower the concentration of HDL cholesterol (a raised level of HDL cholesterol is associated with decreased CHD);
(iii) they may impair glycaemic control in certain individuals.

Low fat diets reduce HDL cholesterol concentrations in both non-diabetic people and NIDDM patients (Garg et al., 1988; Abbot et al., 1990).

Modified fat diets are, therefore, often preferred to low fat diets as cholesterol-lowering diets.

11.2.3 The effect of n-6 polyunsaturated fatty acids on plasma lipids

Oils, such as sunflower and maize oils, which contain a high proportion of PUFA, have traditionally been advocated for the replacement of saturated fatty acids in modified fat diets (FAO/WHO, 1977; European Atherosclerosis Society, 1987). For each 1% energy of saturated fatty acids replaced by linoleic acid, there is, on average, a 5 mg/dl (0.13 mmol/l) reduction in plasma total cholesterol. Intakes of linoleic acid up to 12% of the energy do not lower HDL cholesterol (Mensink and Katan, 1989) but higher intakes do (Mattson and Grundy, 1986). Increasing the proportion of linoleic acid from 4% to 10% energy has little additional benefit in terms of plasma cholesterol reduction if saturated fatty acid intakes are kept constant (Kuusi et al., 1985).

11.2.4 The effect of n-3 polyunsaturated fatty acids on plasma lipids

The n-3 PUFA have different effects on plasma lipids from the n-6 PUFA and the effects observed are strongly dose-related. Certain fatty fish and fish oils containing relatively high proportions of the n-3 fatty acids eicosapentaenoic acid (EPA) and docosahexaenoic acid (DHA) (see Chapter 2) can markedly lower plasma triacylglycerols and VLDL concentrations (Harris, 1989; Sanders et al., 1989). Alpha linolenic acid (18:3 n-3) does not show the same effect at comparable doses (Sanders et al., 1989; McDonald et al., 1989).

Studies in healthy volunteers do not show any significant change in total or LDL cholesterol with moderate intakes of fish oil but very high intakes of fish oil (24 g long chain n-3 PUFA/day) lower the concentration of both LDL cholesterol and apo-protein B – the major protein component in LDL (Harris, 1989). Fish oil can also reduce LDL cholesterol if it is used to substitute for saturated fatty acids in people eating diets which are very rich in saturated fatty acids (Harris et al., 1983). However, if hypercholesterolaemic patients have already reduced their saturated fatty acid intake, then supplementation with n-3 PUFA has inconsistent effects – the average change in plasma cholesterol concentrations in 40 studies was a net decrease of 1.8% (Harris, 1989).

Moderate intakes of fish oils, or the ethyl esters of EPA and DHA, increase HDL_2 cholesterol but high intakes decrease total HDL cholesterol similar to that seen with very high intakes of linoleic acid (Harris, 1989; Sanders et al., 1989).

Dietary EPA and DHA, as found in fish oil supplements, but not alpha linolenic acid as found in linseed oil supplements (Abbey et al., 1990), decrease the activity of lecithin cholesterol acyl transferase (LCAT) and this change is accompanied by an increase in the HDL_2/HDL_3 ratio (see Chapter 6). The increase in the average HDL particle size probably reflects reduced cholesteryl ester acceptor capacity within the smaller pool of VLDL, as well as the decline in lipid transfer activity in plasma involving transfer protein itself, LDL and HDL.

Table 11.1 Summary of some studies on the effects of MUFA on plasma lipoprotein concentrations

Study	Diets	Number of subjects	Effects
Constant fat content			
Mattson and Grundy (1985)	SFA (−17%) replaced by MUFA (+13%) and PUFA (+4%)	20	Decrease in total cholesterol Decrease in LDL cholesterol No change in triacylglycerols No change in HDL cholesterol
Grundy (1986)	SFA (−21%) replaced by MUFA (+20%)	11	Decrease in total cholesterol Decrease in LDL cholesterol No change in triacylglycerols No change in HDL cholesterol
Grundy *et al.* (1989)	SFA (−12%) replaced by MUFA (+12%)	10	Decrease in total cholesterol Decrease in LDL cholesterol No change in triacylglycerols No change in HDL cholesterol
Berry *et al.* (1991)	MUFA compared with PUFA	26	Greater decrease in total cholesterol on PUFA diet
Dreon *et al.* (1991)	MUFA compared with PUFA	39	No difference in the total and LDL cholesterol lowering effects of the two diets
Ginsberg *et al.* (1990)	SFA (−8%) replaced by MUFA (+8%)	36	Decrease in total cholesterol No change in HDL cholesterol No change in triacylglycerols
Modified fat vs low fat			
Mensink and Katan (1987)	CHO (−16%) replaced by MUFA (+15%)	24	Decrease in total cholesterol Increase in HDL cholesterol
Baggio *et al.* (1988)	CHO (−10%) replaced by MUFA (+12%)	11	Decrease in total cholesterol Decrease in LDL cholesterol Decrease in triacylglycerols No change in HDL cholesterol
Mensink and Katan (1989)	CHO (−13%) replaced by MUFA (+10%)	48	Decrease in total cholesterol Decrease in LDL cholesterol Increase in HDL cholesterol
Ginsberg *et al.* (1990)	CHO (−8%) replaced by MUFA (+8%)	36	Decrease in total cholesterol No change in HDL cholesterol No change in triacylglycerols

CHO = carbohydrates; SFA = saturated fatty acids; MUFA = monounsaturated fatty acids; PUFA = polyunsaturated fatty acids

11.2.5 The effect of monounsaturated fatty acids on plasma lipids

The dietary studies carried out in the 1960s on plasma total cholesterol levels led many to believe that monounsaturated fatty acids were neutral with regard to their cholesterol lowering effects. However, the studies of Grundy (1989) and Mensink and Katan (1989) showed that the replacement of fats containing a high proportion of SFA or carbohydrate with oleic acid (*cis*-18:1 n-9) lowered LDL cholesterol by an amount equivalent to that found with linoleic acid (18:2 n-6). These studies used either olive oil or a high oleic safflower oil (see Table 11.1).

More recently, other oils high in monounsaturated fatty acids have been studied. Of particular interest are the studies which showed that high oleic acid rapeseed oil (Macdonald *et al.*, 1989) also lowered LDL cholesterol when substituted for fats rich in saturated fatty acids.

11.2.6 The effect of *trans* unsaturated fatty acids on plasma lipids

Elaidic acid (*trans* 18:1) had a similar effect on plasma total cholesterol to oleic acid in man (Mattson *et al.*, 1975) and a consensus view emerged that *trans* unsaturated fatty acids did not have any consistent effects on plasma cholesterol level (British Nutrition Foundation, 1987). A recent study, though, has shown that a diet containing mixed *trans*-18:1 isomers lowered HDL cholesterol and increased LDL in comparison with a diet rich in oleic acid (*cis*-18:1) (Mensink and Katan, 1990). The amounts of *trans* isomers present in the experimental diets, however, greatly exceeded the amounts normally present in human diets, so the relevance of the findings has been questioned (Grundy, 1990).

11.3 THE INFLUENCE OF UNSATURATED FATTY ACIDS ON POST-PRANDIAL LIPAEMIA

Most people spend about 12 hours per day in the post-prandial phase. The peak in plasma lipids occurs about 5 hours after a meal but about 50% of people show two or three post-prandial peaks. Post-prandial lipoprotein metabolism is now thought by some (Zilversmit, 1979) to offer more clues to atherosclerotic tendency than the measurement of fasting levels of lipoproteins. A persistent and exaggerated post-prandial lipaemia seems to be associated with propensity to atherosclerosis (Simons *et al.*, 1989).

11.3.1 Effect of carbohydrate and fat in mixed diets

Van Amelsvoort *et al.*, (1989) compared the effect of meals with varying amounts of carbohydrate (CHO) and fat on post-prandial hyperglycaemia and hyperlipidaemia. They showed that meals with a high ratio of CHO to fat increased insulin output and hyperglycaemia and that meals with a low ratio of CHO to fat increased post-prandial lipaemia.

11.3.2 Effect of modified fat diets

The nature of the fat intake also modifies post-prandial lipaemia. Chylomicron clearance is increased in subjects following the consumption of fatty fish (Harris, 1989). Fish oil supplementation also tends to lead to less post-prandial hyperlipaemia compared with olive oil supplementation (Sanders *et al.*, 1989; Brown and Roberts, 1991).

Weintraub *et al.* (1988) studied the effect of different types of dietary fat on post-prandial lipoprotein levels. Subjects with normal apoE phenotypes (homozygous for E3/3) received three isocaloric diets for 25 day periods. All the diets provided 42% energy from fat, with 26–28% of fatty acids from monounsaturated fatty acids. The 'saturated fatty acid diet' contained 67% saturated fatty acid, 5% polyunsaturated fatty acid; the 'n-6 diet' contained 30% saturated fatty acids and 42% polyunsaturated (linoleic); the 'n-3 diet' contained 30% saturated fatty acids and 42% polyunsaturated fatty acids (18% long chain n-3 fatty acids, mainly EPA, and 82% linoleic acid). The subjects then received two fat tolerance tests on separate days, one with the mixed fatty acids and the other with pure saturated fatty acids. The degree of post-prandial lipaemia was decreased by a previous dietary regimen of polyunsaturated fatty acids (especially the n-3 diet) compared with the saturated fatty acid pre-treatments. The saturated fatty acids fat tolerance load led to more post-prandial lipaemia than the mixed fatty acids load regardless of previous dietary treatment. The improved clearance of chylomicrons that occurred with the polyunsaturated regimes may have been a result of less competition from endogenously synthesised triacylglycerols because neither the activities of tissue lipoprotein lipase nor hepatic triacylglycerol lipase were affected.

11.4 HYPERLIPOPROTEINAEMIAS

11.4.1 Plasma ranges of cholesterol

Plasma cholesterol concentrations are measured in mmol/l. A desirable concentration as recommended by the European Atherosclerosis Society (1987) is less than 5.2 mmol/l. The majority of the UK

Table 11.2 World Health Organisation (WHO) classification of hyperlipoproteinaemias

Type	Lipoprotein changes	Occurrence
I	CM increased HDL, LDL low	Rare
IIa	LDL increased	Common
IIb	LDL, VLDL increased HDL may be low	Common
III	IDL, CM remnants increased LDL, HDL low	Rare
IV	VLDL increased, HDL low	Common
V	VLDL, CM increased HDL, LDL low	Rare

CM = chylomicrons; HDL = high density lipoproteins; LDL = low density lipoproteins; VLDL = very low density lipoproteins; IDL = intermediate density lipoproteins

Table 11.3 Inherited hyperlipoproteinaemias

Hyperchylomicronaemia (Type I or V)	
Apo C_2 deficiency	Rare
LPL deficiency (homozygotes)	Rare
LPL inhibitor	Rare
Hyper VLDL (or prebeta) lipoproteinaemia (Type IV)	
Familial hypertriglyceridaemia (FHTG)	Uncommon
Familial combined hyperlipidaemia (FCHL)	Common
LPL deficiency (heterozygotes)	Rare
Hyperapobetalipoproteinaemia (hyperapoB)	Common
Familial dyslipidaemic hypertension	Common
Dysbetalipoproteinaemia (Type III, Broad Beta)	
ApoE deficiency	Rare
HTGL deficiency	Rare
Hyper LDL (or beta) lipoproteinaemia with or without high triacylglycerols (Types IIa, IIb)	
Familial hypercholesterolaemia (FH)	1 500
Familial defective ApoB-100	1 500
Familial combined hyperlipidaemia (FCHL)	Common
Beta-Sitosterolaemia	Rare
Hyper HDL (or alpha) lipoproteinaemia (hyperalpha)	
Cholesterol ester transfer protein deficiency	Rare
Hyper Lp(a) lipoproteinaemia	Common
Hyperbetalipoproteinaemia	
Normal sized apoB	?
Truncated apoB	?
Abetalipoproteinaemia	Rare
Hypoalphalipoproteinaemia	Rare
Analphalipoproteinaemia	Rare

middle-aged population have values exceeding this. The value of 6.5 mmol/l is taken to indicate moderate hypercholesterolaemia and a value of 7.8 mmol/l indicates severe hypercholesterolaemia. Individuals with familial hypercholesterolaemia often have plasma cholesterol concentrations in excess of 10 mmol/l. Plasma cholesterol concentrations increase with age, plateauing around the age of 40 in men and approximately 10 years later in women. Plasma cholesterol concentrations tend to be lower in women prior to the menopause and then rapidly overtake the levels found in men.

11.4.2 New knowledge on single gene defects and polygenic inheritance

The genetic disorders of lipid transport were initially classified according to the abnormal concentrations of one or more class of lipoprotein in plasma (Table 11.2). The World Health Organisation (WHO or Fredrickson) phenotype has the advantage of simplicity. Many of the major syndromes are now divisible into more narrowly defined disease entities based on patterns of inheritance or by the detection of aberrant protein molecules, or both (Table 11.3) (see review by Schonfeld, 1990).

Many patients with hyperlipidaemia are hypertensive. In the Lipid Research Clinics study, the prevalence of hypertension in patients with Type IIb and IV hyperlipoproteinaemias was twice as high as in controls. Both these lipid abnormalities are characterised by elevations of VLDL and triacylglycerols. An inherited disorder called 'Familial Dyslipidaemic Hypertension' has been described (Hunt et al., 1989). It is characterised by raised plasma triacylglycerols associated with increased VLDL concentrations and raised plasma insulin. Such individuals are recognised as being at high risk of CHD (Despres, 1991).

Familial hypercholesterolaemia (FH) affects 1 in 500 people. These individuals have defective hepatic receptors to apoprotein B and develop atherosclerosis and CHD prematurely. Hypercholesterolaemia results because of an increased conversion of VLDL to LDL coupled with a low rate of removal of LDL from the blood. FH is not particularly responsive to diet, which means that drug treatment has to be used to reduce LDL levels (Thompson, 1990). However, a modified fat diet (30% energy from fat, less than 10% energy from saturated fatty acids) is usually prescribed as an adjunct to treatment as this helps lower the synthesis of LDL. Hypercholesterolaemia can also result from a defect in apoB which prevents it binding to the receptor (Innerarity et al., 1990). Familial defective apoB-100 also affects about 1 in 500 people but it is uncertain whether it is associated with increased risk of CHD.

Common polygenic hypercholesterolaemia leads

to milder elevations of LDL and is more responsive to changes in diet. It has been found that patients with the apoprotein E4/4 phenotype show a greater response to dietary modification than those with the apoE3/3 phenotype (Tikkanen et al., 1990).

Type III hyperlipoproteinaemia is typically found in patients who are homozygous for the apo E2/2 phenotype, although not all individuals with apo E2/2 have the disorder. Individuals with this disorder have increased intermediate density lipoprotein (IDL) concentrations but low LDL levels. The high IDL levels are thought to arise from poor binding of their defective apoE to the hepatic B/E receptor. This disorder is strongly associated with atherosclerosis of the peripheral vessels. It responds well to a reduction in total fat intake.

Types IV and V hyperlipidaemia respond to a reduction in total fat and saturated fatty acid intake. This is contrary to certain dietetic advice which sometimes advocates a decreased carbohydrate intake. A reduction in fat intake and an increase in carbohydrate intake decreases the conversion of VLDL to LDL (Abbot et al., 1990).

Beta-sitosterolaemia results from the accumulation of plant sterols in plasma. Treatment of this condition involves avoiding vegetable oils such as maize oil which supply these sterols.

11.4.3 The effect of n-6 unsaturated fatty acids in hyperlipidaemic patients

Supplementation of diets with maize oil, safflower oil or evening primrose seed oil has minimal effects on reducing lipid levels in hyperlipidaemic patients (Boberg et al., 1986).

11.4.4 The effect of n-3 unsaturated fatty acids in hyperlipidaemic patients

Several studies have examined the influence of fish oil consumption in patients with hyperlipoproteinaemias. In most studies, the effects of fish oil supplementation have been compared with supplements of maize oil or olive oil. Patients with Type IIa hyperlipoproteinaemia typically have low plasma triacylglycerol and VLDL concentrations which are slightly lowered by fish oil supplements. However, fish oil supplements have no effect on total cholesterol or LDL cholesterol concentrations (Berg-Schmidt et al., 1989); patients with mild hypercholesterolaemia also show no benefit in terms of cholesterol reduction (Wilt et al., 1989).

In patients with the Type IIb phenotype, fish oil supplements or diets providing increased amounts of EPA and DHA lead to a reduction in VLDL triacylglycerols and cholesterol, but no significant change in LDL cholesterol (Phillipson et al., 1985; Dart et al., 1989). HDL cholesterol concentrations tend to increase with fish oil supplements, and increases in LDL have been noted in some, but not all, patients. There also appears to be a sex difference with regard to the influence on HDL (Dart et al., 1989) with HDL in men being more likely to increase.

Patients with the Type III phenotype accumulate high plasma concentrations of IDL which have a late pre-beta electrophoretic mobility. Fish oil supplements have been found to normalise the electrophoretic profile in most patients and this is accompanied by a reduction in plasma triacylglycerols and cholesterol (Molgaard et al., 1990).

Fish oil supplements, but not olive oil, have a marked triacylglycerol lowering effect in patients with type IV phenotype (Saynor et al., 1984; Boberg et al., 1986; Berg-Schmidt et al., 1989; Stacpoole et al., 1989). Total cholesterol concentrations fall or remain unchanged. However, LDL cholesterol concentrations which tend to be low in these patients usually do rise even with low intakes (Radack et al., 1990). This increase in LDL is also seen with lower acyl ester concentrates of EPA and DHA low in cholesterol (Harris et al., 1988). HDL cholesterol concentrations either remain unchanged or increase. Saynor et al. (1984) have shown that fat tolerance is improved in hypertriglyceridaemic patients following ingestion of fish oil. The response of patients with the Type V phenotype (Phillipson et al., 1985) is similar to those with Type IV and the reduction in plasma triacylglycerols can be quite dramatic even when relatively small amounts of certain fish oil supplements are given (10–20 ml/d).

Increases in LDL cholesterol have been observed in patients with insulin dependent diabetes mellitus (IDDM) treated with fish oil supplements (Haines et al., 1986; Vandongen et al., 1988) as well as in non-insulin dependent diabetes mellitus (NIDDM) (Schectman et al., 1989a; Hendra et al., 1990) (see Chapter 17). Glucose control may be impaired in NIDDM given fish oil supplementation as it is with other triacylglycerol lowering agents such as nicotinic acid (Vessby et al., 1990).

As little as 6 g fish oil/day (2 g long chain n-3 fatty acids) has a triacylglycerol lowering effect in hypertriglyceridaemic patients, but the more commonly used dose is 3 g long chain n-3 fatty acids/day. Schectman et al. (1989b) have suggested, on the basis of a study on a small group of patients, that triacylglycerol lowering effects of fish oils cannot be sustained. This suggestion is not supported by other larger controlled trials (Miller et al., 1988b; Saynor et al., 1984).

Most studies investigating the effect of EPA and DHA have used fish oil concentrates containing the free fatty acids. However, some more recent studies have used ester concentrates or re-constituted triacylglycerols. It has been argued that the ability to raise HDL cholesterol and the ability to raise LDL cholesterol observed in some studies with fish oils might be due to cholesterol present in the fish oil. This seems unlikely for the following reasons:

(i) The level of cholesterol in fish oils is too low to elicit an effect.

(ii) Studies using esters or re-constituted fatty acids have also shown that HDL and LDL cholesterol concentrations are increased in some hypertriglyceridaemic subjects (Harris et al., 1988).

11.4.5 The effect of unsaturated fatty acids on lipoprotein (a)

Lipoprotein (a) (Lp(a)) is a lipoprotein of great interest because its concentration in plasma appears to be a better predictor than LDL for CHD events (Scott, 1991). Lp(a) concentrations are not reduced by conventional cholesterol lowering drug therapy. Modification of the total fat, cholesterol and saturated fatty acid intakes only has little influence on Lp(a) levels.

Herrmann et al. (1989) studied patients with elevated levels of Lp(a) and found that fish oil supplementation led to a significant reduction. Similar observations have been made for nicotinic acid, another triacylglycerol lowering agent. However, a subsequent study could not confirm that fish oil decreased Lp(a) concentrations (Gries et al., 1990).

11.5 THE EFFECT OF UNSATURATED FATTY ACIDS ON BILE ACIDS AND STEROL BALANCE

Sterol balance studies have been used to explain why plasma cholesterol concentrations were lowered in subjects replacing dietary saturated with unsaturated fatty acids (Grundy and Denke, 1990) but the results of these studies have been equivocal. In normal and hypertriglyceridaemic subjects, substitution of linoleic acid for saturated fatty acids enhanced faecal steroid excretion. However, patients with hypercholesterolaemia did not show an increase in faecal steroid excretion.

Some experimental studies suggested that high intakes of linoleic acid (about 15% of the dietary energy) were associated with an increased risk of gallstones (Sturdevant et al., 1973). On the basis of acute, rather than chronic, feeding studies,

Beardshall et al. (1989) have suggested that unsaturated fatty acids stimulate the production of cholecystokinin to a greater extent than do saturated fatty acids. Several studies have examined faecal steroid output in vegetarians who usually have slightly higher intakes of linoleic acid compared with the general population. These studies have failed to show that faecal steroid output is greater in vegetarians. Moreover, a lower incidence of gallstones has been reported in vegetarians compared with people eating mixed diets (Pixley and Mann, 1985).

11.6 THE EFFECT OF UNSATURATED FATTY ACIDS ON LIPOPROTEIN METABOLISM AND LIPOPROTEIN RECEPTORS

11.6.1 Background

Hepatic receptors to apoproteins B and E play a key role in regulating plasma cholesterol concentrations (see Figure 11.1). The level of LDL in plasma is determined by the rate of secretion of VLDL and its conversion to LDL, as well as its rate of removal. The rate of conversion of VLDL to LDL is affected by the apoprotein E phenotype as well as the VLDL particle size. As much as 16% of the genetic variance in LDL concentration can be accounted for by allelic difference at the apoE gene locus. The genetic polymorphism of apoE is due to three common alleles e2, e3 and e4 at a single autosomal gene locus. These alleles determine the six phenotypes E2/2, E3/3, E4/4, E4/2, E4/3 and E3/2. Individuals homozygous for E4/4 have higher LDL concentrations than those with E3/3. Individuals with E4/4 also show greater increases in plasma cholesterol in response to dietary cholesterol and saturated fatty acids compared with those with E3/3 (Tikkanen et al., 1990). It has been argued that IDL particles from subjects with E4/4 are more readily taken by the $alpha_2$-macroglobulin receptor in the liver and converted into LDL than those from subjects with other apoE phenotypes. Apoprotein E also binds with high affinity to the apoprotein B receptor and so may influence the removal of LDL by the apoB receptor by competitive binding.

The apoB receptor plays a major role in regulating the rate of removal of LDL as well as its rate of synthesis from VLDL. Hepatic receptors to apoB account for most of the capacity to remove LDL. The binding capacity of the apoB receptor is genetically determined but the number of receptors expressed is influenced by hormonal and dietary factors: e.g. corticosteroids decrease and oestro-

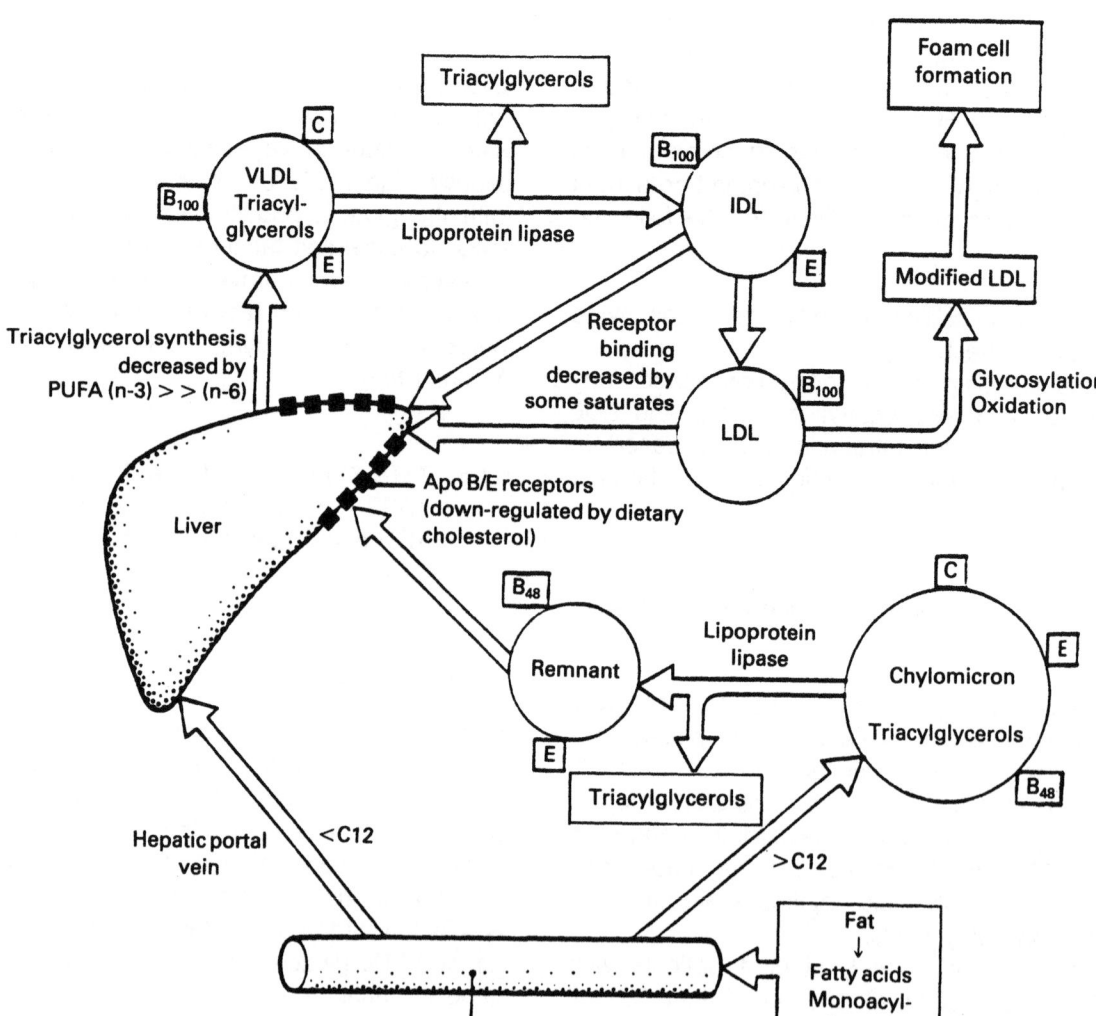

Figure 11.1   The influence of dietary fat on lipoprotein metabolism.

gens increase the number of LDL receptors. Dietary cholesterol decreases expression of LDL receptors. LDL receptor activity is low in man compared with other animals. Consequently, LDL cholesterol concentrations are primarily determined by rates of LDL synthesis rather than by rates of removal. There also appears to be interaction between diet and hormones. Studies with primates have shown that a diet high in saturated fatty acids exacerbates the hypercholesterolaemia induced by prednisone (Ettinger *et al.*, 1989).

11.6.2   The effect of unsaturated fatty acids on lipoprotein synthesis – evidence from protein kinetic studies

Several studies have attempted to explain why LDL cholesterol concentrations are lowered when linoleic acid replaces saturated fatty acids in the diet. One method has been to follow the decay curves of radioactively labelled proteins. These human studies

show a marked reduction in LDL synthetic rate when linoleic acid replaced saturated fatty acids in the diet (Illingworth *et al.*, 1981; Turner *et al.*, 1981; Cortese *et al.*, 1983) and a slight rise in fractional catabolic rate (Illingworth *et al.*, 1981).

There have been few LDL kinetic studies carried out with n-3 fatty acids and all have been with extremely high intakes of fish oil (Illingworth *et al.*, 1981; Nestel *et al.*, 1984; Harris, 1989) . LDL apoB synthesis was markedly inhibited (Illingworth *et al.*, 1981; Harris, 1989), but the reduction in LDL pool size was smaller than would be predicted from the decrease in LDL synthesis. Moreover, despite the reduced LDL pool size, there was no increase in fractional catabolic rate of LDL. This might imply down regulation of the LDL receptors by fish oils.

The kinetic studies carried out so far have suffered from a number of methodological problems, the most important of which has probably been the use of fixed angle ultracentrifuge rotors to prepare lipoprotein fractions. It now appears that only small VLDL particles are converted into LDL and not all

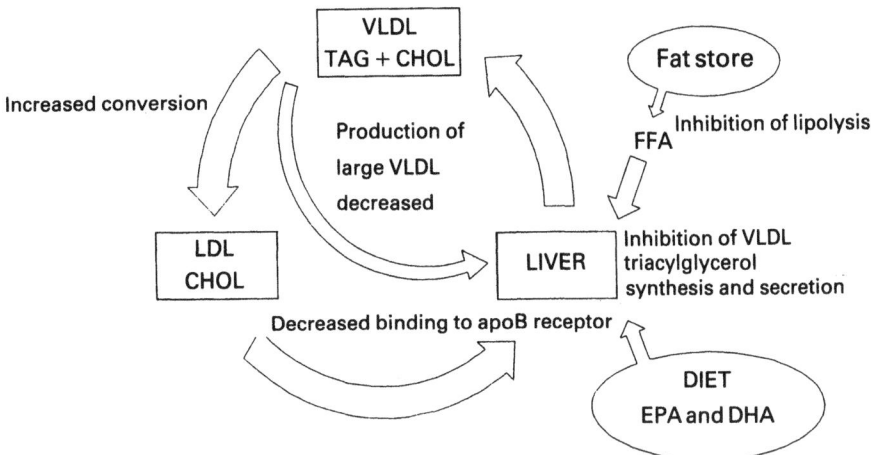

Figure 11.2   The influence of EPA and DHA on lipoprotein metabolism.
FFA = free fatty acids

VLDL as previously proposed (Packard *et al.*, 1986). The uptake and degradation of LDL by isolated monocytes or fibroblasts can also be used to measure the activity of the LDL receptor (Bilheimer *et al.*, 1978). However, no well-controlled studies of dietary unsaturated fatty acids on LDL receptor activity have been carried out in man. The effects of n-3 fatty acids on LDL receptor binding in isolated cells are unclear and deserve further study.

Studies in animals have been of limited value because of between-species differences in lipoprotein metabolism. Another limitation has been the use of levels of linoleic acid in the diet beyond the range of human intakes. Kinetic studies in the hamster, which shows a similar response to dietary lipids as man, suggested that there was an increased rate of removal of LDL by the receptor-dependent pathway when linoleic acid replaced saturated fatty acids (Spady and Dietschy, 1989). Dietary fat composition did not affect the proportion of LDL removed by the non-receptor dependent pathway.

11.6.3   The effect of unsaturated fatty acids on lipoprotein synthesis – evidence from messenger RNA studies

It has been shown that expression of mRNA for the apoB receptor is decreased in primates given dietary cholesterol. However, no influence on mRNA expression for apoB or the apoB receptor was observed in primates fed maize oil compared with those given lard; plasma cholesterol concentrations were nevertheless markedly lower in the group given maize oil (Sorci-Thomas *et al.*, 1989). It is possible that saturated fatty acids interfere with the binding of LDL to the apoB receptor. Wong *et al.* (1989) have carried out

studies on cultured HepG2 cells and claim that EPA and DHA supplementation decrease the synthesis of mRNA for both apoB and the LDL receptor. It appears that n-3 fatty acids may have different effects on LDL receptors.

11.6.4   The effect of unsaturated fatty acids on the physical properties of lipoproteins

Small (1988) has argued that differences in the lipid composition of lipoprotein particles affect their physical properties and thus affect their interaction with receptors. This hypothesis is attractive as it now appears that VLDL particle size affects the extent of conversion to LDL. Influences on VLDL particle size may well explain some of the effects observed for the n-3 fatty acids (Figure 11.2).

11.7   THE POSSIBLE MECHANISMS OF ACTION OF n-3 PUFA ON TRIACYLGLYCEROL AND LIPOPROTEIN METABOLISM

Giving fish oils containing EPA and DHA to human subjects does not affect the activity of lipoprotein lipase or hepatic triacylglycerol lipase (Weintraub, 1988). Consequently, an increased rate of removal of VLDL triacylglycerols from plasma cannot explain the effect of fish oil supplements in reducing triacylglycerol levels. Singer *et al.* (1990) claimed that fish oil decreased free fatty acid release in response to adrenergic stimulation in subjects fed oily fish. This would in turn decrease the availability of substrate for hepatic triacylglycerol synthesis. However, EPA and DHA have also been shown to inhibit directly both triacylglycerol synthesis and apoB synthesis and secretion in cell cultures and in perfused liver (Wong *et al.*, 1989). VLDL turnover studies in man show that fish oil supplementation decreased hepatic triacylglycerol synthesis (Sanders *et al.*, 1985; Harris *et al.*, 1990). Sullivan *et al.* (1986) demonstrated that fish oil supplementation of human diets resulted in an increased proportion of small VLDL particles in circulation. These small particles are known to be more readily converted to LDL than the larger particles which are rich in triacylglycerols and this could explain the increase in fractional catabolic rate of VLDL reported by Harris *et al.* (1990). Huff and Telford (1989) also showed that the conversion of VLDL to LDL was enhanced in mini-pigs whose diet was supplemented with fish oil.

The above findings also explain why fish oil supplements sometimes increase LDL levels in some patients with Type IV and Type V hyperlipoproteinaemias. This is a general phenomenon seen with most forms of therapy aimed at reducing triacylglycerols including caloric restriction as well as fibrate-like drugs (Kesaniemi *et al.*, 1985). Schectman *et al.* (1989a) have tried to explain the heterogeneity of response to EPA and DHA on the basis of the initial LDL particle size; patients with a cholesterol:apoB ratio which is greater than 1.4 were most likely to show a marked rise in LDL apoB levels.

12

UNSATURATED FATTY ACIDS AND ATHEROSCLEROSIS

12.1 RELATIONSHIP BETWEEN PLASMA LIPOPROTEIN LEVELS AND ATHEROSCLEROSIS

Early studies focused on the relationship between diet, plasma total cholesterol and risk of coronary atherosclerosis. Several lines of evidence supported the causative role of elevated plasma cholesterol in atherogenesis. More recent studies have turned to studying the lipoproteins that transport cholesterol around the body, their inter-relationships with diet and how they lead to atherosclerosis. Elevated levels of certain lipoproteins and apoproteins increase the risk of coronary atherosclerosis, in particular high levels of VLDL, IDL, LDL, apoB and Lp(a). On the other hand, low levels of HDL (particularly HDL_2) and apoAI are associated with increased risk. Elevated plasma cholesterol concentrations generally reflect increased concentrations of LDL, VLDL or IDL. Elevated plasma triacylglycerol concentrations may reflect increased VLDL or IDL or chylomicron remnants.

There is unequivocal evidence that certain hyperlipoproteinaemias (IIa, IIb, III, IV) (see Chapter 11) accelerate the development of atherosclerosis. High concentrations of lipoproteins containing apoB are associated with accelerated atherosclerosis, as are defects in the reverse transport of cholesterol. Coronary angiographic studies confirm that high concentrations of LDL and VLDL and low concentrations of HDL are associated with coronary atherosclerosis. Recently it has been shown that reduction of the VLDL and LDL concentrations by effective lipid lowering drug therapy leads to regression of coronary atherosclerosis (Brown *et al.*, 1990). In this study, a reduction in saturated fatty acid intake and an increase in linoleic acid intake was insufficient to cause regression over the period of the study. However, an earlier study (Arntzenius *et al.*, 1985) showed that a reduction in plasma lipid concentrations by strict dietary intervention prevented progression of coronary atherosclerosis.

12.2 THE MECHANISM OF ATHEROSCLEROSIS
(see Figure 12.1)

12.2.1 Uptake of lipoproteins by monocytes

VLDL, LDL, IDL and chylomicrons remnants can be taken up by circulating monocytes which are converted into tissue macrophages. The macrophages become engorged with cholesterol and form foam cells. These cells constitute the 'fatty streak' which many consider to be the earliest recognisable lesion in atherosclerosis (British Nutrition Foundation, 1992). Macrophages may become trapped inside the vascular endothelium and release chemotactic substances, including platelet activating factor (see Chapter 13), that attract other white cells and platelets. These cells interact with the vascular wall and stimulate the production of growth factors that lead to hyperplasia and migration of smooth muscle cells which leads to thickening of the intimal wall (Ross, 1986). Foam cell formation is also associated with inhibition of endothelium-derived relaxing factor (EDRF) from endothelial cells. Decreased EDRF, thought to be nitric oxide, leads to reduced smooth muscle cell relaxation and increased vasoconstriction.

12.2.2 Modification of lipoproteins and uptake by the 'scavenger' receptor

VLDL, IDL, LDL and chylomicron remnants have to be modified before being taken up by macrophages to form foam cells (Steinberg *et al.*, 1989). Native LDL, which can be taken up by the normal LDL receptor on many cells, is not thought to be atherogenic. However, if the protein moiety of LDL is modified, it can then be recognised by the 'scavenger' receptor on macrophages and lead to their conversion to foam cells. Modification can be by oxidation, acetylation or glycosylation. The protein moiety of LDL, like other proteins, can become glycosylated in the presence of high blood

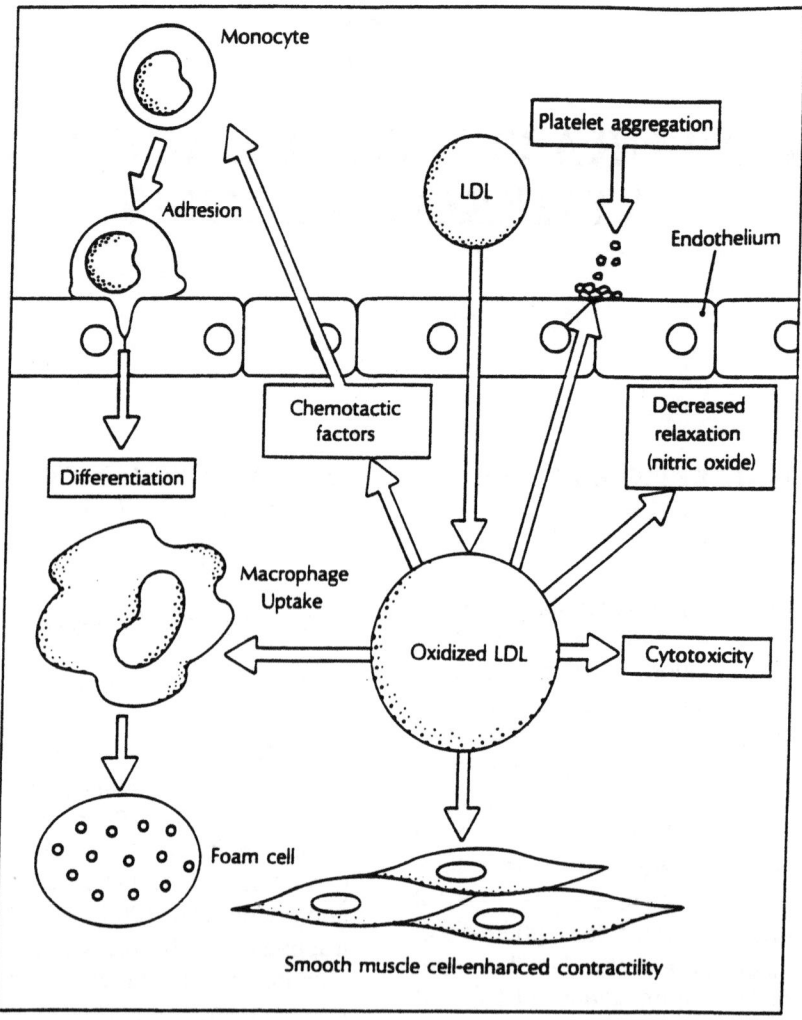

Figure 12.1 The effects of oxidised low density lipoproteins on arterial cells. (From Bruckdorfer, 1990)

glucose concentrations. Glycosylated LDL is then taken up by the 'scavenger' pathway and this could be one of the reasons why diabetes mellitus increases risk of atherosclerosis.

Attention has been focused on lipid oxidation as a major mechanism for the modification of LDL since the discovery that LDL incubated with arterial endothelial cells was oxidatively modified, and was no longer recognised by the LDL receptor but was bound by the 'scavenger' receptor on macrophages (Quinn et al., 1987). These oxidative changes only seemed to occur if the medium contained trace amounts of transition metal ions. Indeed, the effects of the endothelial cells could be simulated by transition metals alone, particularly cupric ions at higher concentrations. The most likely candidates for lipid oxidation would be the PUFA present in the triacylglycerol and cholesterol ester fraction of the lipoprotein. The hypothesis has gained support from the observation that the drug probucol, which has antioxidant properties,

inhibited the development of atherosclerosis in rabbits with genetically high cholesterol levels more than would be predicted from its effects on reducing cholesterol levels alone (Bruckdorfer, 1990).

Hypercholesterolaemic patients with high concentrations of Lp(a) have an increased risk of coronary disease compared with those having normal concentrations (Seed et al., 1990; Wiklund et al., 1990). The atherogenicity of Lp(a) may be a consequence of its uptake by the 'scavenger' pathway as it is not taken up by the normal apoB receptor. Furthermore, high local concentrations of Lp(a) in the arterial wall might inhibit the breakdown of mural thrombi. Lp(a) is believed to be an inhibitor of tissue plasminogen activator (tPA) because of structural homology between its constituent apoprotein apo(a) and plasminogen (Scott, 1989). However, there is no evidence that fibrinolytic activity (see Chapter 13) is impaired in patients with elevated concentrations of Lp(a).

12.2.3 The role of unsaturated fatty acids in lipid oxidation and lipoprotein modification

Oxidation of the lipids in lipoproteins is thought to take place in the arterial wall and not in the plasma although the constituents of the plasma lipoproteins may be of relevance to their subsequent fate. Linoleic and arachidonic acid are the principal polyunsaturated fatty acids in LDL lipids. Parthasarathy et al. (1990) found that the fatty acid composition of the LDL particle affected its susceptibility to subsequent oxidative modification; LDL particles rich in oleic acid were not modified to the same extent as those rich in linoleic acid.

Oxidation of these polyunsaturated fatty acids can be catalysed either by free radicals produced by cells or by transition metal ions. The lipid peroxides, and their breakdown products such as 4-hydroxynonal and malonaldehyde (see Chapter 7), can react with cholesterol to form oxidation products and modify the structure of apoB causing it to crosslink and become insoluble (Carpenter et al., 1990; Bruckdorfer, 1990). The modified apoB will then not be recognised by the LDL receptor but will be recognised by the 'scavenger' receptor on macrophages.

Furthermore, certain oxidation products of cholesterol have long been known to cause damage to the artery wall and stimulate smooth muscle growth.

Endothelial cells and macrophages appear to play a role in the oxidation of LDL and inhibitors of lipoxygenase prevent modification of LDL (Esterbauer et al., 1990). Vitamin E, beta-carotene and ascorbic acid in vitro inhibit the oxidative modification of LDL (Jilial et al., 1990). However, rabbits given cholesterol supplemented with high intakes of vitamin E showed increased rates of atherosclerosis (Godfried et al., 1989).

12.3 THE EFFECT OF UNSATURATED FATTY ACIDS ON THE DEVELOPMENT OF ATHEROSCLEROSIS

12.3.1 Experimental models of atherosclerosis

It is possible to induce a form of atherosclerosis in animals by providing large amounts of cholesterol in the diet, usually 1–2% by weight, a concentration 50–100 times greater than in human diets. Human hypercholesterolaemia is characterised by increased concentrations of LDL, whereas cholesterol-induced hypercholesterolaemia in experimental animals is associated with the production of cholesteryl ester rich lipoproteins with a density of 1.006 g/ml with beta-electrophoretic mobility (Beta-VLDL) (Mahley, 1982). Beta-VLDL may represent remnants of the intestinal lipoproteins or may be synthesised by the liver in response to the delivery of excessive amounts of cholesterol to this organ. Beta-VLDL will cause macrophages to accumulate massive amounts of cholesteryl esters, which resemble foam cells, in the arterial wall (Parthasarathy et al., 1990).

Rabbits, pigs and certain primates when fed on diets containing a relatively high proportion of saturated fatty acids over a long period of time develop lesions that are not dissimilar to those in man (Wissler and Vesselinovich, 1988). The development of genetic animal models for hyperlipidaemia such as the Watanabe hereditary hyperlipidaemic (WHHL) rabbit and the JCR-corpulent rat have also increased understanding of the relationship between hyperlipidaemia and atherosclerosis. These animal models clearly demonstrate the interaction between diet and genetics (Reue et al., 1990). For example, some species of primates are particularly prone to atherosclerosis induced by diets high in saturated fatty acids, whereas others are not (Hayes, 1991).

12.3.2 The effect of n-6 unsaturated fatty acids

The severity of experimental atherosclerosis is influenced by the type rather than the quantity of fat in the diet. When vegetable oils rich in unsaturated fatty acids, such as sunflower seed oil, are substituted for butter fat, this is accompanied by less atherosclerosis (Mendelson and Mendelson, 1989). Generally, saturated fatty acids such as those found in butter, tallow, lard and hardened coconut oil have been found to be atherogenic. Cocoa butter appears to be less atherogenic than expected (Kritchevsky, 1988). Monounsaturated fatty acids such as oleic acid (cis 18:1) and elaidic acid (trans 18:1) do not appear to be atherogenic in primates (Kritchevsky et al., 1984). Leth-Espensen et al. (1988) found lower rates of cholesterol accumulation in the aortas of rabbits given olive oil compared with control animals. No effect was found with maize oil even though plasma cholesterol concentrations were similar between the olive oil and maize oil groups. This could possibly reflect differences between oleic and linoleic acid (18:2 n-6) in reducing the uptake of cholesterol by the arterial wall.

Many of the animal models have used extreme hypercholesterolaemia to induce atherosclerosis. Masuda and Ross (1990a,b) have shown that moderate hypercholesterolaemia (5.2–10.4 mmol/l) in primates leads to fatty streak formation followed by conversion of the fatty streak to a fibrous plaque. The atherogenic diet used in these studies provided 42% of the dietary energy as fat, 300 mg cholesterol/1000 kcal, and the fat was provided as a mixture of lard, butter and beef fat. The lesions observed were similar to those seen in man and were typically found at the bifurcations of vessels.

12.3.3   The effect of n-3 unsaturated fatty acids

12.3.3.1   Effect on lipoproteins

EPA and DHA decrease the plasma concentration of several atherogenic lipoproteins (VLDL, IDL, chylomicron remnants) and, at moderate doses, may increase the ratio of HDL_2/HDL_3 cholesterol.

On the other hand, they lead to an increase in LDL levels in some patients. This observation, coupled with the knowledge that EPA and DHA are particularly prone to oxidation, might indicate that fish oil supplementation could increase the risk of atherosclerosis. However, atherosclerosis was reported to be rare among North American Inuits (Eskimos) on their traditional diet of fish, seal and caribou (Sinclair, 1953; Schaeffer, 1959). Moreover, the consumption of fish oil inhibits the development of atherosclerosis in pigs, dogs and primates (Davis et al., 1987; Leaf and Weber, 1988; Wiener et al., 1986; Shimokawa and Vanhoute, 1988; Parks et al., 1990). The inhibition occurs even in the presence of hypercholesterolaemia, but not in the hereditary hyperlipidaemic rabbit or the rabbit

fed cholesterol (Rich et al., 1989). It seems likely therefore that the protection afforded by EPA and DHA is mediated by mechanisms independent of plasma LDL concentrations and oxidative modifications of LDL particles.

12.3.3.2   Effect on mitogenic factors and intimal thickening

A concomitant of fibrous plaque formation in atherosclerosis is the thickening of the arterial wall which is brought about by mitogenic factors from the arterial endothelial cells which stimulate smooth muscle cell proliferation and migration from deeper layers in the arterial wall (see British Nutrition Foundation, 1992).

EPA and DHA have been shown to inhibit the production of mitogenic factors by endothelial cells in vitro (Fox and DiCorletto, 1988; Smith et al., 1989). There is some evidence that fish oil supplements decrease the production of tumour necrosis factor and interleukin-1 in man (Endres et al., 1989). Landymore et al. (1985) observed that dietary cod liver oil markedly reduced intimal thickening in autogenous vein grafts implanted as arterial bypasses in cholesterol-fed dogs. These studies were confirmed by Cahill et al. (1988) using a fish oil concentrate low in vitamins A and D. Sarris et al. (1989) showed that fish oil concentrates, with or without aspirin, were more effective than aspirin alone in reducing intimal thickening. In the pig, the luminal encroachment of coronary arteries was inhibited by dietary cod-liver

Table 12.1   A meta-analysis of controlled trials of fish oil supplements on risk of restenosis, according to intention to treat

Treatment	Daily dose of EPA + DHA	Duration	No. of patients with restenosis		Odds Ratio (95% confidence limits)	Reference
			Treatment group	Control group		
18 g MaxEPA/d vs untreated group.	5.5 g	6 months	8/43	18/39	0.27 (0.10–0.73)	Dehmer et al. (1988)
9 g SuperEPA/d vs untreated group.	4.5 g	6 months	21/95	35/99	0.52 (0.28–0.98)	Milner et al. (1989)
10 g MaxEPA/d vs aspirin. (Patients with single vessel disease.)	3.1 g	6 months	5/20	8/23	0.63 (0.17–2.37)	Cheng et al. (1990)
12 g SuperEPA/d vs Promega vs olive oil placebo.	6 g	6 months	44/150	15/72	1.58 (0.81–3.08)	Reiss et al. (1989)
10 g MaxEPA/d vs olive/ maize oil placebo.	3.1 g	4 months	16/52	19/56	0.87 (0.38–1.95)	Grigg et al. (1989)
10 g unspecified fish oil vs olive oil.	3.1 g	4 months	25/67	21/62	1.16 (0.56–2.38)	Franzen et al. (1990)
All trials			119/427	116/351	0.80 (0.58–1.10)	

MaxEPA = 17% EPA; 11.4% DHA
SuperEPA = 30% EPA; 20% DHA
Promega = 28% EPA; 12% DHA

oil (Wiener *et al.*, 1986; Shimokawa and Van Houte, 1988) and mackerel oil supplements (Hartog *et al.*, 1989).

12.3.3.3 Effect on restenosis

Percutaneous transluminal coronary angioplasty (PTCA) is a surgical procedure used to relieve myocardial ischaemia by widening the coronary arteries. It can lead to marked symptomatic improvement, but re-narrowing (restenosis) occurs in about 30% of patients. As yet, there is no way of preventing restenosis; it is believed to result from thickening of the intimal layer. Several clinical trials have examined the effects of fish oil supplementation on restenosis following coronary angioplasty (Table 12.1). Although the first three studies showed benefit from fish oil supplementation, three others showed no significant benefit. Overall, there is a non-significant trend for the restenosis rate to be about 20% lower with the fish oil treatment. All studies except Cheng *et al.* (1990) and Franzen

et al. (1990) have compared the influence of fish oil in addition to aspirin treatment. Consequently any effects mediated by prostaglandins that might be exerted by fish oil may well have been masked by the aspirin. The studies used different doses of n-3 fatty acids and not all of the studies initiated fish oil supplementation at least two weeks before carrying out the angioplasty. Furthermore, the studies lacked sufficient statistical power to detect small changes in restenosis rate. For a 20% reduction in the restenosis rate where the usual rate is 30%, it would require more than 1,000 patients to have a 90% power of detecting a statistically significant difference. Further studies are in progress to evaluate whether fish oil supplementation is of benefit.

In summary, while it is uncertain whether fish oil supplements inhibit the development of atherosclerosis in man, there is at least no evidence that they accelerate its development.

13

UNSATURATED FATTY ACIDS, THROMBOGENESIS AND FIBRINOLYSIS

Interest in possible effects of dietary fat composition on the haemostatic and fibrinolytic systems has arisen following recognition of the role of thrombosis in coronary heart disease, stroke and peripheral vascular disease.

13.1 THE THROMBOGENIC AND THE FIBRINOLYTIC PATHWAYS

13.1.1 Normal role in vascular repair

The walls of the blood vessels are subject to continual wear and tear, for which a complex system of repair has evolved. Microscopic injuries by chemical, biological or mechanical agents trigger a set of local responses involving three main components: blood platelets, plasma coagulation factors and the fibrinolytic system. They act as follows:

(i)   Cellular adhesive reactions lead to the attachment of platelets to the damaged surface and the recruitment of inflammatory cells.

(ii) Activation of platelets by their attachment to collagen fibrils and their exposure to adenosine diphosphate (ADP) which is released from the damaged endothelial cells, leads to platelet aggregation (see Section 13.2.4).

(iii) Growth factors, secreted by activated platelets, stimulate reparative growth and mitosis of fibroblasts and smooth muscle cells in the subendothelial tissue, as well as in the endothelial cell layer.

(iv) Exposure of a specific cell-membrane protein, known as tissue factor, which is normally 'hidden' from the blood, permits its binding to the circulating plasma coagulant factor VII (see Section 13.9).

(v)   Creation of the tissue factor–factor VII complex at the site of injury triggers the cascade of precursor-to-enzyme transformations that culminates in the local deposition of fibrin (see

Section 13.9). These humoral reactions are localised partly by the requirement for assembly of the enzyme, its co-factor and its precursor substrate on the phospholipid surface of the endothelial cell or the attached platelet, and partly by the presence of circulating protease inhibitors, various co-factor neutralising reactions, and platelet damping activities.

(vi) The fibrinolytic system is also activated (see Section 13.11), with the generation of the thrombolytic enzyme, plasmin.

This reparative role of the thrombogenic/fibrinolytic system is likely to represent its primary function and probably accounts for its basal and continual level of activity in normal healthy individuals.

13.1.2 Role in more serious damage

More serious disruption of the endothelium prompts an explosive reaction, with deposition of a mass of fibrin and aggregated platelets to staunch bleeding (blood clot formation). When this process creates a mass which blocks the vessel lumen, rather than just repairing a breach in the vessel wall, a thrombus forms. If this is large enough, it can have pathological consequences. A thrombotic event may arise:

(i)   through exposure of normal blood to a strongly pro-coagulant and pro-aggregatory surface (e.g. a ruptured atheromatous plaque); or

(ii) as a consequence of disorders of the systems which respond to vascular injury.

The role of unsaturated fatty acids in the prevention, or correction, of a pro-aggregatory state is now considered.

13.2 THE PLATELET RESPONSE TO INJURY

13.2.1 Resting platelets

Blood platelets are formed in the bone marrow from their precursor cell, the megakaryocyte. The normal platelet count varies widely in healthy people between 150 and 440×10^9 per litre of blood. The platelet is the smallest cellular element in the blood. It lacks a nucleus and has a limited life-span of 8–11 days. Until called on for haemostasis and repair, the platelet circulates in a 'resting' state. Once activated, it reveals a remarkably complex and sophisticated metabolic armoury which subserves its roles in haemostasis, inflammatory cell recruitment, initiation of tissue repair and fibrinolysis.

The resting platelet can be thought of as continually scanning the vascular surface for biochemical evidence of local injury. Its surveillance equipment is based on a system of protein receptors on its plasma membrane surface which recognise and adhere to collagen and other adhesive glycoproteins in the exposed subendothelium. Other receptor proteins serve to detect the presence of ADP released from damaged cells, and thrombin generated by the coagulant system (see Table 13.1).

Platelet adhesion is followed by:

(i)   shape changes as the discoid cell spreads itself over the site of injury;
(ii) secretion of the contents of its storage granules;

(iii) assembly of functional fibrinogen receptors on the plasma surface membrane;
(iv) recruitment of more platelets with subsequent platelet aggregation.

The platelet surface membrane is a typical bilayer of phospholipids, unesterified cholesterol, glycoproteins, glycolipids and proteins. The proteins are responsible for many metabolic roles of the membrane and are held in position by interactions with each other, with the hydrophobic domains of the surrounding lipid matrix, and sometimes by attachment to the intracellular constituents of the cytoskeleton, such as actin fibrils.

The glycoproteins in the surface membrane which function as receptors for collagen, ADP and other agonists are functional in the resting state and they protrude like antennae from the platelet into the surrounding media. Other receptor proteins are non-functional until the platelet has been activated, and are the proteins responsible for aggregation. Others are wholly embedded in the membrane or protrude partly through its cytoplasmic surface. These proteins tend to be concerned with the generation of second messenger molecules and trigger the platelet's response when the receptor complex is linked to its ligand. The second messengers are responsible for the initiation of reactions in the cytoskeleton and other intracellular structures which culminate in:

(i)   shape change;
(ii) the release reaction;
(iii) adherence to neighbouring platelets;
(iv) further anchoring to the adhesive proteins of the subendothelium;

Table 13.1 Platelet receptors

Receptor	Ligand	Result of binding	Reference
GPVI	Collagen	Activates PLA_2	Moroi et al. (1989)
GPIa	Collagen	Generates TXA_2 platelet shape change	Nieuwenhuis et al. (1985)
57-kD glycoprotein	TXA_2/PGH_2	Platelet shape change. Promotes secretion of ADP	Arita et al. (1989)
Aggregin	ADP	Platelet shape change. Permits assembly of fibrinogen receptor Promotes aggregation	Colman (1990)
Inhibitory receptor	ADP	Inhibits adenylate cyclase; leads to decreased cAMP	Colman (1990)
GPIb	Thrombin	Stimulates PLA_2 and releases ADP Activates PLC Permits fibronectin-GPIIb/IIIa binding	Harmon and Jamieson (1986)
Alpha$_2$ adrenoreceptor	Adrenalin	Potentiates ADP effects on the platelet	Regan et al. (1986)
GPIIb/GPIIIa	Fibrinogen/fibronectin/ von Willebrand's factor	Platelet adhesion	Plow and Ginsberg (1989)

(v) counter-regulatory signals which limit these responses.

13.2.2 Collagen-stimulated activation of resting platelets

13.2.2.1 Attachment to collagen fibrils

The attachment of the platelet to subendothelial collagen (step 1 in Figure 13.1) is mediated by von Willebrand's factor, a protein present in the platelet protein storage granule, and also synthesised by the endothelial cell (see Section 13.2.3). The complex of receptor proteins that is responsible for this adherence appears to include two specific glycoproteins, GP1a and GPVI (Table 13.1).

13.2.2.2 Release of platelet arachidonic acid

Arachidonic acid comprises about 30% of the fatty acids of human platelet phospholipids (Broekman et al., 1976). The most important initial platelet response to collagen binding is activation of the enzyme, phospholipase A_2 (PLA_2) (step 2 in diagram). This enzyme is found mainly on intracel-lular membranes. It releases arachidonic acid or other PUFA esterified at position 2 of the glycerol backbone of phosphatidylcholine (PtdCho) and phosphatidylethanolamine (PtdEtn) (Blackwell et al., 1977).

The initial release of arachidonic acid following adhesion to collagen depends preferentially upon the PLA_2 pathway described above but it can also be released by a second pathway requiring phospholipase C (PLC) (not shown in Figure 13.1). PtdIns 4,5 bisphosphate (PIP_2) is first converted to 1,2-diacylglycerol (DAG) and inositol 1,4,5, trisphosphate (IP_3). Arachidonic acid is then released by the action of diacylglycerol lipase on DAG (Rittenhouse-Simmons, 1979). Once arachidonic acid is liberated by either of these methods, it is available for eicosanoid synthesis. Oxygenation and cyclisation generates the 2-series prostaglandin E (PGE_2) (see Chapter 8).

With appropriate dietary adjustment, the phospholipid synthetic processes will substitute dihomogamma linolenic acid (DGLA) (20:3 n-6) or eicosapentaenoic acid (EPA) (20:5 n-3) for a proportion of arachidonic acid (20:4 n-6) at position 2 of the glycerol molecule. Liberation, oxygenation and cyclisation of these 20 carbon PUFAs leads to the generation of prostaglandins PGE_1 and PGE_3. As a rule, more PGE_2 is produced than PGE_1 or PGE_3, a simple reflection of the relative abundance of arachidonic acid as substrate.

The conversion of arachidonic acid to PGE_2 involves the preliminary formation of a number of unstable, but biologically important, intermediate compounds (see Chapter 8). In the platelet, the unstable intermediates PGG_2 and PGH_2 are alternatively metabolised by the enzyme thromboxane synthetase to form the key product thromboxane (TXA_2) (Needleman et al., 1976) (step 3 in Figure 13.1). The action of this extremely potent substance is confined to the site of vascular injury by its instability. Within about 30 seconds, TXA_2 breaks down spontaneously to the inactive stable product TXB_2.

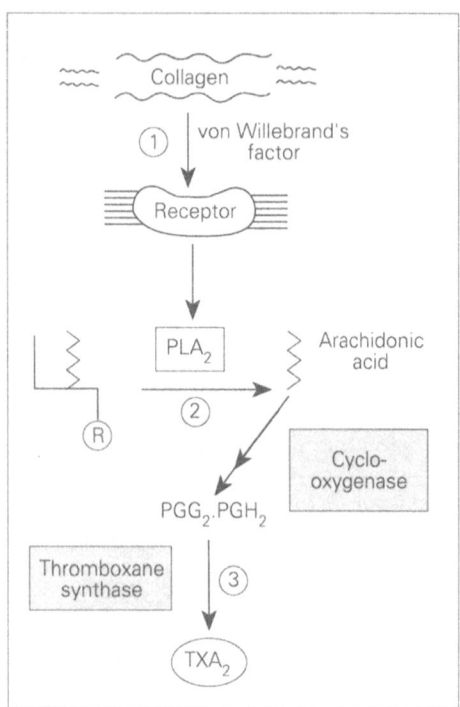

Figure 13.1 Collagen binding leads to thromboxane formation: **1** Collagen binds to the platelet via von Willebrand's factor; **2** Collagen binding activates phospholipase A_2 (PLA_2) which releases arachidonic acid from membrane phospholipids; **3** Arachidonic acid is metabolised via cyclo-oxygenase to unstable prostaglandin intermediates PGG_2 and PGH_2. These are metabolised by thromboxane synthetase to form thromboxane A_2 (TXA_2).

13.2.2.3 Changes involving the platelet thromboxane receptor

Another of the plasma membrane proteins partially exposed on the outer surface of the platelet is the thromboxane (TXA_2) receptor (Hanasaki et al., 1989), which is also able to bind the unstable intermediate, PGH_2. Both TXA_2 and PGH_2 are generated in the dense tubular system of the platelet (see Figure 13.1), and must travel from this intracellular site to the platelet surface by an unknown route to bind to their receptor. Morinelli et al. (1987) have proposed that the TXA_2/PGH_2 receptor complex contains several binding sites

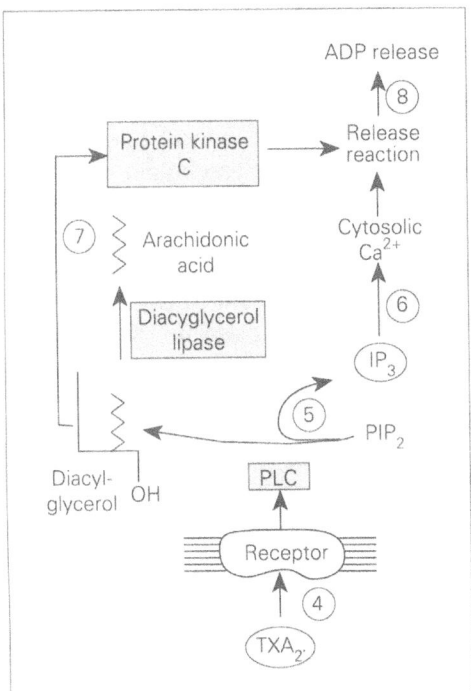

Figure 13.2 Thromboxane/receptor interactions lead to the platelet release reaction: **4** Thromboxane (TXA_2) binds to platelet receptor; **5** Phospholipase C (PLC) acts on phosphoinositol 4,5 biphosphate (PIP_2) to produce the second messenger inositol 1,4,5 tri-phosphate (IP_3) and diacylglycerol (DAG); **6** IP_3 mediates calcium ion (CA^{2+}) mobilisation; **7** DAG activates protein kinase C (PKC); **8** PKC releases contents of amine storage granules including adenosine diphosphate (ADP).

with different affinities for the prostaglandin ligands (step 4 in Figure 13.2).

The binding of TXA_2 to high-affinity sites on its receptor brings about changes in platelet shape. Further TXA_2-receptor binding on lower-affinity sites appears to induce an increase in cytosol calcium (Ca^{2+}) concentration which is required for secretion of granule contents. This mobilisation of Ca^{2+} involves a number of second messenger compounds as intermediaries. Activation of PLC (step 5 in Figure 13.2) leads to the generation of inositol 1,4,5 trisphosphate (IP_3) and diacylglycerol (DAG) from phosphoinositol 4,5 bisphosphate (PIP_2). In the subsequent step (step 6 in Figure 13.2), IP_3 mediates a transient Ca^{2+} mobilisation from membrane stores (Streb *et al.*, 1983). The other messenger, DAG, triggers activation of a further membrane enzyme, protein kinase C (PKC) (Nishizuka, 1986), which promotes phosphorylation of a number of substrate proteins as part of the activation process (step 7 in Figure 13.2).

Thus, collagen-stimulated activation of the resting platelet seems to proceed by stages which reinforce each other (Arita *et al.*, 1989). The initial wave of TXA_2 generation by PLA_2 leads to stimulation of the TXA_2/PGH_2 receptor and change

in platelet shape. The formation of the TXA_2-receptor complex triggers PLC activation, IP_3-mediated Ca^{2+} mobilisation, DAG-mediated activation of PKC, and the release of the contents of the intracellular storage granules.

13.2.3 The platelet release reaction

Platelets contain protein storage granules (alpha granules) and amine storage granules (dense granules). Several of the proteins stored in the alpha granules are exclusive to the platelet including platelet factor 4, beta-thromboglobulin, thrombospondin and platelet-derived growth factor. Other stored proteins, not specific to the platelet, include fibrinogen, fibronectin, von Willebrand's factor, factor V, tissue-plasminogen activator (t-PA), plasminogen-activator inhibitor, plasminogen and albumin. The membrane of the alpha granule also contains specific receptor proteins, especially the GPIIb-IIIa adhesive receptor (Wencel-Drake *et al.*, 1986) (see Section 13.2.5).

The amine storage granule contains adenosine diphosphate (ADP) as the dominant nucleotide, as well as ATP, serotonin and Ca^{2+}. When activated by secretory agonists such as PKC, the contents of the granules are released from the platelet (step 8 in Figure 13.2), and the granule membrane appears to fuse with the surface membrane (Wencel-Drake *et al.*, 1986). Released ADP is immediately coupled to its receptor protein on the platelet surface (aggregin), and the ligand–receptor complex induces exposure of fibrinogen-binding sites and platelet aggregation.

13.2.4 Platelet aggregation

13.2.4.1 Collagen-induced ADP-dependent platelet aggregation

Collagen-induced platelet aggregation proceeds via TXA_2 production, the release reaction, and ADP-binding to its aggregin receptor. Thus it can be inhibited by TXA_2- and ADP-receptor blockers. Eicosanoid-induced aggregation is dependent upon the release of ADP and ADP-receptor coupling (Morinelli *et al.*, 1983). ADP-induced aggregation is reversible at low concentration (<0.5 µmol/l), but irreversible at higher concentration (secondary aggregation) as a result of stimulation of TXA_2-mediated secretion of platelet-granule contents.

13.2.4.2 Thrombin-induced ADP-independent platelet aggregation

Unlike other agonists, thrombin's ability to induce aggregation is independent of ADP. Binding of

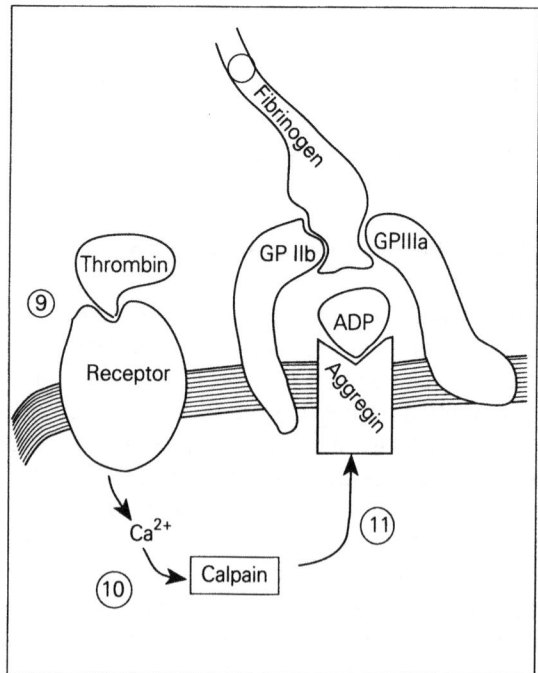

Figure 13.3 Role of thrombin and aggregin in platelet aggregation: 9 Thrombin binds to its receptor; 10 Raised intracellular Ca^{2+} activates calpain; 11 Calpain causes the proteolysis of aggregin, permitting exposure of the fibrinogen receptor.

thrombin to its receptor (step 9 in Figure 13.3) appears to stimulate phospholipase C (PLC) activity and exposure of fibrinogen receptors by a Ca^{2+}-dependent pathway. One possibility is that a raised intracellular Ca^{2+} induces the localisation and activation of the enzyme calpain (step 10 in Figure 13.3) (Sakai et al., 1989) which by proteolysis of aggregin (step 11 in Figure 13.3), permits exposure of the fibrinogen receptor (Colman, 1990).

13.2.4.3 Role of aggregin and fibrinogen in platelet aggregation (see Figure 13.4)

The binding of ADP to its surface receptor, aggregin, permits changes in the fibrinogen receptor, fibrinogen binding, and platelet aggregation. This direct effect of ADP on aggregin-mediated changes in the fibrinogen receptor means that

ADP-induced aggregation can occur prior to ADP-stimulated TXA_2 synthesis and the secretory response (Charo et al., 1977). The interaction of aggregin with ADP, or possibly its proteolysis by calpain, is thought to permit conformational re-alignment of the two glycoprotein sub-units of the fibrinogen receptor, the complex of GPIIb and GPIIIa, so that it can bind fibrinogen, von Willebrand's factor and fibronectin. The dimeric structure of fibrinogen allows it to link the GPIIb-GPIIIa receptors of adjacent platelets, thereby promoting aggregation. The ability of the receptor to bind multiple proteins resides in its recognition of the sequence arginine–glycine–aspartic acid which is common to many adhesive proteins (Plow et al., 1985) (see Table 13.1).

13.2.5 Platelet inhibitory mechanisms

As outlined above, platelet adhesion and activation triggers shape changes, the release reaction, agonist induction of the GPIIb-IIIa receptor, platelet aggregation and reinforcement of platelet adhesion to the site of injury. It is paramount that the birth, growth and maturation of this acute process must be followed by its in-built senescence so that it does not become over-extensive and threaten platelet thrombus and embolism. This final step in the sequence appears to be achieved partly by inhibitory agonists, and partly by the phenomenon of platelet desensitisation.

13.2.5.1 Inhibitory agonists

The principal inhibitory agonists are adenosine, the prostaglandin PGD_2 and prostacyclin (PGI_2). The coupling of the inhibitory agonist to its platelet surface receptor is thought to activate the inhibitory second messenger system, in which the membrane-bound enzyme adenylate cyclase plays a key role. Activation of this enzyme leads to an increase in intra-platelet levels of cyclic adenosine monophosphate (cAMP).

The inhibitory second messenger system may also act in part by stimulating Ca^{2+} pump systems, thereby lowering cytosolic Ca^{2+} levels and reducing the rate of calcium dependent reactions which

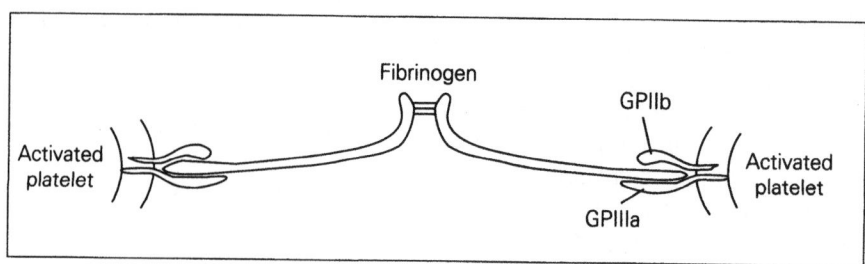

Figure 13.4 Role of fibrinogen in platelet aggregation.

effect secretory and aggregatory responses. The local concentrations of these inhibitory agonists would presumably increase until they are sufficient to down-regulate the processes of activation and aggregation occurring within the platelets.

13.2.5.2 Platelet desensitisation by low levels of excitatory agonists

Low levels of excitatory agonists, which presumably arise some distance from the centre of the acute process, appear to desensitise the platelet to higher concentrations of the agonist, though the mechanism by which this happens is unclear. This ability may serve to confine platelet recruitment to within spatial limits consonant with the extent of injury.

The complex interplay of the excitatory and inhibitory platelet pathways is illustrated by the presence on the platelet surface of a distinct ADP receptor which, when coupled by ADP, counteracts the accumulation of cAMP which has been induced by prostacyclin (PGI_2) (see Table 13.1) (Mills et al., 1985).

13.2.6 Platelet effect on vasoconstriction

Platelet-derived thromboxane (TXA_2) not only induces platelet aggregation, but also acts as a potent local vasoconstrictor (Bunting et al., 1976) thereby restricting blood flow and promoting haemostasis.

13.3 VESSEL WALL RESPONSES TO INJURY

13.3.1 Prostacyclin formation

Several substances released from platelets (ADP and ATP, serotonin, platelet-derived growth factor (PDGF)) or generated by the coagulant response to injury (e.g. thrombin) act as ligands for endothelial cell surface receptors in the walls of blood vessels (see steps 1 and 2 in Figure 13.5).

Formation of the ligand-receptor complex leads to liberation of arachidonic acid and synthesis of the unstable prostaglandin intermediates PGG_2 and PGH_2, as in the platelet (steps 2 and 3 in Figure 13.5). In the endothelial cell, these products are converted to prostacyclin (PGI_2) by the enzyme prostacyclin synthetase (Moncada et al., 1977) rather than being converted to thromboxane as they are in the platelet. Prostacyclin is a potent inhibitor of platelet TXA_2 synthesis (Lefer et al., 1978) and platelet aggregation (step 7 in Figure 13.5). It is also a powerful vasodilator via its action on smooth muscle cells (step 5 in Figure 13.5) (Moncada et al., 1976).

Prostacyclin has a very short lifespan, breaking down spontaneously into a stable, but inactive, metabolite. As in the platelet, however, the consequences of this short burst of eicosanoid synthesis and metabolic activity depend upon the half-lives of the products of this activity, which may be more sustained.

13.3.2 Endothelium-derived relaxing factor (EDRF)

Various ligands, including ATP, stimulate the generation of another substance from endothelial cells called endothelium-derived relaxing factor (EDRF) (step 1 in Figure 13.5). This promotes relaxation of the underlying smooth muscle and vasodilation (Furchgott et al., 1980). Recent evidence suggests that the nature of EDRF may, in fact, be nitric oxide (Palmer et al., 1987). The generation of nitric oxide appears to activate the enzyme guanylate cyclase and the resultant production of cyclic guanosine monophosphate (cGMP) induces relaxation of smooth muscle cells (step 4 in diagram). Like prostacyclin, EDRF appears also to act as a platelet inhibitory agonist, raising platelet cGMP level and thereby down-regulating aggregatory activity (Radomski et al., 1987) (step 6 in Figure 13.5). The initial step in both prostacyclin and EDRF production seems to involve activation of phospholipase C (PLC) in the endothelial cell.

13.3.3 A balance of opposing factors?

The opposing effects of TXA_2 and prostacyclin/EDRF have led to the concept that the extent of platelet recruitment after vessel-wall injury is a function of the ratio of the concentrations of the opposing factors. Thus interventions that change the balance are conceived as altering thrombosis risk. If the ratio of TXA_2/prostacyclin-EDRF is raised, the risk of thrombosis is higher and vice versa.

13.4 UNSATURATED FATTY ACIDS AND CELL MEMBRANE FUNCTION

13.4.1 Background

Many of the cellular responses to vessel wall injury are mediated by receptors, signal transducer systems and effector systems located within the surface and intracellular membranes. The func-

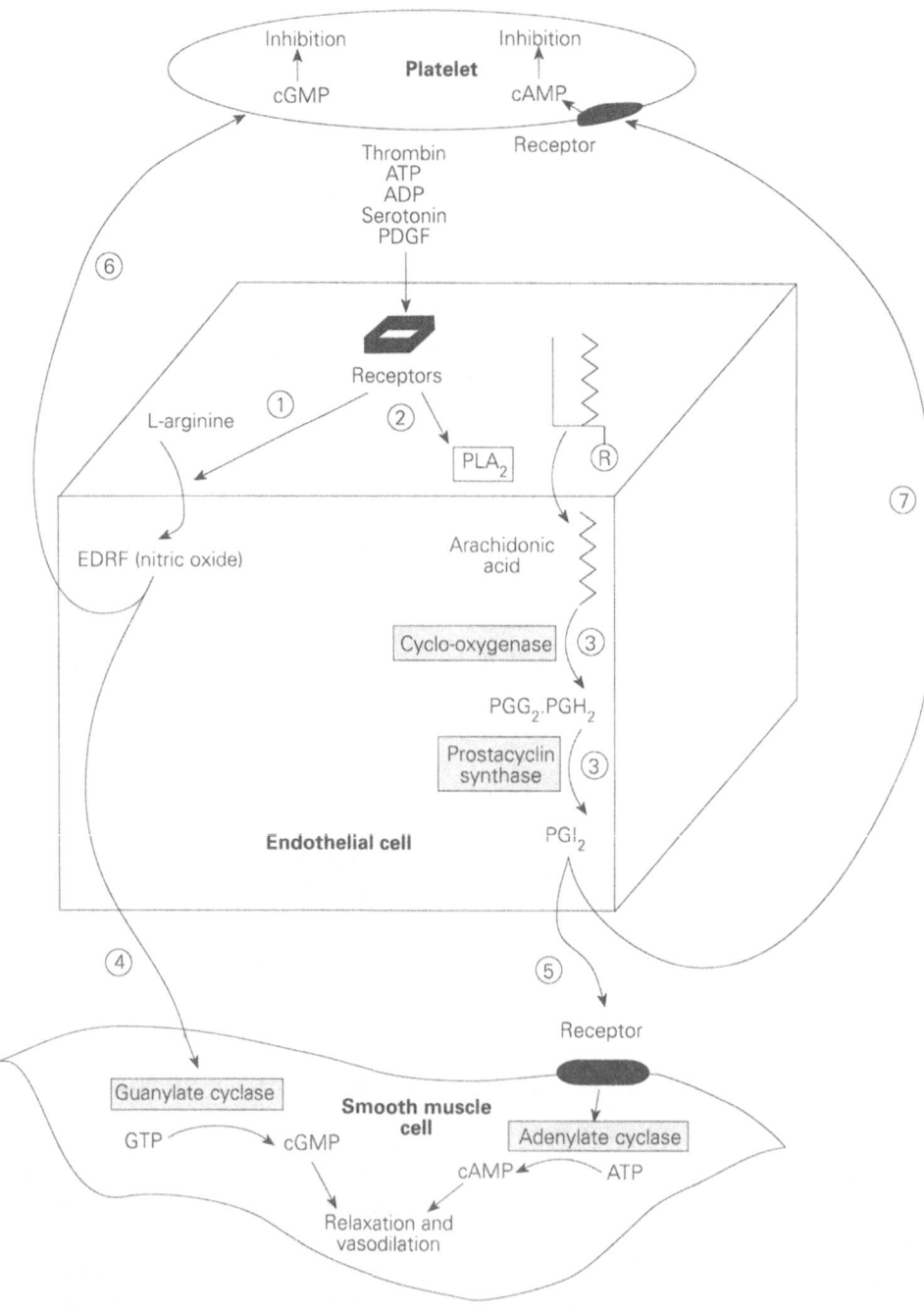

Figure 13.5 Vessel wall responses: **1** Various ligand–receptor complex formations on the endothelial cell surface membrane trigger the generation of endothelium derived relaxing factor (EDRF); **2** Various ligand–receptor complex formations on the endothelial cell surface membrane trigger phospholipase A$_2$ (PLA$_2$) formation which releases arachidonic acid from phospholipids; **3** Arachidonic acid is converted via prostaglandin intermediates to prostacyclin I$_2$ (PGI$_2$); **4** EDRF stimulates smooth muscle cell relaxation via cyclic GMP (cGMP) formation; **5** PGI$_2$ stimulates smooth muscle cell relaxation via cyclic AMP (cAMP) formation; **6** EDRF inhibits platelet aggregation via cGMP; PGI$_2$ inhibits platelet aggregation via interaction with receptor and subsequent cAMP formation.

tional alignment of the protein components, both in their resting and activated state, is partly governed by interactions with surrounding phospholipid molecules. Furthermore, phospholipids serve as substrates for enzymes such as phospholipase A (PLA_2) and phospholipase C (PLC). Two questions arise:

(i) Is the efficiency of these systems modulated in any important way by their interactions with the surrounding lipid?
(ii) Does the fatty acid composition of the diet influence membrane lipid composition in ways that have consequences for the metabolic processes underlying vessel wall repair, haemostasis and thrombosis?

The complex sequences of molecular events which constitute effective performance of the numerous protein systems in cell membranes call for a certain amount of rapid, short-range, ordered molecular motion and conformational change. Such movements are permitted by the liquid-crystalline state of the lipid bilayer (Singer and Nicholson, 1972). The fluidity of the bilayer can be conceived in terms of the amplitudes and rates of motion of the molecules in the lipid matrix, both with respect to rotations about their axes and diffusion. These parameters vary with the physico-chemical properties of the constituent phospholipids and with the cholesterol/phospholipid ratio. In general (but see Chapter 8), unsaturated fatty acids increase membrane fluidity in ways which depend upon the number and position of double bonds in the acyl chains (Hague, 1988). Cholesterol enrichment, on the other hand, decreases membrane fluidity (Shattil and Cooper, 1976).

13.4.2 Some examples of membrane lipid effects on enzyme *in vitro* systems

In human platelets, basal and PGE_1-induced adenylate cyclase activity are raised when platelets are enriched *in vitro* with phosphatidylcholine with linoleic acid at position 2 (Chambaz *et al.*, 1983). Salesse and Garnier (1984) have proposed that the PGE_1 receptor and its membrane regulatory complex normally repress the activity of adenylate cyclase. Ligand binding would therefore lead to some dissociation and de-repression, allowing increased generation of cAMP. According to the hypothesis, enrichment of membrane phospholipid with an unsaturated fatty acid would create a shift in the equilibrium of the receptor/regulatory unit/ catalytic unit complex towards dissociation, elevation in intra-platelet cAMP, reduction in cytosolic Ca^{2+}, and a lowering of platelet responsiveness to

excitatory agonists, i.e. a reduction in platelet aggregation.

In one study (Sinha *et al.*, 1977), but not in another (Insel *et al.*, 1978), cholesterol enrichment of the platelet membrane reduced the ability of PGE_1 binding to stimulate adenylate cyclase activity. In other studies, the platelet aggregatory responses to ADP (Shattil *et al.*, 1975) and thrombin (Stuart *et al.*, 1980) were enhanced by cholesterol enrichment of membranes, with liberation of arachidonic acid and conversion to prostaglandins.

Whether changes such as these are of value in survival terms, or in prevention of morbidity due to thrombosis, is not clear at present.

13.4.3 Membrane fatty acids and eicosanoid metabolism (Figure 13.6)

The preceding sections have described how, in human platelets, arachidonic acid at position 2 of the glycerol backbone of phospholipid serves as the normal substrate for TXA_2 synthesis. In the endothelial cell, arachidonic acid also acts as the substrate for eicosanoid metabolism, with PGE_2 and prostacyclin (PGI_2) as the major end-products. However, other UFA, particularly dihomogamma linolenic acid (20:3 n-6) and eicosapentaenoic acid (EPA) (20:5 n-3), can compete with arachidonic acid for esterification at position 2 of the phospholipid. Once esterified, they can also be liberated by phospholipase A_2 (PLA_2) (Needleman *et al.*, 1979) and can then act as alternative substrates for cyclo-oxygenase.

The conversion of EPA to PGH_3 and then to TXA_3 in the platelet is of particular significance (Needleman *et al.*, 1979; Hamberg, 1980). In contrast to PGH_2 and TXA_2, PGH_3 and TXA_3 are very poor inducers of platelet activation. The enrichment of platelets with EPA might therefore be expected to lower platelet responsiveness to excitatory agonists because the ratio of TXA_2:TXA_3 would be lowered.

The effect of EPA in the endothelial cell is to produce PGI_3 from PGH_3. This has inhibitory effects similar to those of prostacyclin PGI_2 (Needleman *et al.*, 1979).

Considering the combined role of platelets and the endothelial cells of the vessel wall, the overall effect of substituting EPA for arachidonic acid would be to reduce the likelihood of platelet aggregation because the balance of opposing factors (section 13.3.3) would be shifted so that the important ratio of TXA_2 : prostacyclin-EDRF is lowered.

In summary, the results of studies considered here provide ample reason to believe that the lipid

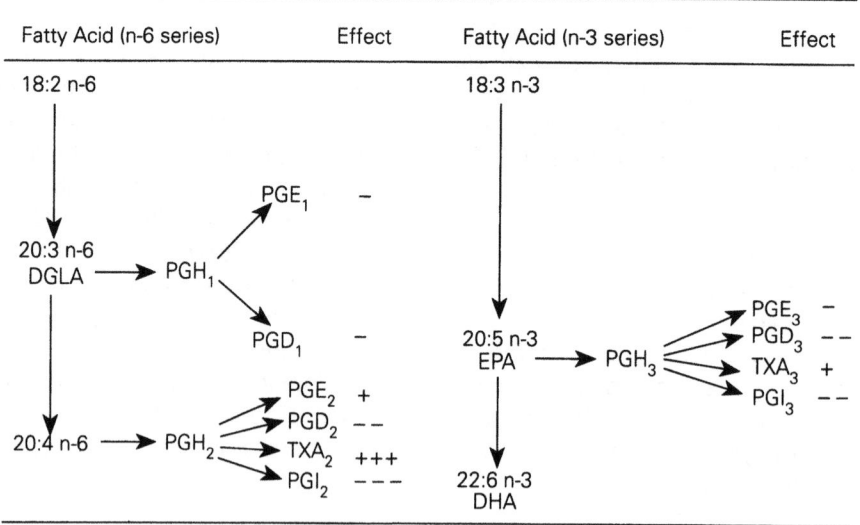

Figure 13.6 Effect of different prostaglandins on platelet aggregation. PGH = prostaglandin endoperoxides; TXA = thromboxanes; PGI = prostacyclins; PGD/E = types of prostaglandin (subscripts denote number of double bonds). +, weakly stimulating; +++, strongly stimulating; −, weakly inhibitory; − −, moderately inhibitory; − − −, strongly inhibitory.

composition of cell membranes could modulate many of the processes concerned with vessel wall repair and haemostasis. The rest of this Chapter will discuss whether these changes could be protective against thrombosis and whether they could be of survival value.

13.5 DIETARY MANIPULATION OF PLATELET CELL MEMBRANE LIPID COMPOSITION

13.5.1 Background

There is much evidence for changes in the fatty acid composition of many cell membranes with alteration in the composition of dietary fat. There is less evidence, however, for dietary-induced change in the cholesterol/phospholipid ratio of cell membranes. A diet with a high content of margarine rich in PUFA led to both linoleate and cholesterol enrichment of the cell membranes of guinea pig red blood cells, thereby minimising changes in membrane fluidity (Edwards-Webb and Gurr, 1988). Dietary-induced changes in fatty acid phospholipid composition in membranes appear to be limited to exchange of one species of UFA (principally arachidonic acid) for another at the position 2 of the glycerol backbone. Furthermore, elongation, desaturation and retroconversion enzymes subject these fatty acids to continual re-modelling processes, which seem to operate preferentially on the n-3 series rather than the n-6 series (see Chapter 7).

13.5.2 Dietary monounsaturated fatty acids

The consequences of dietary enrichment with oleic acid (18:1) on platelet membrane composition have not been studied extensively. Barradas et al. (1990) gave 21 g of olive oil to 7 healthy adults daily for 8 weeks, and noted a small, but statistically significant, increase in platelet-membrane oleic acid, together with a small reduction in membrane arachidonic acid content. The proportions of other fatty acids in membrane total fatty acid were unaffected.

13.5.3 Dietary n-6 PUFA

A diet rich in n-6 PUFA, particularly linoleic acid, has no effect on the phospholipid composition or the cholesterol/phospholipid ratio of the platelet. It does, however, induce small increases in the linoleic acid content of phosphoglycerides PtdEtn, PtdSer and PtdCho (Nordoy and Rodset, 1971). Vegetarians have been found to have a significantly higher proportion of linoleic acid in their platelet phospholipids than adults eating a mixed diet (Fisher et al., 1986). The presence of highly unsaturated fatty acids with more than 20 carbon atoms in platelet phospholipids does not appear to be influenced by dietary linoleic acid (Nordoy and Rodset, 1971).

The two initial products of desaturation and elongation of linoleic acid are gamma linolenic acid and dihomogamma linolenic acid (DGLA); these are present in human diets only in trace quantities and can therefore be neglected in the present context. The desaturation product of DGLA, arachidonic acid, is found only in quantities of about 1.0 g/day in the mixed diet and most membrane arachidonic acid is derived from the metabolism of dietary linoleic acid. Significant increases in the proportion of arachidonic acid in platelet phospholipid fatty acids can be achieved only by pharmacological dosages of the purified acid (Seyberth et al., 1975) and decreases achieved only by feeding diets deficient in essential fatty acids (EFA) (Loiacono et al., 1987). Thus, of the dietary n-6 fatty acids, only the content of linoleic acid is of potential relevance for the function of cell membranes and platelet and endothelial cell responses to vessel wall injury.

13.5.4 Dietary n-3 PUFA

Current intense interest in the relation between marine oils, thrombosis tendency and cardiovascular disease stems from the proposal that the high content of these oils in the traditional diet of the Inuit (Eskimo) (Bang et al., 1976) may be antithrombotic and may contribute to their low incidence of coronary heart disease (CHD) (Dyerberg et al., 1978) (see Chapter 10). Whereas in most 'Western' diets, the long chain highly unsaturated oils of marine origin amount to less than 0.2 g/day, Inuits consume them in quantities of about 9–13 g/day. Because of this, eicosapen-

taenoic acid (EPA) accounts for about 8% of platelet lipid in the Inuit and about 0.5% of platelet lipid in his Danish counterpart. There are compensatory reductions in arachidonic acid and linoleic acid in platelet lipids (8% and 4% respectively in the Inuit, as compared with 22% and 8% in the Dane) (Dyerberg and Bang, 1979).

Numerous dietary feeding studies, in which fish isocalorically replaced other foods or in which cod liver oil, sardine oil or other commercial fish oil concentrates were given as dietary supplements, have all confirmed the prompt, sustained and dose-dependent enrichment in cell membrane phospholipid EPA content induced by such interventions (see Table 13.2). Membrane phospholipid arachidonic acid content is simultaneously reduced in a dose-dependent manner, but because the decrease is proportionally smaller than that in EPA, it frequently fails to achieve statistical significance in small studies.

13.6 INDICES OF THE RISK OF THROMBOSIS AND THE LIMITATIONS OF THE METHODOLOGY

In view of the ability of EPA to compete with arachidonic acid as a substrate for cyclo-oxygenase (Needleman et al., 1979) and the fact that PGH_3 and TXA_3 produced by processing of EPA are

Table 13.2 The effect of dietary n-3 PUFA on EPA and arachidonic acid changes in platelet membrane phospholipids

Study	Intervention	Duration of trial (weeks)	Dose	Membrane phospholipids	
				EPA	Arachidonic acid
Siess et al. (1980)	Fish	1	7–11 g/day EPA	Increase	Decrease
Thorngren and Gustafsen (1981)	Fish	11	3 g/day EPA	Increase	Non-significant decrease
Thorngren et al. (1984)	Fish	6	3 g/day EPA	Increase	Decrease
Sanders et al. (1981)	Cod liver oil	3	1.8 g/day EPA	Increase	Decrease
Von Schacky (1985)	Cod liver oil	4–8	1–4 g/day EPA	Increase	Dose-dependent decrease
Brox et al. (1981)	Cod liver oil	4	2 g/day EPA	Increase	Decrease in PtdCho only
Hirai et al. (1982)	Sardine oil	4	1.4 g/day EPA	Increase	Non-significant decrease
Goodnight et al. (1981)	Sardine oil	4	10 g/day n-3	Increase	Decrease
Schmidt et al. (1990)	Fish oil concentrate	6	1.3–9 g/day n-3	Dose-dependent increase	Non-significant decrease on 9 g/day
Sanders and Roshanai (1983)	Fish oil concentrate	3	3.3 g/day EPA	Increase	Decrease
Terano et al. (1983)	Fish oil concentrate	4	3.6 g/day EPA	Increase	No change

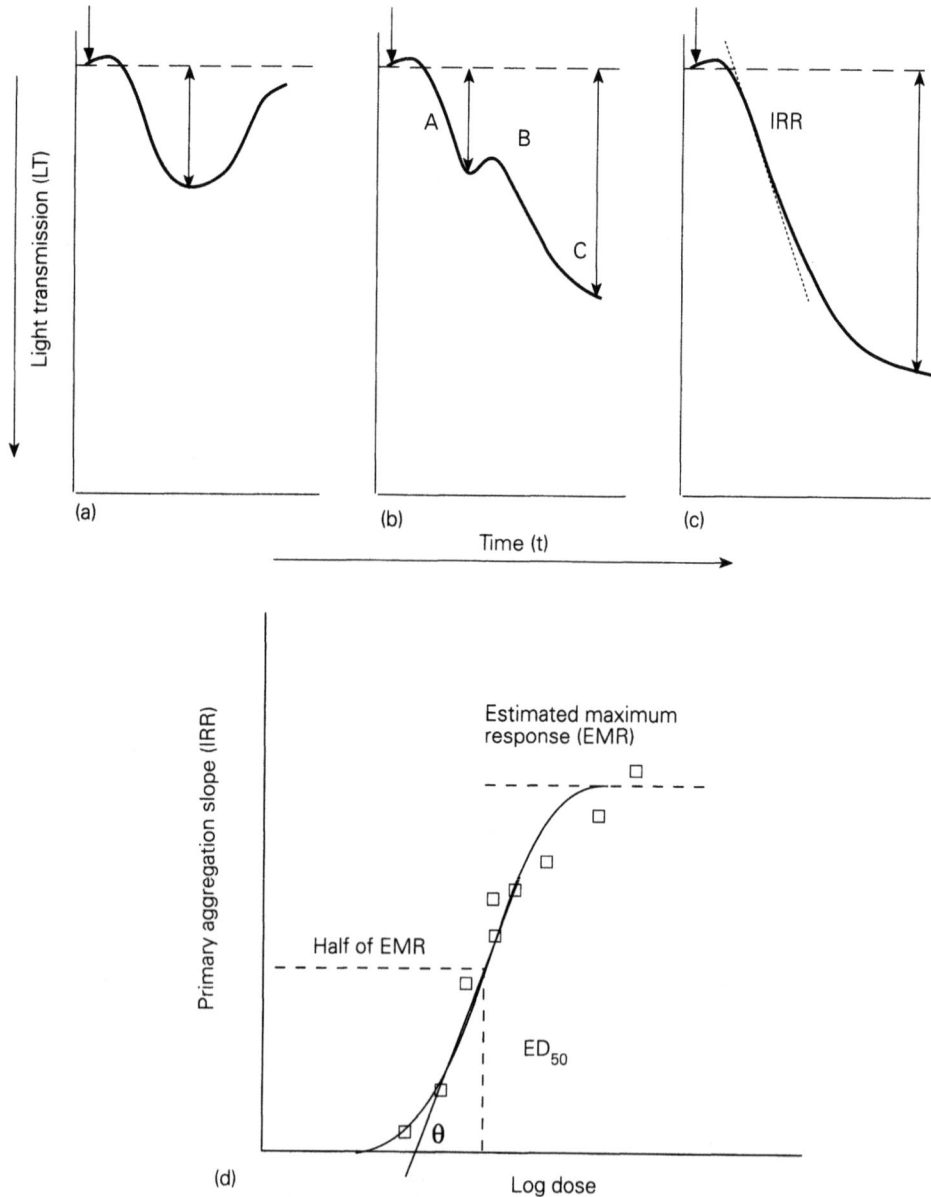

Figure 13.7   Optical densitometry response patterns when agonists are added to platelet-rich plasma: (a) reversible aggregation response – aggregation followed by disaggregation; (b) primary and secondary aggregation responses (A primary aggregation, B release reaction, C secondary reaction); (c) irreversible aggregation (IRR = initial rate of reaction); (d) an estimated platelet aggregation dose–response curve (EMR = estimated maximum response; ED_{50} = estimated dose giving 50% of the EMR; θ = scaled angle; $\downarrow$ = time at which agonist is added to stirred platelet suspension; $\updownarrow$ = maximal response expressed as change in light transmission).

relatively poor inducers of platelet activation (Needleman *et al.*, 1979), the question arises as to whether dietary fish oils are able to reduce thrombotic tendency. This problem has been addressed primarily by examination of the effect of marine n-3 fatty acid intake upon platelet responsiveness to excitatory agonists *in vitro*, and on the bleeding time.

13.6.1   Platelet studies *in vitro* (see Figure 13.7)

The responses of platelets to aggregating agents are currently assessed by three techniques:

(i)   Optical densitometry is a technique where the changes in light transmission through platelet-rich plasma are measured as the platelets are induced to aggregate. As the platelets aggregate, the transmission approaches that of platelet-poor plasma (Born, 1962). This is the original and most popular procedure and will be discussed in detail below.

(ii) Single platelets can be counted in whole blood before and after agonist exposure (Frojmoric et al., 1989)

(iii) The electrical impedence of whole blood can be measured before and after agonist exposure (Cardinal and Flower, 1980).

13.6.1.1 Problems with optical densitometry

The conditions under which platelet responsiveness is assessed by this technique differ considerably from those in which platelets probably function *in vivo*. Some of the differences are:

(i) The preparation of platelet-rich plasma removes red and white blood cells, which may normally influence the performance of platelets *in vivo*.

(ii) There may be a loss of platelets which spontaneously aggregate during preparation. Prostacyclin is occasionally added during preparation to prevent this problem (von Schacky et al., 1985).

(iii) The platelet count is standardised (usually at 300×10^9 per litre) by addition of platelet-poor plasma to platelet-rich plasma derived from the same patient. There is an inverse relation between platelet count and platelet size, so any potential effect of this interplay *in vivo* on overall function might not be observed *in vitro*.

Further problems with optical densitometry arise from the plethora of indices used to describe platelet responsiveness to a variety of agonists. All of the following have been measured and used as indices:

(i) the decrease in optical density upon exposure to one or more standardised concentrations of agonist, within either a fixed time or an unspecified period (see Figure 13.7a-c);

(ii) the 'lag period', i.e. the time elapsed between the addition of agonist and the start of a detectable response;

(iii) maximum velocity of response;

(iv) the integrated velocity of response or rate of change of light transmission within a fixed time period;

(v) the concentration of agonist needed to obtain a prespecified proportion of the maximum response;

(vi) the concentration of agonist needed to elicit a detectable response;

(vii) the minimum concentration needed to produce

secondary phase aggregation (i.e. to induce the platelet release reaction) (see Figure 13.7b).

The assumption is that platelets with a low tendency to aggregate will need higher doses of agonist to elicit minimal, half-maximal and maximal responses and will show a small change in optical density in standard time with standard exposure to agonist. They will also be less likely to show secondary aggregation. There appear to be only limited data to confirm or refute these assumed inter-relations.

13.6.1.2 Improvements in optical densitometry

The assessment of aggregability in terms of a dose–response curve is more time-consuming but a more reliable procedure (Sanders et al., 1981; Thompson and Vickers, 1985). Different final concentrations of agonist are created sequentially in aliquots of platelet-rich plasma, and the rate of primary aggregation is measured as the change in light transmittance per unit time over the steepest initial sector of the response curve. The relation between initial rate of response (IRR) and log dose of reagent is then plotted, generally revealing a sigmoid curve. Several indices are taken off this curve to describe aggregability, namely the maximal velocity of response (the upper asymptote of the curve (EMR)), the estimated log dose giving half maximal velocity of response (ED_{50}) and the scaled angle (the angle in degrees which the tangent to the curve at ED_{50} describes with the horizontal axis) (see Figure 13.7d).

The natural assumption in dose–response aggregometry is that a decrease in maximal response is associated with an increase in ED_{50} and a decline in the scaled angle. This is sometimes, but not always, the case. Furthermore, in one study, correlations between corresponding indices for ADP and adrenaline, although of expected sign, were generally weak (Thompson and Vickers, 1985).

It is also important to determine the relationship between the dose of agonist and the eicosanoid response in platelet studies. There is clearly a maximal dose of agonist above which no further platelet aggregation can be achieved (Thompson and Vickers, 1985). Lands et al. (1985) showed that thromboxane (TXA_2) formation continued to increase with increasing amounts of agonist above the dose which produces a maximal aggregatory response. Platelet aggregation studies using high doses of agonist are therefore likely to be insensitive to small decreases in eicosanoid production. Although reduced, these may still remain above that needed to induce maximal aggregation.

13.6.1.3 Other sources of error in platelet aggregation studies

The spontaneous within-person variability in platelet indices of aggregability is large, even in dose–response aggregometry. In a study of platelet response to the agonists ADP and adrenaline in 15 healthy adults over an 18-month period, within-person variability accounted for as much as about 50% of the total variability in the ED_{50} estimates and method error amounted to about 40% of total variance observed in the estimated maximal response (Vickers and Thomson, 1985).

Method error and spontaneous within-individual variability in measures of platelet aggregometry appear to be so large that the risk of false conclusions from dietary feeding studies seeking small effects is likely to be considerable, even when the individual is used as his or her own control. The reliability of the results would be somewhat improved if the average of repeated studies was used, but this precaution is not often taken.

Other weaknesses in study design and analysis include the lack of assurance that the dietary treatments being compared are given in randomised alternate order, and that those making measurements have no knowledge of the origins of the samples tested, and the failure to appreciate the effects of multiple comparisons on statistical analyses.

Many studies have been too small for reliable detection of small dietary-induced changes in platelet behaviour. Large numbers of subjects are needed to provide valid results at the 5% level of statistical significance. A study of 12 people has only a limited chance of detection of a shift in platelet aggregability of less than one standard deviation (SD) of the within-person variability in the index chosen. 468 subjects must be studied to provide a 90% probability of detection of a change in platelet sensitivity of 0.15 within-person SD units (Meade et al., 1985a). Few data are available for the inherent within-person variability in platelet behaviour as assessed by single-dose optical densitometry (Yarnell et al., 1987).

Newer techniques of instrumentation such as whole blood aggregometry are only part of the solution to these design weaknesses. In one study, the simple correlation between responsiveness to ADP given by single-dose optical densitometry in platelet-rich plasma and by impedance changes in whole blood was 0.22, indicating considerable discrepancy (Sharp et al., 1990).

13.6.2 The bleeding time

13.6.2.1 Origins

The bleeding time was first described by Milian in 1901. Within a few years, reports appeared of an increased bleeding time in Inuits (Trebitsch, 1907). Since then, bleeding time has been used clinically as an index of bleeding tendency and as a measure of the efficacy of therapy for bleeding disorders. Its use as an index of thrombotic risk has stemmed from the observation that Inuits have a traditionally low mortality from coronary heart disease, and hence a low risk of coronary thrombosis.

13.6.2.2 Improvements

In an attempt to improve reproducibility, numerous modifications to the original procedure have been introduced. Most modern versions are forms of the test proposed by Ivy et al. (1941). The incision is made on the volar aspect of the forearm using a sphygmomanometer cuff on the upper arm to maintain a standardised venous blood pressure. Channing et al. (1990) estimated the reliability of the bleeding time as an indicator of aspirin therapy. (Aspirin irreversibly inactivates platelet and endothelial cyclo-oxygenase, and thus reduces the aggregability of platelets in vitro.) They concluded that the bleeding time was very unreliable because it had low sensitivity, low specificity and was almost certainly prone to many errors.

As with tests of platelet aggregability, use of the individual as his or her own control will obviate but not completely control for these weaknesses in the test. Apparent discrepancies between studies, or between the results of bleeding time and platelet aggregometry in a single study, may arise simply because of the limited reliability of the methods employed.

These inherent weaknesses in the methods of assessment must always be remembered in assessing the evidence for an effect of dietary unsaturated fatty acids on thrombotic tendency.

13.7 DIETARY UNSATURATED FATTY ACIDS AND PLATELET RESPONSIVENESS

13.7.1 n-3 PUFA and platelet aggregability

13.7.1.1 Large amounts

Dyerberg and Bang (1979) examined the bleeding time and platelet responsiveness to ADP in 14 men and 7 women with typical Inuit diets, and compared the findings with those for age and sex-matched

Danish adults. Mean bleeding times were 8.05 and 4.76 min respectively. In 10 Inuits, but in none of the Danes, secondary-phase aggregation (i.e. a platelet release reaction) could not be induced with the highest doses of agonist employed (collagen and ADP). However, marine oil n-3 fatty acid intake is exceptionally high in these people, and is derived from marine mammals rather than fish. Hirai et al. (1982) compared platelet performance in people belonging to a Japanese fishing village (average individual fish intake, 250 g/day) and a farming village (fish intake, 90 g/day). The concentration of ADP needed to produce 50% of maximum platelet activation was reported to be higher in the people living in the fishing villages.

The amount of fish consumed by Japanese fishermen (Hirai et al., 1982) was estimated to have supplied about 2.5 g of EPA daily. This amount is contained in approximately 25 ml of cod liver oil or 12.5 g of EPA capsules (these oils are also rich in oleic acid, palmitic acid and DHA). Table 13.3 summarises the reported effects of this intake of EPA in 10 experimental feeding studies.

There is little doubt that bleeding time is prolonged, an impressive finding considering the questionable reproducibility of the test.

Platelet responsiveness to collagen in vitro (at various doses between 0.25 and 4.0 µg/ml platelet-rich plasma) was depressed in most studies. Platelet responsiveness to ADP gave inconsistent results. The aggregation response to collagen is dependent on TXA_2 and ADP, whilst ADP can induce aggregation by a direct effect upon

aggregin. The balance of the evidence therefore suggests an effect of EPA on aggregatory mechanisms that are TXA_2-dependent (see Section 13.2.2.3). However, in two of the studies, a reduced sensitivity of platelets to ADP was observed and it persisted for several weeks after feeding EPA, by which time platelet phospholipids had returned to normal. These observations led to the suggestion that fish oils may also influence platelet aggregation by other means.

One problem that seems not to have been considered sufficiently is the likely effect of dietary-induced changes in platelet number and size on the results of optical densitometry on platelet-rich plasma. There is little doubt that the platelet count frequently decreases when the diet is enriched with EPA or linoleic acid and that this reduction is accompanied by an increase in mean platelet volume (Goodnight et al., 1981; Li and Steiner, 1990). On account of the routine standardisation of platelet count in platelet-rich plasma prepared for optical densitometry (see Section 13.6.1.1), post-dietary test plasmas are likely to contain more platelet material than pre-dietary plasmas. This difference may well tend to obscure any suppressive effects of UFA on platelet responsiveness. Another factor not apparently taken into account is the possible effect of dietary fat composition on plasma fibrinogen concentration (see Section 13.10.4). Platelet aggregometry results are undoubtedly influenced by changes in plasma fibrinogen within the physiological range (Bloom and Evans, 1969; Meade et al., 1985a).

13.7.1.2 Normal dietary amounts

While large amounts of dietary fatty fish (about 250 g/day) or large amounts of orally administered EPA appear to prolong bleeding times and diminish TXA_2-dependent aggregation, intakes of EPA consistent with lower levels of fish consumption which are acceptable to communities eating a mixed diet (say, 30 g/day) appear less likely to have effects on platelet aggregability (Rogers et al., 1987). In the Zutphen study (Van Houwelingen et al., 1989), 40 healthy elderly men from the study population were selected on the basis of their habitual consumption of fish. The high-fish eating group took an average of 33 g/person/day, whereas the low-fish eating group took 2 g/person/day. In spite of the expected differences in serum phospholipid composition, no significant differences were observed in bleeding time or collagen-induced platelet aggregation, although trends towards those anticipated were found.

In one study, however, it was reported that a reduction took place in the aggregability of platelets to collagen in diabetic patients after 50 mg of EPA (Velardo et al., 1982), and in another study it was

Table 13.3 Effects of 2–3 g EPA/day on platelet aggregability and bleeding time

Study	Bleeding Time	Platelet responsiveness to Collagen	Platelet responsiveness to ADP
Thorngren and Gustafsen (1981)	Increased	Decreased	Decreased
Von Schacky et al. (1985)	–	Decreased	–
Brox et al. (1981)	No effect	Decreased	No effect
Hirai et al. (1982)	–	Decreased	No effect
Sanders and Hochland (1983)	–	Decreased	No effect
Thorngren et al. (1984)	Increased	No effect	Decreased
Schmidt et al. (1990)	Increased	No effect	–
Sanders and Roshanai (1983)	Increased	No effect	–
Atkinson et al. (1987)	Increased	Increased	No effect
Sanders et al. (1981)	Increased	–	Increased

–, not measured

found that reductions occurred in platelet responsiveness to collagen, adrenaline and low-dose ADP in the elderly after the consumption of 150 mg of EPA daily for one month, even though no change in platelet fatty acids was discerned (Driss et al., 1984).

Together, these studies suggest that marine n-3 fatty acids may well lower platelet aggregability, but that large numbers are necessary to demonstrate the small reductions that may accompany nutritionally practicable increases in fish intake in Westernised communities.

13.7.2 n-3 PUFA and eicosanoid metabolism

13.7.2.1 Effects on thromboxanes

Several investigators have tried to measure effects of dietary n-3 PUFA on eicosanoid metabolism *in vivo*. Thorngren et al. (1984) could not demonstrate any effect of fish oils on the concentration of TXB_2 (the principal metabolite of TXA_2) in blood *in vivo*. The measurement of TXA_2 released in platelets exposed to collagen *in vitro* yielded inconsistent findings (Brox et al., 1981; Hirai et al., 1982; Sanders and Hochland, 1983). Dose–response studies of TXA_2 formation upon agonist exposure would probably be more informative than fixed-dose procedures.

13.7.2.2 Effects on urinary metabolites of eicosanoids

Von Schacky et al. (1985) showed that endogenous PGI_3 production, as indicated by urinary excretion of its metabolites, was raised when 6 volunteers consumed a 10–40 ml cod liver oil supplement daily for 5 months. In contrast, prostacyclin (PGI_2) production was not affected by treatment, so that the total turnover of anti-aggregatory and vasodilatory prostaglandins (i.e. PGI_3 plus PGI_2) appeared to be increased by the cod liver oil treatment. TXA_2 and TXA_3 metabolites assayed in the urine did not change when basal pre-treatment output was low, but were significantly decreased with therapy in two subjects with high basal outputs. A similar effect on PGI_2 and PGI_3 was reported by Fischer and Weber (1984) when cod liver oil supplied about 4 g of EPA daily, and by Hornstra et al. (1990) who gave a dietary supplement of mackerel paste (135 g) for 6 weeks.

These studies raise the possibility that inducement of PGI_3 synthesis without disturbance of PGI_2 production may represent one mechanism whereby fish oils induce prolonged bleeding times. Certainly, the balance of TXA_2 and TXA_3 to PGI_2 and PGI_3 as assessed from urinary metabolites appears shifted to a potentially antithrombotic state in the Inuit

(Fischer and Weber, 1986), and in patients given as much as 10 g of EPA daily (Knapp et al., 1986).

A serious weakness of studies involving urinary PGI metabolites is that they do not necessarily reflect vessel-wall production of PGI_2 and PGI_3. The gastrointestinal tract and the kidneys are important sites of prostanoid production from dietary EFAs (Pace-Asciak and Wolfe, 1971; Grant et al., 1988; Ramesha et al., 1985; Levenson et al., 1982). Studies of urinary metabolites need to be interpreted with considerable caution.

13.7.3 n-3 PUFA and platelet adhesiveness

Platelet responsiveness consists of two components: aggregability and adhesiveness. Platelet adhesion can be measured by recirculating a perfusate of fresh, anticoagulated whole blood over a small section of everted rabbit aorta at shear rates approximating those found in medium-sized arteries (shear rate has an important influence on platelet adhesion). Segments of the vessel wall can then be examined under light microscopy and the platelet attachments can be categorised as 'contact alone', 'spread' (evidence of shape change over the endothelial surface), and thrombus (aggregates greater than 5 μm in height). Owens and Cave (1990) gave 2.7 g of EPA/day to 6 healthy adults for 4 weeks and showed that all three indices tended to increase (rather than the anticipated decrease) after EPA treatment, but not to a statistically significant extent.

Li and Steiner (1990) gave a fish-oil equivalent of 6 g EPA daily to 8 healthy adults for 25 days, and somewhat different findings emerged. Platelet adhesion to fibrinogen and collagen, studied at low shear rates in the laminar flow chamber, was markedly reduced after the EPA supplement. The ability of the platelet to spread over the surface was limited, thereby reducing its adhesiveness.

In summary, high doses of n-3 PUFA, as found in fish oil concentrates may interfere with platelet pseudopodia formation, and increase bleeding times by lowering platelet adhesiveness.

13.7.4 n-6 PUFA and platelet aggregability (see Table 13.4)

Observational and intervention studies indicate that dietary linoleic acid (18:2 n-6) has no effect on platelet responsiveness to collagen.

Sirtori et al. (1986) and Sanders and Hochland (1983) found the amount of TXA_2 formed by collagen-stimulated platelets was not influenced significantly by dietary linoleate. Platelet sensitivity to adrenaline (another ADP-dependent response)

Table 13.4 Effect of n-6 unsaturated fatty acids on platelet aggregability

Study	Design	Platelet responsiveness to			
		Collagen	Adrenaline	ADP	Thrombin
Renaud *et al.* (1978)	Observational P/S ratio increased	No effect	No effect	Reduced	Reduced
Fisher *et al.* (1986)	Observational	No effect	No effect	No effect	–
Nordoy and Rodset (1971)	Intervention	No effect	–	No effect	No effect
Sanders and Hochland (1983)	Intervention	Reduced at low dose	–	No effect	
		No effect at high dose	–	–	–
Renaud *et al.* (1986)	Intervention P/S ratio increased	No effect	No effect	Increased	Reduced
Sirtori *et al.* (1986)	Intervention	No effect	–	–	–
Macdonald *et al.* (1989)	Intervention	No effect	–	–	–
Beswick *et al.* (1991)	Intervention	No effect	–	No effect	No effect

–, not measured

was also unaffected by dietary linoleate.

With respect to primary ADP-responsiveness, the data are not entirely consistent. No effect of dietary linoleate was observed in four studies. However, in a comparison of farmers consuming diets of differing fat composition, those whose diets had a relatively high ratio of PUFA to saturated fatty acids (P/S ratio) were found to have a reduced platelet ADP-responsiveness (Renaud *et al.*, 1978). Increasing the ratio to about 1:0 in those with a baseline value of 0.3 seemed to increase responsiveness rather than to depress it, the latter being achieved when the ratio was raised only to 0.7–0.8 rather than 1.0 (Renaud *et al.*, 1986).

Increasing the P/S ratio (>1.5) has been reported to increase ADP-responsiveness of platelets in the African Green Monkey (Lewis and Taylor, 1989). Further evidence for an effect of linoleate on platelet behaviour comes from Beswick *et al.* (1991), who found that the secondary (release-reaction dependent) response to ADP on optical densitometry and the response to ADP by impedence aggregometry on whole blood were both lower in men with a P/S ratio above 0.5, as compared to men with a lower ratio.

Dietary linoleate did not influence platelet responsiveness to thrombin in two studies but thrombin-induced aggregation was reported to be depressed in farmers with a raised dietary P/S ratio.

Two studies have reported apparent effects of a diet, rich in linoleic acid and depleted in saturated and monounsaturated fatty acids, on the time taken for platelets to aggregate in fresh heparinised blood drawn through a filter, without exposure to agonists (Hornstra *et al.*, 1973; Fleischman *et al.*, 1975). The rate of aggregation appeared to be prolonged by the increase in linoleate and reduction in saturated fatty acids.

In one study of dietary linoleate enrichment and the bleeding time, a tendency towards its prolongation was not statistically significant, but there was a significant reduction in TXA_2 metabolite in the bleeding time blood (McDonald *et al.*, 1989).

13.7.5 MUFA and platelet aggregability

A high oleic acid (18:1) diet appears to lead to some enrichment of platelet membranes with this fatty acid at the expense of arachidonic acid (Barradas *et al.*, 1990). Such diets do not influence platelet responsiveness to high doses of collagen (1 μg/ml or more) in platelet-rich plasma, although aggregation induced by lower doses may be depressed (Sirtori *et al.*, 1986; McDonald *et al.*, 1989; Barradas *et al.*, 1990). Collagen-induced aggregation in whole blood is unaffected by dietary oleic acid content (Barradas *et al.*, 1990).

On the other hand, TXA_2 formation in platelet-rich plasma after collagen exposure is reduced by an increased intake of oleic acid (Sirtori *et al.*, 1986; Barradas *et al.*, 1990). In the two studies that have examined responsiveness to ADP and adrenaline, platelet sensitivity after a high oleic acid diet is reduced at some doses of agonist but not at others (Sirtori *et al.*, 1986; Barradas *et al.*, 1990).

McDonald *et al.* (1989) noted a prolonged bleeding time and an apparent reduction in PGI_2 formation following dietary enrichment with canola oil (which has a high content of oleic acid and alpha linolenic acid).

13.8 PLATELET AGGREGABILITY *IN VITRO* AND THE RISK OF THROMBUS FORMATION – IMPORTANT CONSIDERATIONS

Attempts to reduce platelet responsiveness to agonists *in vitro* by dietary enrichment with unsaturated fatty acids implicitly assume that the selected indices of aggregability will provide information about risk of clinical thrombosis. For primary prevention, this remains to be established. The reasons for doubt can be summarised as follows.

13.8.1 Inconsistency of *in vitro* results with observational studies

Epidemiological studies of the characteristics associated with platelet responsiveness in healthy adults have found that aggregability was relatively high in some groups known to be at relatively low risk of acute coronary thrombosis. For example, responsiveness to ADP was less in men than in women, and lower in male cigarette smokers than in non-smokers. This is the opposite of what might have been anticipated. On the other hand, aggregability increased with age in both sexes, and tended to be increased in men with a history of coronary heart disease (Meade *et al.*, 1985b).

Furthermore, Elwood *et al.* (1991) noted a significant positive association between platelet responsiveness to ADP and thrombin, but not collagen, and the frequency of a history of myocardial infarction in a study of more than 1800 middle-aged men. Since the primary aggregation response to ADP, and platelet responsiveness to thrombin, do not depend upon cyclo-oxygenase activity and TXA_2 formation, these results suggest that the associations with myocardial infarction may have arisen through other mechanisms not necessarily related to diet, although a dietary influence cannot be excluded.

13.8.2 Dose–response considerations

The dose of agonist that platelets are exposed to upon vessel wall damage *in vivo* is not known. One might expect exposure to collagen on breach of the endothelium, or to thrombin on coagulation, to be supra-maximal for aggregatory responses. During the haemostatic response, platelets are exposed to multiple agonists, and such exposures induce a greater TXA_2 response than a single agonist (Di Minno *et al.*, 1982). Only about 10% of the platelet's capacity for TXA_2 production is needed for a maximal aggregatory response (Di Minno *et al.*, 1983). Therefore, for dietary fish oils to be effective as anti-aggregatory agents, they would need to reduce platelet TXA_2 production upon supra-maximal and multiple agonist exposure to below 10% of maximal. This would appear to be unlikely.

13.8.3 Mechanistic considerations

n-3 PUFA might impair platelet-vessel wall interac-

tions by several mechanisms:

 (i) by reducing the platelet adhesive response and ability to spread over a damaged surface;
 (ii) by suppressing platelet aggregatory responses that are TXA_2-independent (e.g. the effect of ADP on aggregin); or
 (iii) by raising the anti-aggregatory properties of the vessel wall by increasing PGI_2, PGI_3 and EDRF production.

Each of these possibilities requires further study, preferably using doses of fish oil that are nutritionally acceptable to communities not habitually eating large amounts of fish. On present evidence, intakes of this level seem unlikely to have the potential to induce changes likely to reduce the risk of thrombosis.

13.8.4 Problems with mixtures of fatty acids

The interpretation and comparison of studies would be enhanced if they were performed with pure fatty acids. Conflicting findings between studies may have arisen because of the mixed and varying composition of the fish oils or n-6 preparations used (Tichelaar, 1990). Ackman (1988) has pointed out that the long chain n-3 unsaturated fatty acids of polar bear and seal depot fat (the component of main interest in the Inuit diet) are mainly found at the 1 and 3 positions of triacylglycerols, whereas they occur mainly at the 2 position in fish oil. Intestinal lipases preferentially hydrolyse triacylglycerol at positions 1 and 3 on glycerol. Therefore, the proportion of n-3 fatty acid delivered to the liver is likely to be considerably greater in diets of marine mammal oils than in diets of marine fish oil.

13.9 THE COAGULATION PATHWAYS (see Figure 13.8)

13.9.1 Background

The coagulation pathway comprises a series of linked proteolytic reactions which culminates in the generation of thrombin activity. Thrombin converts soluble circulating fibrinogen to fibrin, which along with platelet aggregates, forms the haemostatic plug.

Blood coagulation is initiated:

 (i) when the specific lipoprotein tissue factor is exposed to blood (extrinsic pathway activation) – see Section 13.9.3;
 (ii) when blood contacts a negatively charged

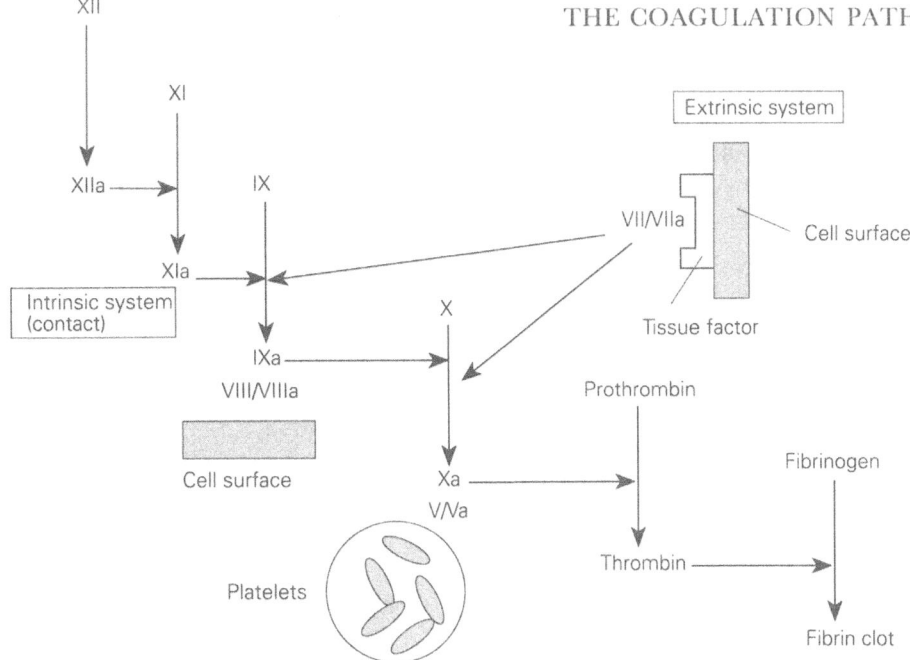

Figure 13.8 The coagulation pathway. Zymogens and pro-cofactors are designated by Roman numerals. The enzyme and co-factor derivatives have the subscript a.

surface (intrinsic or contact activation) leading to factor X1a generation (see Section 13.9.2).

Several different clotting factors participate in blood coagulation. They all circulate in the blood as zymogens or proenzymes which are activated to produce the active enzyme. The precursor (or zymogen) forms of the active enzymes are given Roman numerals, and the enzyme itself is designated by the letter a.

The products of both the intrinsic and extrinsic systems catalyse the conversion of factor IX to IXa. Factor IXa converts factor X to Xa, which in turn generates thrombin from prothrombin. Most steps take place on platelet and endothelial surfaces, and involve cofactors which possess the ability to bind the reactants in an appropriate conformation. The reactions are confined to the area of damage by the anticoagulant properties of surrounding healthy endothelial cells, the need for a phospholipid surface, and the presence of circulating inhibitors such an antithrombin III.

Fibrin, deposited by haemostatic activity, is eventually degraded during clot lysis by the enzyme plasmin. Fibrinolysis (see Section 13.11) consists of a series of reactions located on the fibrin clot or thrombus, and like haemostatic activity is regulated by an interplay of activators and inhibitors of enzyme activity.

13.9.2 The intrinsic (or contact) system

This system consists of factor XII, factor XI, prekallikrein and high molecular weight kininogen

(HMWK). When blood is exposed to a surface with an appropriate negative charge, such as that provided by collagen fibrils or platelet membranes, these components are assembled on the surface and undergo a complex series of reactions leading to the generation of factor XIIa, kallikrein, and factor XIa (Revak *et al.*, 1977, 1978; Dunn *et al.*, 1982). Factor XIIa activates factor XI. Then factor XIa and kallikrein activate factor IX (Osterud *et al.*, 1978). Factor IXa is the enzyme which catalyses factor X activation. Factor XIIa is also able to cleave the native single-chain factor VII of the extrinsic system to its fully-activated form, VIIa (Radcliffe *et al.*, 1977; Seligsohn *et al.*, 1979).

13.9.3 The extrinsic system

Factor VII(a) denotes factor VII activity due to either factor VII, VIIa, or a mixture of both species.

The essential co-factor needed for the expression of factor VII(a) coagulant activity is a specific protein called tissue factor. This protein is associated with phospholipid in the membranes of many tissue cells such as fibroblasts that are only exposed to blood upon physical disruption or in response to inflammatory changes induced by complement activation (Carson and Johnson, 1990). Additionally, endothelial cells and monocytes express tissue factor on their surfaces when exposed to cytokines such as tumour necrosis factor and interleukin I (Bevilacqua *et al.*, 1985).

The tissue factor–VII(a) complex cleaves factor IX to IXa, and is also able to cleave factor X to Xa. There is no naturally occurring circulating inhibitor of free factor VII(a). A lipoprotein-associated coagulant inhibitor (LACI) can bind to factor Xa to form a complex which has the capacity to inhibit surface-bound tissue factor-VII(a) complex (Novotny et al., 1989).

13.9.4 The 'Tenase' complex

Factor IXa generated either by the intrinsic or extrinsic systems forms a surface-assembled macromolecular complex with its co-factor VIIIa and its substrate factor X, leading to the generation of factor Xa (Lollar et al., 1984). Factor Xa then enters the surface-bound prothrombinase complex. Any 'excess' Xa appears to bind instead to the coagulation inhibitor, LACI. In turn, the resultant complex suppresses the Xa generating activity of the tissue factor–VII(a) complex. It is important to note that the tissue factor–VII(a) – Xa – LACI complex is Ca^{2+} dependent, so that any such complexes on the surfaces of circulating cells will dissociate in plasma collected in tubes coated with citrate. Factor VII/VIIa is then free for assessment in assays in vitro.

13.9.5 The 'Prothrombinase' complex

This is a surface-bound macromolecular complex comprising factor Xa, its co-factor Va and its substrate prothrombin. Procofactor V is present in platelets and in plasma, and is converted to factor Va by thrombin. Receptors for factor Va exist on the surfaces of platelets, monocytes and endothelial cells. Thrombin is the product of the prothrombinase complex. It has three important properties:

(i) It binds to specific receptors on platelets and endothelial cells, stimulating platelet aggregation and secretion, and synthesis and secretion of plasminogen activator and plasminogen activator inhibitor respectively.
(ii) It acts upon fibrinogen to produce the fibrin clot.
(iii) It binds to specific receptor proteins and heparin-like molecules on nearby normal endothelial cells to induce the release of prostacyclin (PGI_2) and the conversion of protein C to its active anticoagulant derivative.

Excess thrombin binds to its surface-bound inhibitor, antithrombin-III.

13.9.6 Activation peptides

Although not shown in Figure 13.8, cleavage of factor IX, factor X and prothrombin to their active enzymes releases specific fragments into the circulation which do not have coagulant properties. These are called activation peptides. The plasma concentrations of these peptides are proving useful markers of the in vivo rate of cleavage of their parent precursors, and therefore of 'steady-state' procoagulant activity.

13.9.7 Assessment of procoagulant activity

Methods of assessment of procoagulant activity in vivo have usually relied on measurement of the plasma concentrations of the precursors (e.g. factor X), the inhibitors (e.g. antithrombin III), the co-factors (e.g. factor V) or the substrates (e.g. fibrinogen). This approach has serious limitations:

(i) All components circulate in very considerable excess of requirements, and small changes in concentration induced by dietary or other means are likely to have no influence on reaction rates.
(ii) Coagulant activity normally occurs on vessel walls rather than in blood itself.
(iii) Any free enzyme released from surfaces is rapidly bound to its inhibitors, so that its activity is neutralised.

These difficulties have been overcome to some extent by the introduction of assays for activation peptide levels (Bauer and Rosenberg, 1987).

13.9.8 Factor VII as a predictor of CHD

Factor VII coagulant activity (VIIc), as determined by bioassay, is strongly predictive of coronary heart disease (CHD) in middle-aged men (Meade et al., 1986). The basis of the high VIIc in high-risk individuals is uncertain, but that associated with hypertriglyceridaemia and a fat-rich diet appears to be due to a combination of an increased factor VII concentration and an increased proportion circulating as the fully activated two-chain species (Miller et al., 1986, 1991). Whatever the basis, a raised VIIc appears to serve as another marker of a hypercoagulable or pre-morbid coronary thrombotic state.

13.10 UNSATURATED FATTY ACIDS AND THE COAGULATION PATHWAY

13.10.1 Unsaturated fatty acids and the intrinsic system

Early studies demonstrating the dependency of plasma coagulation on its lipid content (MacFarlane et al., 1941) are now known to have reflected the need for assembly of the prothrombinase complex on a phospholipid surface. Coagulant activity in lipid-free plasma can be restored by addition of lipoproteins rich in triacylglycerols. This demonstrates that platelets, chylomicrons and very low density lipoproteins (VLDL) can provide the necessary surface. When the coagulant activity was restored by lipid replenishment of plasma, it often failed to match the changes in blood coagulability observed after a fatty meal. This led to suspicion that free fatty acids may also influence coagulability. Subsequently, sodium stearate and palmitate, but not sodium linoleate or the sodium salts of short chain fatty acids, were found to increase coagulability in vitro (Poole, 1955).

Conner (1962) was the first to propose that plasma free fatty acids may in some way activate the contact system. He found that stearate failed to stimulate thrombus formation in blood which was deficient in factor XII. More recent research indicates that the more relevant fatty acid pool is possibly the one which exists in liquid-crystalline phase on the surface of large lipoprotein particles. Such surfaces appear to provide the necessary negatively charged groups for factor XII activation (Mitropoulos et al., 1991).

13.10.1.1 Effect of P/S ratio

While total dietary fat intake is important for factor VII coagulant activity, dietary fat composition appears to have no significant influence, at least in the short term. Thus, in 9 healthy adults, factor VII activity was very similar after two 7-day diets with standardised total fat and energy content, one with a P/S ratio of less than 0.3 and the other with a ratio of greater than 3.0 (Miller et al., 1991). Markmann et al. (1990) fed diets with P/S ratios of 0.28 and 0.89 respectively to 11 adults for 2 weeks, after which factor VII activity differed by only 1%.

13.10.1.2 Effects of fish oils

Several studies have not demonstrated any effect of fish oils on factor VII coagulant activity (Sanders et al., 1981; Berg Schmidt et al., 1989, 1990; Hansen, 1989; Haines et al., 1986; Muller et al., 1989; Brox et al., 1983). The one exception is the report of Hornstra (1992), who noted an increase in factor VII coagulant activity and a reduction in factor VII concentration after feeding a fish oil concentrate for 6 weeks. This would indicate that fish oil can activate factor VII.

13.10.2 Unsaturated fatty acids and the extrinsic system

Hansen et al. (1989) showed that tissue factor activity of isolated mononuclear cells, before and after stimulation by lipopolysaccharide (an inducer of tissue factor), was reduced when 40 healthy adults were supplemented with 25 ml cod liver oil each day for 8 weeks. However, this study and that of Muller et al. (1989) found no effect of fish oils on the tissue factor activity of unstimulated monocytes.

13.10.3 Unsaturated fatty acids and the tenase and prothrombinase complexes

Reports of an increase in factor X activity after fish oil dietary supplements (Haines et al., 1986; Muller et al., 1989) were not confirmed in a third study (Sanders et al., 1981). There have been no studies of the effects of UFA on turnover rates of factor IX, X or prothrombin as determined by the plasma concentrations of their activation peptides.

13.10.4 Unsaturated fatty acids and fibrinogen concentration

Plasma fibrinogen concentration is a strong and positive predictor of risk of CHD (Meade et al., 1986). Most studies have found fibrinogen concentration to be unchanged by dietary fish oil enrichment, although a small number have observed a reduction (see Table 13.5). In one study of insulin-dependent diabetic patients, an increase in plasma fibrinogen was observed after treatment with fish oils.

13.11 THE FIBRINOLYTIC PATHWAY

13.11.1 Plasmin generation

The dissolution of fibrin in clot or thrombus is mediated by the enzyme plasmin. Fibrin, the substrate for plasmin, serves as a co-factor for tissue plasminogen activator (tPA) which is secreted by the endothelial cells of the arteries and veins (Kristensen et al., 1984) (step 1 in Figure 13.9). Plasmin is formed locally by the proteolytic action

Table 13.5 Effect of unsaturated fatty acids on fibrinogen concentration

Study	Dose (g/day)	Duration of trial (weeks)	Effect on plasma fibrinogen
Sanders et al. (1981)	4	6	No effect
Brox et al. (1983)	4.7	6	No effect
Mortensen et al. (1983)	4	4	No effect
Haines et al. (1986)	3.6	6	Increased[a] in insulin-dependent diabetics
Norris et al. (1986)	2.3	6	No effect
Rogers et al. (1987)	2.9/1.6	1 1.5–5	No effect
Hostmark et al. (1988)	6.5	6	Reduced
Saynor and Gillot (1988)	3	4 years	Reduced
Berg Schmidt et al. (1989)	5.1	6	No effect
Hansen et al. (1989)	4	8	No effect
Muller et al. (1989)	4.7	6	No effect
Radack et al. (1989)	1.1/2.2	20	Reduced
Berg Schmidt et al. (1990)	1.3	6	Reduced
Gans et al. (1990)	3	16	No effect
Haglund et al. (1990)	4.5/9 then 4.5	4 24	Reduced

[a] In insulin-dependent diabetics

of plasminogen activators on the precursor plasminogen (PLG) (step 2 in Figure 13.9).

Intracellular tPA synthesis is stimulated by thrombin (Levin et al., 1984) and suppressed by interleukin-1 (Stern et al., 1985). Hormonally induced increases in tPA synthesis appear to be mediated by the cyclic adenosine monophosphate (cAMP) second messenger (O'Connell et al., 1987). Thrombin-induced synthesis seems to involve activation of inositol-specific phospholipase C generation of diacylglycerol and activation of protein kinase C (Hatzakis et al., 1987).

Native tPA is secreted as a single chain protein. The native species appears to bind to fibrin, where it cleaves a small amount of plasminogen (PLG) to plasmin. This plasmin acts on plasminogen to convert it to a more catalytically susceptible species, PLG* (step 3 in Figure 13.9). It also cleaves single-chain tPA to the more potent two-chain tPA species (step 4 in Figure 13.9), thus stimulating a rapid rise in fibrinolytic activity at the site of tPA binding (step 5 in Figure 13.9).

13.11.2 Fibrinolytic inhibitors

13.11.2.1 Alpha$_2$-antiplasmin

Alpha$_2$-antiplasmin is the main physiological plasmin inhibitor in plasma. It is a single chain

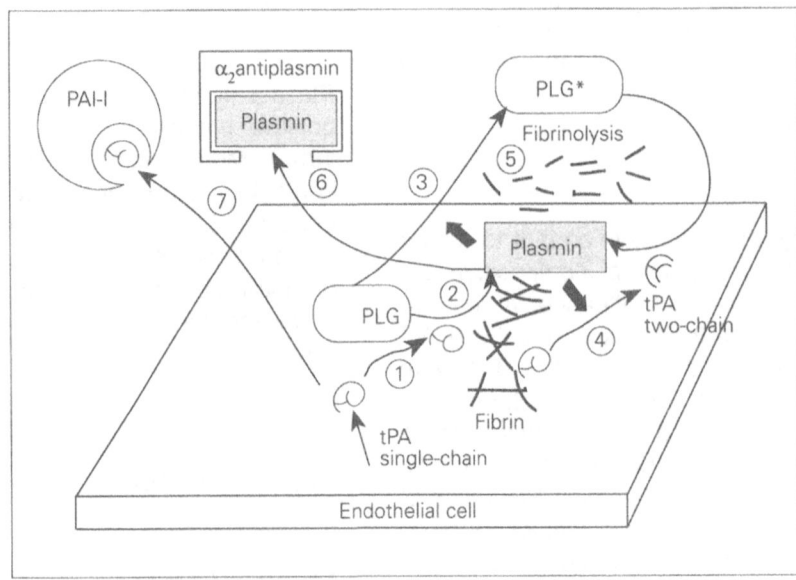

Figure 13.9 The mechanism and control of fibrinolysis: **1** Single chain tissue plasminogen activator (tPA) is secreted by the endothelial cells and binds to fibrin fibrils; **2** Plasminogen (PLG) is converted to plasmin by tPA; **3** Plasmin acts on plasminogen to convert it to a more catalytically susceptible species (PLG*); **4** Plasmin converts single chain tPA to activated two-chain tPA, accelerating the conversion of PLG and PLG* to plasmin; **5** Plasmin degrades fibrin fibres (fibrinolysis); **6** Free plasmin is neutralised by alpha$_2$ anti-plasmin; **7** Any excess tPA in the circulation is neutralised by plasminogen activator inhibitor (PAI-1).

glycoprotein which neutralises free plasmin by forming a 1:1 stoichiometric complex (step 6 in Figure 13.9).

13.11.2.2 Plasminogen activator inhibitor (PAI-1)

There are many types of circulating proteinase inhibitors accounting for about 10% of plasma total protein. Most of them lack specificity and have neutralising effects on several proteinases. Fast-acting endothelial cell type plasminogen activator inhibitor (PAI-1) is the most important specific inhibitor of plasminogen activator (Sprengers and Kluft, 1987). Plasma contains up to 1.3 nmol of PAI-1 per litre. It is present in platelets and is synthesised by endothelial cells and hepatocytes. More than 90% of circulating PAI-1 is in an inactive or latent form. This is converted to the active form by conformational changes induced by its binding to a negatively charged phospholipid surface (Lambers et al., 1987).

The release of PAI-1 by platelets, and its abundance in sub-endothelial sites, can prevent the premature action of plasminogen on newly formed fibrin. The neutralising effect of PAI-1 on the single chain tPA cleavage of plasminogen is so rapid that the half-life of tPA in plasma is only a matter of minutes (step 7 in Figure 13.9). As soon as it is secreted, tPA will bind to fibrin or endothelial-cell surface receptors (Hajjar et al., 1987). Any excess tPA in the circulation will be neutralised by PAI-1 and the complex will then be rapidly cleared by the liver.

13.12 UNSATURATED FATTY ACIDS AND FIBRINOLYSIS

13.12.1 n-3 PUFA

Because of the complex control mechanisms operating in the fibrinolytic pathway, some studies have reported an increase in PAI-1 activity after dietary fish oil supplementation (Berg-Schmidt et al., 1989, 1990; Emeis et al., 1989) whereas others have reported a reduction in PAI-1 (Barcelli et al., 1985; Mehta et al., 1988) or no change (Radack et al., 1989; Takimoto et al., 1989). The reasons for these discrepant results are unclear, but because of the method of measurement, the results obtained in vitro will depend upon the balance of tPA and its inhibitors in vivo, rather than specifically on PAI-1 levels.

In most studies, fish oil has failed to induce any significant change in tPA levels (Radack et al., 1989; Mehta et al., 1988; Hansen et al., 1989; Emeis et al., 1989; Takimoto et al., 1989), although two studies reported an increase (Barcelli et al., 1985; Berg-Schmidt et al., 1990) and one a reduction (Berg-Schmidt et al., 1989) after a fish-oil supplement to the diet.

13.12.2 n-6 PUFA

There appears to be little evidence to show whether n-6 unsaturated fatty acids influence fibrinolysis directly. Takimoto et al. (1989) found no effects of wheat germ oil (supplying 10 g of linoleic acid daily) on either tPA or PAI-1 levels in healthy adults. Mehrabian et al. (1990) studied the effects of a 3 week low fat diet with a P/S ratio of 2.4, on indices of the fibrinolytic pathway in middle-aged and elderly adults. Highly significant reductions in plasminogen, tPA and PAI-1 activity were found.

14

UNSATURATED FATTY ACIDS AND BLOOD PRESSURE

14.1 INTRODUCTION

Raised blood pressure (BP) is known to be a major risk factor in common cardiovascular and renal diseases of adulthood, including coronary heart disease (CHD), haemorrhagic and atherothrombotic stroke, aortic aneurysm and peripheral vascular disease (Shurtleff, 1974; Pooling Project Research Group, 1978; Kannel et al., 1986; Mac Mahon et al., 1990). High BP can be reduced by:

(i) drugs which decrease cardiac output (e.g. beta-adrenoceptor antagonists); or
(ii) agents which reduce peripheral vascular resistance (e.g. alpha-adrenoceptor antagonists, calcium antagonists, angiotensin-converting enzyme inhibitors).

Several large-scale clinical trials have been undertaken to assess the benefits of reducing BP in people who are above the normotensive range (i.e. having a resting diastolic pressure consistently more than 90 mmHg). Even in those people with mild hypertension (diastolic pressure between 90 and 109 mmHg) or moderate hypertension (diastolic pressure between 110 and 115 mmHg), stroke mortality during the first 5 years of treatment was reduced by about 40%, and CHD mortality was reduced by about 15% (Collins et al., 1990).

14.2 EFFECT OF MONOUNSATURATED FATTY ACIDS ON BLOOD PRESSURE

Mensink et al. (1988) gave two diets of constant saturated fatty acid and unsaturated fatty acid contents to healthy normotensive adults. One diet was rich in carbohydrates and the other was rich in olive oil. During the 55 days of study, and at 7 weeks after return to the usual diet, changes in blood pressure did not differ significantly between the two dietary groups. Thus the monounsaturated fatty acid content of the diet appears to have no appreciable influence on blood pressure, at least in normotensive individuals.

14.3 EFFECT OF n-6 PUFA ON BLOOD PRESSURE

Daily dietary supplements of 4 g or more of n-6 PUFA for 2 to 10 weeks had no effect on resting BP (Mortenson et al., 1983; Knapp and Fitzgerald, 1989; Bonaa et al., 1990; Singer et al., 1990).

Margetts et al. (1985) randomised 54 healthy normotensive men and women to either a control diet with a P/S ratio of 0.3, or to one of two groups who changed to a diet with a P/S ratio of 1.0 for 6 weeks in a cross-over design. The high P/S ratio diet did not reduce BP. These results indicate that the reduction of BP associated with a vegetarian diet (Sachs et al., 1974; Haines et al., 1980; Rouse et al., 1983) is unrelated to its high linoleate content.

Iacono et al. (1983) and Puska et al. (1983) have presented evidence to show that an increase in the dietary P/S ratio to about 1.0 can reduce BP in mild hypertensives when given for 6 weeks. In both of these studies, however, the consumption of total dietary fat was also reduced.

14.4 EFFECT OF n-3 PUFA ON BLOOD PRESSURE

14.4.1 Effect on hypertensive patients

Most studies on the effects of eicosapentaenoic acid (EPA) and docosahexaenoic acid (DHA) dietary supplements on BP have shown modest reductions in both normotensive and mildly hypertensive people (see Table 14.1).

14.4.2 Effect on normotensive people

There appears to be a small, but inconsistent, effect of n-3 fatty acids on the blood pressure of normotensive patients (see Table 14.2).

Cobiac et al. (1991) have recently shown that a

Table 14.1 The effects of n-3 unsaturated fatty acids on blood pressure of hypertensive patients

Study	No. of patients (controls)	Supplement and dose	Duration of trial	Change in blood pressure
Singer et al. (1986)	12(11)	Supplements of mackerel giving 2.2 g EPA and 2.8 g DHA; then supplements giving 0.5 g EPA and 0.7 g DHA	2 weeks 8 months	Systolic: 13 mmHg fall Diastolic: 11 mmHg fall
Norris et al. (1986)	16	2.3 g/day mixed n-3 fatty acids	cross-over trial; 6 weeks on each diet	Systolic: 6% fall Diastolic: no change
Knapp and Fitzgerald (1989)	24(8)	(i) 3 g n-3 fatty acids (ii) 15 g n-3 fatty acids (iii) 39 g n-6 fatty acids (iv) control	4 weeks	(i) No change (ii) 5% fall in mean BP (iii) No change (iv) No change
Bonaa et al. (1990)	78(78)	(i) Fish oil supplements; total dietary n-3 fatty acids 6.1 g/day or (ii) Maize oil supplements; total dietary n-6 fatty acids 12.3 g/day	10 weeks	(i) 3% fall in systolic and diastolic BP (ii) No change

4.2 g/day fish oil supplementation can enhance the effect of dietary sodium restriction on blood pressure in elderly people. Therefore the interactive effects of fish oils and sodium restriction might produce greater reductions on blood pressure than either factor acting alone.

These studies indicate that moderate intakes of n-3 unsaturated fatty acids (about 5 g/day) can slightly reduce blood pressure by about 3 to 5 mmHg in systolic and diastolic pressure. The effect is greater in subjects with mild to moderate hypertension than in normotensive individuals and it appears to persist as long as intake levels are maintained. Larger reductions cannot be expected unless, perhaps, other measures such as dietary sodium restriction are also employed.

Table 14.2 The effects of n-3 unsaturated fatty acids on blood pressure of normotensive people

Study	No. of subjects (controls)	Supplement and dose	Duration	Change in blood pressure
Lorenz et al. (1983)	8	Cod-liver oil 10 g/day mixed n-3 fatty acids	25 days	Systolic: 10% fall Diastolic: no change
Mortensen et al. (1983)	20	(i) 4 g/day mixed n-3 fatty acids or (ii) 4 g/day mixed n-6 fatty acids	4 weeks cross-over	(i) Systolic: 4 mmHg fall Diastolic: no change (ii) No change in either BP
Rogers et al. (1987)	30(30)	(i) 16 g/day fish oil concentrate then 9 g/day fish oil or (ii) 9 g/day olive oil	7 days 10–35 days	(i) Systolic: no change Diastolic: 5 mmHg fall attributable to fish oil (ii) No significant change on olive oil
Van Houwelingen et al. (1987)	40(42)	(i) Multi-centre study; fish paste supplements providing 1.7 g EPA and 3.0 g DHA/day (ii) meat paste supplements	6 weeks	No significant difference between groups
Kestin et al. (1990)	33	(i) Fish oil concentrate 3.4 g mixed EPA + DHA (ii) 14.3 g/day linoleic acid (iii) 9.2 g/day alpha linoleic acid	6 weeks	(i) Systolic: 5 mmHg fall (ii) No change (iii) No change

14.5 POSSIBLE MECHANISMS FOR THE EFFECTS OF n-3 PUFA

The basis for the small effect of the n-3 PUFA in fish oils on blood pressure is not known for certain. There are four possible mechanisms (Sections 14.5.1–14.5.4).

14.5.1 Prostacyclin effects within blood vessels

When the diet is enriched with fish oil, eicosapentaenoic acid (EPA) partially replaces arachidonic acid at position 2 of membrane phospholipids. As explained in Chapter 8, the action of cyclo-oxygenase on EPA gives rise to prostacyclins with three double-bonds in their side chains (the 3-series) rather than the 2-series which would be formed from arachidonic acid. The endothelial cells of the blood vessels produce mainly prostacyclins (PGI) but both PGI_2 and PGI_3 have similar vasodilatory properties. Total urinary excretion of prostacyclin increases during the initial period of fish oil supplementation but returns to baseline levels before the end of the supplementation period. The blood pressure reduction, however, is sustained for the whole length of the supplementation period (Knapp et al., 1989).

The effect of unsaturated fatty acids seems unlikely to be mediated solely via the prostacyclins released from blood vessels.

14.5.2 Prostaglandins from other tissues

Many tissue cells, apart from vascular endothelial cells, secrete prostaglandins, including the vasodilatory prostacyclins. Both the gastro-intestinal tract (Pace-Asciak and Wolfe, 1971) and the kidney (Levenson et al., 1982) produce increased amounts of prostaglandins when fish oil supplements are given. Renal phospholipase A_2 responds to many hormones, including angiotensin II, noradrenaline, arginine vasopressin and bradykinin and thus releases unsaturated fatty acids from membrane phospholipids. Renal prostaglandins, PGE_2 and PGF_2, are the major products of renal endoperoxide metabolism of these unsaturated fatty acids (Hassid and Dunn, 1980). One of the physiological effects of the renal prostaglandins is to counteract the actions of vasoconstrictor substances, such as angiotensin II and noradrenaline, in certain situations and thus help to regulate renal blood flow. Renal prostaglandins produced from arachidonic acid can stimulate renin release from the juxtaglomerular apparatus (Seymour et al., 1979), and are involved in the regulation of salt excretion. Renin promotes the formation of angiotensin which causes the kidneys to retain salt and water, thus increasing the extracellular fluid volume and therefore increasing blood pressure. A reduction in urinary PGE excretion appears to be a feature of essential hypertension; whether this reduction is cause or effect is uncertain, however (Tan et al., 1978).

The EPA metabolites, unlike the arachidonic acid metabolites, cannot stimulate renin release, at least in animals (Gerkens et al., 1981; Weber et al., 1976) and so fish oil supplementation may reduce blood pressure by decreasing renin activity in plasma. Studies in man do not, however, show consistent effects. Lorenz et al. (1983) showed an appreciable (though not statistically significant) fall in renin activity in plasma following cod-liver oil supplementation. Mortensen et al., (1983), however, showed that renin levels in plasma were similar after supplements of either fish oil or vegetable oil.

Since fish oils do not appear to influence urinary sodium and potassium excretion (Singer et al., 1986; Mortensen et al., 1983; Van Houwelingen et al., 1987), the effect of unsaturated fatty acids seems unlikely to be mediated solely via renal prostaglandins.

14.5.3 Effects via endothelium-derived relaxing factor

The n-3 PUFA in fish oils may play a role in correcting the impaired release of endothelium-derived relaxing factor (EDRF) in atherosclerotic vessels exposed to increased blood pressure (Shimokawa et al., 1987).

Healthy endothelium modulates the activity of vascular smooth muscle and thereby vessel wall tone, including that of the resistance vessels (Vallance et al., 1989). Damage to the endothelium reduces the vasodilator response to acetylcholine and other substances (Furchgott and Zawadzki, 1980). Panza et al. (1990) found that the blood flow and vascular resistance response to acetylcholine were reduced in untreated hypertensive patients. When exposed to acetylcholine, damaged endothelium fails to release normal amounts of EDRF (Palmer et al., 1987), which normally modulates the smooth muscle response to vasodilator substances. This impairment of endothelial function may be secondary to atherosclerotic damage, or a consequence of a raised blood pressure per se (Miller et al., 1987), but nevertheless serves to increase blood pressure further.

Shimokawa et al. (1987) examined the effect of a dietary cod liver oil supplement on EDRF responses in porcine coronary arteries and found that EDRF release in response to bradykinin, serotonin and thrombin was augmented when fish oil was added to the diet. EPA incorporation into phospholipids could therefore alter membrane fluidity in ways which promote the synthesis and/or the release of EDRF in response to agonists.

14.5.4 Effects via ion transport systems

Another possible mechanism for the influence of n-3 PUFA on blood pressure might be through changes in membrane fluidity which in turn produce alterations in ion transport via effects on second messengers.

Changes in membrane fluidity induced by incorporation of unsaturated fatty acids have been shown to stimulate chloride ion transport in corneal epithelial cells (Schaeffer and Zadunaisky, 1979). Increased fluidity might permit increased interaction between ligand–receptor complexes and the effector enzyme, adenylate cyclase. Increased cyclic adenosine monophosphate (cAMP) generation then stimulates the membrane transport system for the chloride ion. Ion transport systems could be similarly influenced in other cell types.

Lorenz et al. (1983) measured the co-transport of sodium and potassium and the countertransport of sodium and lithium across red cell membranes in 8 healthy men before and after they were given a 10 g n-3 PUFA supplement for 25 days. Despite an increased n-3 PUFA incorporation into red cell membranes, there was no change in arachidonic acid content and no evidence of any effect on ion fluxes.

Animal studies have shown that the activity of the membrane ion Na^+K^+-ATPase is influenced by membrane lipid composition (Sinensky et al., 1979) and have found that linoleic acid, arachidonic acid, alpha linolenic acid and DHA inhibit Na^+K^+-ATPase activity (Bidard et al., 1984).

However, although sodium–lithium countertransport is increased in essential hypertension (Canessa et al., 1980), and unsaturated fatty acids appear to play a regulatory role in ion transport, there is no good evidence, as yet, in humans for an effect of n-3 PUFA reducing blood pressure through an influence on ion-transport systems.

15

UNSATURATED FATTY ACIDS AND CARDIAC ARRHYTHMIAS

15.1 INTRODUCTION

Diseased heart muscle is susceptible to bouts of irregular electrical activity (arrhythmias) which are potentially lethal and often cause sudden cardiac death. In an American population-based study, the cumulative 6-year incidence of sudden death was about 1% in adults over 30 years of age (Chiang et al., 1970), while in a follow-up of 1739 male survivors of acute myocardial infarction, the 2-year mortality from sudden death was 7.3% (Ruberman et al., 1977). The presence of premature ventricular beats in the resting electrocardiogram (ECG) was a strong predictor of sudden cardiac death.

Many factors have a role in precipitating arrhythmias, including the extent of the previous infarct, sympathetic nervous activity, a low concentration of potassium ions, and a low glycogen content in heart tissue. Experimental studies in animals have shown that lethal arrhythmias may occur either:

(i) during the period when the heart muscle is starved of blood (myocardial ischaemia) induced by coronary occlusion; or

(ii) during the onset of reperfusion as the coronary blood flow is established.

The underlying mechanisms of these two forms of arrhythmia may differ (Manning and Hearse, 1984).

15.2 ROLE OF PROSTAGLANDINS IN ARRHYTHMIA

15.2.1 Balance between prostacyclin and thromboxane

Parratt and Coker (1985) suggested that the prostaglandins thromboxane (TXA_2) and prostacyclin (PGI_2) play an important role in the risk of both forms of experimentally induced ventricular arrhythmias. In the greyhound, ventricular arrhyth-

mias were observed most frequently between 10 and 12 minutes after onset of occlusion, and within the first minute of reperfusion. The risk of ventricular arrhythmias induced by ischaemia was related to the balance between TXA_2 (causing local platelet aggregation and vasoconstriction) and PGI_2 (opposing the TXA_2 effects). TXA_2 was released as an early response to occlusion whilst PGI_2 was released after the onset of ischaemia. Drugs which inhibited TXA_2 synthesis or blocked the TXA_2 receptor reduced the incidence not only of arrhythmias induced by ischaemia but also arrhythmias induced by reperfusion. On the other hand, when local PGI_2 activity was raised, this also reduced the frequency and severity of both types of arrhythmias.

Sivakoff et al. (1979) showed that the major site of prostaglandin synthesis in the rabbit heart was the vessel wall rather than the muscle-cells, and that prostacyclin PGI_2 was the principal eicosanoid product of arachidonic acid metabolism.

15.2.2 Balance between types of prostacyclins and types of thromboxanes

Gudbjarnason (1989) reviewed the dynamics of n-3 and n-6 PUFA in the major phospholipids of heart muscle, and concluded that the balance between these two types appears to influence the risk of sudden cardiac death. A reduced ratio of arachidonic acid/DHA shifted the spectrum of eicosanoid production (i.e. an increase in TXA_3 and PGI_3 at the expense of TXA_2 and PGI_2), and reduced the risk of ventricular fibrillation and sudden cardiac death in a diseased heart.

15.3 ANIMAL STUDIES

Nestel (1991) has concluded that the effects of PUFA in reducing cardiac arrythmias are most likely to be related to eicosanoid metabolism, but that the leukotrienes, as well as the prostaglandins, might be involved. The formation of potentially injurious leukotrienes is also inhibited by n-3 fatty acids, although not equally by EPA and DHA (see Chapter 19). The compounds may also act directly on cardiac electrophysiological function or indirectly by reducing coronary blood flow.

15.3.1 Effect of unsaturated fatty acids on myocardial phospholipids

Hock et al. (1987) examined the effects of feeding either dietary fish oil or maize oil on the fatty acid composition of phospholipids in the rat heart. The diet rich in fish oil reduced the level of n-6 PUFA in the heart muscle cells, whilst markedly increasing the n-3 PUFA content, relative to the maize oil diet. These differences were observed to different degrees in both phosphatidylcholine (PtdCho) and phosphatidylethanolamine (PtdEtn). Cardiolipin is the phospholipid which is specific to heart cell mitochondrial membranes. The linoleic acid (18:2 n-6) content of cardiolipin was much higher in the diet rich in maize oil, whereas the n-3 PUFA content of cardiolipin was increased when the diet was rich in fish oil. The changes in phospholipid composition induced by manipulations of dietary triacylglycerols were similar in all intracellular structures, i.e. sarcolemma, sarcoplasmic reticulum and mitochondria (Gudbjarnason et al., 1983).

15.3.2 Effect of PUFA on ischaemic damage and cardiac arrhythmias

Hock et al. (1987) also studied the frequency and severity of experimentally induced ventricular arrhythmias in rats given dietary supplements of fish oil or maize oil for 4 weeks. No significant difference was noted between the two groups. However, the rats given fish oil lost significantly less creatine kinase from the left ventricular wall (i.e. the ischaemic zone) than those given maize oil suggesting that the group given fish oil experienced less ischaemic damage to the heart muscle. The mechanism of the effect is unknown.

McLennan et al. (1990) gave rats long-term diets containing 35% of the total energy as fat, which were either rich in saturated fatty acids, n-3 PUFA or n-6 PUFA. Rats on the diets rich in both types of PUFA showed a reduction in the severity of arrhythmias induced during ischaemia, while those on the n-3 PUFA rich diet also showed a relative reduction of arrythymias induced during reperfusion. PUFA of the n-3 and n-6 series, but particularly the former, seem therefore to protect against experimentally induced ventricular arrhythmias.

Riemersma and Sargent (1989) examined the effects of small amounts of a dietary fish oil supplement on ischaemic arrhythmias in a strain of rat which was particularly susceptible to arrythmias. The incorporation of EPA in total myocardial phospholipids was increased from 0.03% to 0.13%. The frequency of induced ventricular arrythmias was reduced from 61% to 50%, but this was not statistically significant. The effect of feeding a diet rich in linoleic acid also reduced arrhythmias but the frequency of arrhythmias was increased by a diet rich in saturated fatty acids (Riemersma et al., 1988).

15.4 CLINICAL STUDIES

The saturated fatty acid content of subcutaneous adipose tissue tends to reflect that of cardiac phospholipids in men who die suddenly from non-cardiac causes (Riemersma and Sargent, 1989). Abraham et al. (1989) explored the relation between the fatty acid composition of subcutaneous adipose tissue and serious ventricular arrhythmias during or after acute myocardial infarction in hospitalised patients. The linoleic acid content (as a percentage of total fatty acid in adipose tissue) was not related to a history of ventricular arrhythmias, but the saturated fatty acid content of adipose tissue was significantly higher in those patients with a positive history of arrhythmias. As expected, vulnerability to ischaemic arrhythmias was related to the extent of myocardial damage as indicated by cardiac enzyme concentrations in the serum. More extensive athero-thrombotic disease existed in men with the highest levels of saturated fatty acids in adipose tissue who presumably had the highest levels of saturated fatty acids in their diets. These patients were drawn from a Scottish population that traditionally consumes a high fat, high saturated fatty acid diet in which average fish intake amounts only to about 2 to 3 g/day (Thomson et al., 1985).

In a study of two groups of Icelandic patients dying of either heart attacks or following accidents, Skuladottir et al. (1988) noted a reduction in the amount of phospholipid in the non-infarcted cardiac muscle of heart attack victims. There was also a significant increase in the percentage of phospholipid fatty acids as stearic acid (18:0) and a significant reduction in the percentage found as arachidonic acid. No differences were observed in the n-3 PUFA content of cardiac phospholipid between the two groups of patients.

Thus the experimental and limited clinical studies indicate that saturated fatty acid enrichment of cardiac phospholipids may be pro-arrhythmic when the heart is diseased, whereas unsaturated fatty acids (particularly n-3 fatty acids) may be anti-arrhythmic and protective against ischaemic damage.

16

UNSATURATED FATTY ACIDS AND CANCER

16.1 BACKGROUND

Cancer is second only to heart disease as the leading cause of death in the UK. For certain groups of people, such as children and women aged between 30 and 54, it is the primary cause of death. A cancerous growth (tumour or neoplasm) is initiated either by a physical or chemical change to the DNA, or by the insertion into the genome of an appropriate DNA sequence from a retrovirus. The ultimate formation of a tumour after this initiation event depends on proliferation of the damaged cells under the influence of a tumour promoter. While some cancers can be traced back to a known carcinogen, the cause of most cancers is not known and attention has focused on dietary components as initiators or promoters of carcinogenesis (see British Nutrition Foundation, 1988).

16.2 ROLE OF POLYUNSATURATED FATTY ACIDS IN TUMOUR PROMOTION

16.2.1 Epidemiological studies

16.2.1.1 Colon cancer

Several studies have suggested that dietary lipids may be involved in the development of cancer at several sites. The strongest evidence relates to fat and colon cancer. Cross-country comparisons indicate that populations which consume diets high in fat are at a greater risk of colon cancer than those which consume less fat (Committee on Diet, 1982). Willet et al. (1990) have provided evidence from prospective studies in women to confirm this association with total fat and saturates. There was no association with linoleic acid intake.

16.2.1.2 Breast cancer

A high breast cancer incidence and mortality has also been shown to correlate strongly with fat 'disappearance' per capita in several populations (Figure 16.1). In contrast to these surveys of whole populations, a study based on 89,538 US nurses aged between 34 and 59 years, who obtained between 32 and 44% of their dietary energy from fat, gave no evidence of a positive relationship between the consumption of saturated fatty acids, linoleic acid (18:2 n-6) or cholesterol and the risk of breast cancer (Willett et al., 1987). Since the minimum level of dietary fat consumed comprised 32% of the total energy, the effect of reducing the fat intake below this level could not be addressed, nor the possible influence of fat intake early in life.

The difference between this study and that derived from per capita fat 'disappearance' may be due to differences in the intake of alcohol, vegetables, minerals or fish oil. A low incidence of breast cancer was observed in Greenland Inuits (Stefansson, 1960) and also in Japanese women, although in both populations the situation has changed in recent years. In Japan, the number of annual deaths from breast cancer doubled between 1955 and 1975 (Hirayama, 1978). Over the same period, the Japanese diet became more like that of Western countries, especially among the younger people in the urban population. The traditional Japanese diet contains primarily fish and some seaweed, so that the principal components are long chain polyunsaturated fatty acids (PUFA) of the n-3 series. This was also true of the Inuits, but with modernisation there has been a change in dietary habits, which had led to an increased consumption of saturated fatty acids and PUFA of the n-6 series.

A comparison of breast cancer mortality and food consumption trends in the USA since the 1920s showed that breast cancer mortality had risen with little change in the total fat consumption (Enig et al., 1978). In that period, however, there has been a change in the type of fats consumed with animal and dairy fats being replaced by vegetable fats. However, other studies have shown no correlation between overall cancer mortality and PUFA in the

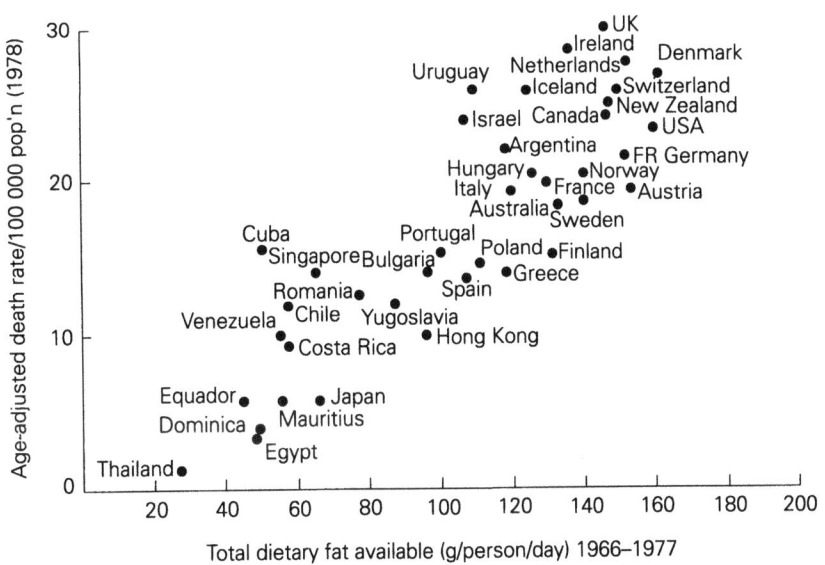

Figure 16.1   The relationship between available dietary fat and age-adjusted death rate from breast cancer in various countries. (From Carroll *et al.*, 1986)

diet or for individual PUFA such as linoleic acid (Carroll *et al.*, 1986). A group of men in Los Angeles, receiving a high PUFA diet because of a risk of developing coronary heart disease, showed an increase in colon cancer (Pearce and Dayton, 1971), but five other studies showed no such increase in patients on cholesterol-lowering diets (Ederer *et al.*, 1971). Re-analysis of the original Los Angeles data showed that PUFA could not have been responsible for the raised incidence, since this also occurred in men who did not receive the PUFA diet (Heyden, 1974).

16.2.1.3   Prostate cancer

Even if the overall PUFA intake may not be correlated with the incidence of cancer, the n-6/n-3 ratio might be important. In a retrospective case-control study of 100 patients with prostate cancer in Japan, it was found that they had consumed less seafood and green/yellow vegetables than had controls (Mishira *et al.*, 1985) which suggests a reduced intake of both beta-carotene and n-3 PUFA. Since the general trend is towards increased consumption of PUFA, a decreased n-3 intake might be expected to be correlated with an increased n-6 intake. An increase in the dietary n-6/n-3 ratio might therefore be a risk factor for cancer.

16.2.2   Experimental studies

While the epidemiological studies give rise to much debate as to the role of PUFA in cancer, the experimental studies leave little doubt that PUFA,

especially those of the n-6 series, can act as promoters. Most experimental studies have concentrated on colon, breast and pancreatic cancer.

16.2.2.1   Colon cancer

Several animal studies have demonstrated that high fat diets containing maize oil, sunflower oil, lard or beef tallow increased the incidence of chemically induced colonic tumours in rats. Diets containing high proportions of coconut oil, olive oil or *trans* fatty acids had no such effects (Bull *et al.*, 1981; Reddy and Maeura, 1984; Reddy *et al.*, 1985). These studies suggest that the fatty acid composition of the fat is an important factor in colon carcinogenesis. Furthermore, fish oils rich in n-3 PUFA have been shown not to increase the yield of azoxymethane-induced tumours in rats (Reddy and Maruyama, 1986). The excretory pattern of the faecal secondary bile acids, deoxycholic and lithocholic acids, which can act as tumour promoters in the large bowel, is positively correlated with the tumour incidence in animal models given various types and amounts of dietary fat (Reddy, 1983). It is possible that fish oils have an inhibitory effect on the colonic concentration of these secondary bile acids.

16.2.2.2   Breast cancer

Experimental studies have shown that high levels of dietary fat promote the development of mammary tumours, with PUFA being more effective than saturated fatty acids (Rogers and Wetsel, 1981; Braden and Carroll, 1986). Induction of mammary tumours in both rats and mice have generally been shown to be positively correlated with the content of linoleic acid (18:2 n-6) in the diet (Ip *et al.*, 1985). It is important, however to ensure that the control groups in such experiments are not deficient in essential fatty acids (EFA) since such a state has been shown to be associated with reduced tumour growth rate and to increase the latent period after transplantation (Sauer and Dauchy, 1988). Carroll (1980) reported that di-methylbenz-[a]anthracene (DMBA)-induced mammary carcinogenesis was similar in rats fed on a 20% fat diet containing predominantly either oleic acid (olive oil) or linoleic acid (sunflower seed oil). Dayton *et al.* (1977) also found no marked differences in the ability of diets containing high oleic and high linoleic acid to promote DMBA-induced mammary tumours in the rat.

Other studies have indicated that total energy intake is a more stringent determinant of tumour growth than fat intake, because energy reduction significantly inhibits tumour growth even when the diets are relatively high in fat (Kritchevsky *et al.*, 1986). Thus the promotional effect of dietary fat has recently been attributed to an *ad libitum* feeding protocol (Welsch *et al.*, 1990) since a slight restriction in the amount consumed (12% less than *ad libitum*) abolished the differential between a high fat and a low fat diet. The importance of unsaturated fatty acids in the promotion of mammary tumours may therefore be limited and, once a minimal requirement for unsaturated fatty acids is met, the total amount of fat, whether saturated or unsaturated, determines the influence of high fat levels (see Figure 16.2). A recent review has analysed the data from 100 animal experiments which studied the effects of different levels of dietary fat and/or calorie intake on the development of mammary tumours (Freedman *et al.*, 1990). The authors concluded that, in mammary tumour development, there is a specific enhancing effect of

dietary fat, and a general enhancing effect of total energy. This suggests that future studies to determine the effect of dietary modification on the incidence of human breast cancer should include reduction of both total fat and total energy.

The type of fat may also be important since fatty acids of the n-3 series always have an inhibitory effect (Braden and Carroll, 1986). Furthermore, the geometric configuration of the double bonds and the chain length have an influence. Diets with *trans* unsaturated fatty acids are similar to saturated fatty acids in their mammary-tumour promoting capability (Selenskas *et al.*, 1984).

Although it may appear that the animal and human studies are in conflict with regard to PUFA intake, in that the human data indicate that total fat and saturated fat in the diet are important whereas the PUFA content is not, it may be that PUFA content of the diet is critical only below a certain caloric intake. Thus the lack of evidence for a role of PUFA in human breast cancer arises because the normal intake exceeds the threshold for promotional effects. The observation that n-3 fatty acids inhibit tumour induction suggests that Inuits may have a low incidence of breast cancer despite eating a relatively high fat diet because their dietary fat is derived from marine sources rich in n-3 fatty acids (Neilsen and Hansen, 1980).

16.2.2.3 Pancreatic cancer

Epidemiological investigations have suggested that there may be an increased risk of pancreatic cancer in the presence of biliary tract disease and/or cholecystomy (Fraumeni, 1975). The latter increases the flow of secondary bile salts and so enhances tumour formation in both the colon (Narisawa *et al.*, 1974) and the pancreas (Ura *et al.*, 1986).

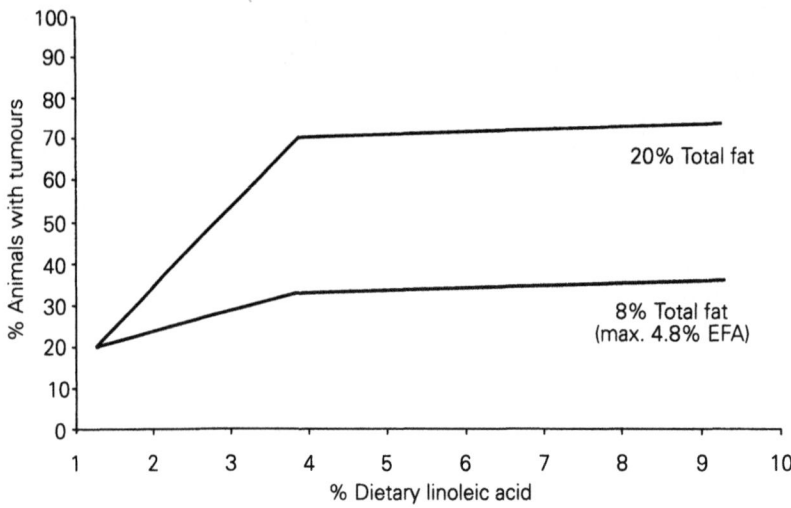

Figure 16.2 Tumour responses to linoleic acid. (From Ip *et al.*, 1987)

PUFA, particularly linoleic acid, have been shown to promote pancreatic carcinogenesis in experimental animals (Roebuck et al., 1985). High fat diets increase the yield of azaserine-induced pancreatic tumours in the rat, provided that the minimum EFA content is between 4% to 8% of dietary energy. This is not due to a general, non-specific stimulation of the rat pancreas, but is focal in nature.

As with mammary carcinogenesis, n-3 PUFA tend to have an antipromotional effect. Thus, feeding a diet containing 20% menhaden oil rich in n-3 PUFA produced a significant decrease in the number and size of pre-neoplastic lesions of the rat pancreas, when compared with a diet containing 20% maize oil (O'Connor et al., 1985).

16.2.3 Mechanism of action

Several hypotheses have been advanced to explain the promotional role of dietary fat in experimental cancer. In breast cancer, dietary fat was initially thought to act by enhancing prolactin secretion, but in subsequent studies it was shown that the effect is independent of any change in circulatory prolactin or oestrogen (see review by Welsch and Aylsworth, 1983).

16.2.3.1 Role of prostaglandins

Several studies have suggested that the effects of the n-6 PUFA may be mediated by modulation of prostaglandin synthesis and metabolism (see Chapter 8). Increased production of prostaglandin E_2 (PGE_2) has been reported in a number of human cancers (Balch et al., 1982). The possible role of prostaglandins in tumour promotion by PUFA is suggested by the observation that such effects can be prevented or reduced by the addition of inhibitors of eicosanoid biosynthesis in the diet (Hillyard and Abraham, 1979). Both eicosapentaenoic acid (EPA) (20:5 n-3) and docosahexaenoic acid (DHA) (22:6 n-3) may exert their effects through competition with arachidonic acid (20:4 n-6) for the enzyme cyclo-oxygenase, thereby curtailing the formation of arachidonic acid metabolites (Needleman et al., 1980). Diets rich in n-3 PUFA have been shown to suppress the excessive production of PGE_2 which accompanies colonic tumours, whereas a diet rich in n-6 PUFA causes further increases (Minoura et al., 1988). In pancreatic tumours induced by azaserine, the serum levels of thromboxane B_2 (TXB_2), PGE_2 and 6-keto-PGF_1alpha are decreased as the ratio of dietary n-3 to n-6 fatty acids is increased, again suggesting that the anti-promotional effects of n-3 fats may be mediated through inhibition of the formation of the 2-series of prostaglandins (O'Connor et al., 1985).

16.2.3.2 Role of immunosuppression

Prostaglandins, particularly PGE_2, have a profound immunosuppressive effect. Such immunosuppression, together with increased PGE_2 production, has been reported in a number of human cancers (Balch et al., 1982). In addition, dietary fat appears to play a role in modulating immune function. Thus diets high in PUFA are more immunosuppressive and are more effective promoters of carcinogenesis than are diets high in saturated fatty acids (Vitale and Broitman, 1981). The effect of PUFA on immune function appears to be mediated by a change in cell membrane structure leading to increased 'fluidity'. The more fluid the membrane, the less responsive it is to stimuli (see Chapter 8).

16.2.3.3 Modification of metabolism of carcinogens

Another mechanism by which PUFA of the n-6 series may influence carcinogenesis is by increasing the metabolism of xenobiotics. Wade et al. (1982) have shown that liver microsomes from rats fed on a 10% maize oil diet metabolise N-nitrosodimethylamine to mutagenic metabolites more readily than do microsomes from rats fed on a similar diet devoid of maize oil, suggesting that the activity of the mixed-function oxidase system is increased. The effect may not be related to the overall promotional effect however, since it is the dietary level of fat after administration of the carcinogen that appears to be the important factor. Increased tumour development is still observed when rats are fed on a high fat diet for up to 20 weeks after administration of the carcinogen (Ip, 1980).

16.2.3.4 Inhibition of gap junction-mediated communication

The possible role of inhibition of gap junction-mediated intercellular communication in tumour promotion by PUFA has been considered (Aylsworth et al., 1987). PUFA inhibit metabolic co-operation at non-cytotoxic concentrations, whereas saturated fatty acids fail to do so at any concentration. In addition, the long chain PUFA of the cis configuration are more efficacious than short chain fatty acids or trans-fatty acids.

16.2.3.5 Role of lipid peroxidation

Metabolism of PUFA to eicosanoids is accompanied by the generation of free radicals which may initiate tumorigenesis via lipid peroxidation or by direct reaction with DNA. However, since most experimental diets are adequately supplemented with antioxidants such as vitamin E, the importance of this mechanism to tumour induction is uncertain. The possible adverse health implications of high n-3 PUFA diets with inadequate vitamin E content has recently been discussed (Odeleye and Watson, 1991).

16.2.4 Conclusions

On the basis of the evidence available from the experimental and epidemiological studies, it appears that a reduction in total fat intake may decrease the risk of developing cancer. It would also seem prudent to modify the intake of n-6 and n-3 PUFA in favour of the latter.

16.3 ROLE OF POLYUNSATURATED FATTY ACIDS IN TUMOUR CELL GROWTH

In addition to a promotional role of carcinogenesis, PUFA have profound effects on the growth of established tumours.

16.3.1 Effects on tumour growth *in vitro*

Unsaturated fatty acids such as palmitoleic, oleic, linoleic and arachidonic acid are more effective destroyers of tumour cells *in vitro* than the corresponding saturated fatty acids of the same chain length (Siegel *et al.*, 1987). In addition, PUFA destroy human breast, lung and prostate cancer cells *in vitro* at concentrations which have no adverse effects on normal human fibroblasts or normal animal kidney cell lines (Begin *et al.*, 1985a). The most consistent and selective effects are obtained with fatty acids containing 3, 4 and 5 double bonds (GLA 18:3 n-6, arachidonic acid 20:4 n-6, EPA 20:5 n-3). Linoleic acid is inactive which suggests that GLA has a specific action against human hepatoma cells.

In contrast to these results, complexes of linoleic acid with bovine serum albumin (BSA) stimulated the growth of human breast cancer cells *in vitro* (Rose and Connolly, 1989). The use of a linoleic acid-BSA complex approximates more closely to physiological conditions than does the addition of fatty acid dissolved in ethanol and these results correlate more strongly with the effects *in vivo* of linoleic acid. Other studies have shown that linoleic acid also stimulates the growth of mouse melanoma and Chinese hamster cells *in vitro*.

16.3.2 Effects of tumour growth *in vivo*

16.3.2.1 n-6 fatty acids

Linoleic acid, administered as a 10% maize oil diet, has been shown to enhance significantly the growth of transplantable mammary adenocarcinomas in mice, when compared with a 10% hydrogenated oil diet (without linoleic acid) (Abraham and Hillyard, 1983). The other 18-carbon fatty acids (stearic, oleic and columbinic) did not increase tumour growth, suggesting that both double bonds are required for this effect. Stimulation of tumour growth in adult rats after acute starvation appears to be due to the uptake of linoleic acid and arachidonic acid present in hyperlipaemic blood (Sauer and Dauchy, 1988); these PUFAs appear to be rate limiting for tumour growth *in vivo*.

In contrast to these results in which the fatty acids were present in triacylglycerols and were administered orally, both oleic and linoleic acid administered separately, intraperitoneally as an emulsion, significantly prolonged the life spans of mice bearing the Ehrlich carcinoma (Zhu *et al.*, 1989). The methyl esters of the fatty acids were less effective than were the free acids in prolonging survival, suggesting that the free carboxyl group of the free fatty acid (FFA) plays a role in destroying tumour cells. Although evening primrose seed oil, which contains about 9% GLA, has been shown to inhibit the growth of a rat mammary adenocarcinoma, the effect was not dose-related, and was not reproducible with pure GLA (Karmali *et al.*, 1985). These results suggest that the effects observed with the oil may be related to constituents other than GLA.

Clinical studies with high doses of GLA (18–27 GLA capsules per day) plus vitamin C (6 g per day) appeared to show a prolongation of survival of patients with primary liver cancer in comparison with historical controls (van der Merwe, 1984) but a larger study, using a placebo control group, showed no significant effect of 36 GLA capsules per day on either survival time or liver size (van der Merwe *et al.*, 1990). The results in the latter study were attributed to a low GLA intake (1.44 g per day) and it was suggested that further studies should be carried out with pure GLA. However, in both studies, GLA was administered as evening primrose seed oil which contains about 70% linoleic acid and only about 9% GLA. In view of the possible antagonistic actions of these two fatty acids on tumour growth rate, care should be taken when attributing an effect to a single fatty acid when complex mixtures of oils are administered.

16.3.2.2   n-3 fatty acids

In contrast to these disparate effects with n-6 fatty acids, most studies with n-3 fatty acids have shown an inhibitory effect on tumour growth. Thus Kort *et al.* (1987) showed a significant reduction in the growth of a transplantable rat mammary adenocarcinoma when animals were fed on a diet rich in n-3 fatty acids compared with a diet rich in n-6 fatty acids. Similar results were obtained with another rat mammary adenocarcinoma (Karmali *et al.*, 1984). In addition to these effects on syngenic rat mammary tumours, the growth rate of a human mammary carcinoma was significantly reduced in athymic 'nude' mice fed on a diet rich in fish oil (Borgeson *et al.*, 1989). Furthermore, the tumours in the animals fed fish oil showed a greater sensitivity to two anti-neoplastic agents, mitomycin C and doxorubicin. A similar reduction in tumour growth rate has been observed with human prostatic tumour cells in nude mice, when the animals were fed on diets containing 20% fish oil (capsules containing 18.7% EPA and 10.9% DHA) (Karmali *et al.*, 1987).

The level of n-3 fatty acids required to produce an antitumour effect is much higher than would normally be achieved from the diet. Thus, while a 9% fish oil diet (containing 6.8% EPA and 11.2% DHA) has been reported as having no effect on the growth of a colon tumour in rats (Fady *et al.*, 1988), a significant reduction in the growth of a mouse colon tumour was reported, when the fish oil (capsules containing 18.7% EPA and 10.9% DHA) comprised 25 to 50% of the total energy (Tisdale and Dhesi, 1990). In this study, the fish oil diet had a more profound antitumour effect than conventional cytotoxic drugs, without the attendant toxicity, thus promising a more specific antitumour effect. The antitumour effect of the fish oil diet has been attributed to its content of EPA rather than DHA (Tisdale and Beck, 1991) which could also explain the discrepancy in the two studies cited above.

So far, there has been only one clinical evaluation of n-3 PUFA and no evaluation of the individual fatty acids. In this study 12 patients with breast cancer who had received substantial therapy previously were given 20 EPA capsules per day, delivering 3.6 g EPA and 2.4 g DHA (Holroyde *et al.*, 1988). A measurable clinical response was observed in two patients and no toxic side effects were observed. This result would warrant further evaluation, using pure fatty acids, enabling higher doses to be achieved.

16.3.3   Mechanism of the cytotoxic effect

16.3.3.1   Role of lipid peroxidation

The cytotoxicity *in vitro* of PUFA has been attributed to the generation of peroxides resulting in extensive lipid peroxidation (Begin *et al.*, 1985b). Further studies confirmed that the cytotoxic potential of PUFA varied with the ability of the free fatty acid to stimulate the production of superoxide radicals. Neither hydrogen peroxide nor hydroxy radicals appear to be significantly involved (Begin *et al.*, 1988). Thus GLA and arachidonic acid with three and four double bonds generated the most hydroperoxide breakdown products (as measured by thiobarbituric acid-reactive material, TBARM) and were the most cytotoxic, whereas DHA with six double bonds was the least effective in raising TBARM and in destroying the malignant cells. The formation of superoxide anion in response to PUFA appears to occur in tumour cells, but not in normal cells, perhaps because of a decreased capacity of the former to detoxify superoxide radicals (Das *et al.*, 1987). Lipid peroxidation appears to be only partly responsible however since the antioxidant alpha-tocopherol, while completely suppressing lipid peroxidation, only partially reduced the growth suppression by PUFA (Chow *et al.*, 1989). This effect of alpha-tocopherol was more pronounced with PUFA of the n-6 series than with the n-3 series, suggesting that lipid peroxidation may not be the primary mechanism of action of n-3 acids.

16.3.3.2   Role of prostaglandins

Although there have been suggestions that prostaglandins can influence the effects of PUFA on tumour growth, the relationship is somewhat tenuous. Although linoleic acid produced a stimulation of mammary tumour growth, this was not apparent when pure arachidonic acid was added to the diet (Hillyard and Abraham, 1979). It is possible that either dietary arachidonic acid does not enter the same pool as arachidonic acid derived from linoleic acid, or that prostaglandins of the 1-series derived from dihomogamma linoleic acid (DGLA) are involved, rather than prostaglandins of the 2-series.

It appears, however, that a variety of cell lines transformed by both chemicals and viruses have lost the enzyme 6-desaturase, which converts linoleic acid to GLA (Horrobin, 1980) (see Chapter 7). Stores of both GLA and DGLA are strictly limited in most cells and very little is transported in the plasma. Thus cells which have lost their 6-desaturase also lose their ability to make PGE_1 from DGLA, once the limited stores of GLA and DGLA have been utilised. Horrobin (1980) has argued that loss of the ability to make PGE_1 and/or

TXA_2 may be the critical change in many forms of cancer and that restoration of normal PGE_1 synthesis, by providing GLA or DGLA and thereby by-passing the desaturase, should normalise malignant cells and reverse cancer growth. This hypothesis has received support from the observation that GLA significantly reduced the growth rate of murine and human tumour cells while having no effect on the growth of normal epithelial cells (Dippenaar et al., 1982). The antiproliferative effect of DHA however, suggests that mechanisms other than eicosanoid activity may be involved.

16.3.3.3 Effect of oncogene products

Several studies have suggested a role for growth factors or oncogene products in the growth-regulating effects of PUFA. For example, linoleic acid enhances the proliferation of mouse mammary epithelial cells only in the presence of other growth factors such as the marnmogenic hormones (progesterone and prolactin) (Bandyopadhyay et al., 1987), i.e. PUFA are incomplete mitogens. In contrast, synthetic phospholipids added as liposomes stimulated the proliferation of mouse mammary epithelial cells in a basal medium which contained only insulin as a supplement (Imagawa et al., 1989). Dilinoleoyl phosphatidic acid or phosphatidylserine (PtdSer) were most effective, whereas dilinoleoyl phosphatidylcholine (PtdCho) was less so, and phosphatidylethanolamine (PtdEt) did not stimulate growth at all. The optimal mitogenic effect of the phospholipids was dependent on the presence of a PUFA esterified at the 2-position of the glycerol moiety. The presence of a basic group seemed to inhibit this activity. The relative effects of the phospholipids on growth did not correlate directly with their incorporation into cell lipids, indicating that free phospholipid was the most important determinant of mitogenesis.

The mechanism by which phosphatidic acid and PtdSer stimulate proliferation remains obscure, but it is probable that several growth-regulatory pathways are activated. Phosphatidic acid may affect proliferation by the generation of diacylglycerol (DAG), a known activator of protein kinase C (see Chapter 8). In addition, those phospholipids containing arachidonic acid and linoleic acid can indirectly enhance the activity of the 21-K Dalton gene product of the Harvey ras proto-oncogene (p21) by decreasing the conversion of bound GTP to GDP (Tsai et al., 1989). In addition to regulating the activity of oncogene products, PUFA may also regulate their expression. Thus the expression of the oncogenic ras p21 protein was shown to be increased after culturing murine hyperplastic alveolar nodules, a preneoplastic lesion of the mammary gland, with linoleic or arachidonic acid,

whereas expression was decreased after treatment with EPA (Telang et al., 1988). Such an effect could explain the apparently disparate effects of PUFA of the n-6 and n-3 series.

16.3.3.4 Other effects

There is a greater infiltration of mast cells into tumours in mice which are fed on high fat diets. It has been postulated that these cells stimulate the migration of capillaries within the tumour (Hubbard and Erickson, 1987a) and this could be a factor in the decreased latent period of tumour onset and the increased growth of the tumours.

16.4 ROLE OF POLYUNSATURATED FATTY ACIDS IN TUMOUR METASTASIS

16.4.1 Experimental models

Metastasis is a complex, multistep process, by which malignant cells from the primary tumour are disseminated to other parts of the body to form secondary growths. PUFA have a role in this process but its nature is not clear. The effect obtained depends on the tumour model and the type of fat tested. Basically, two types of models are used to study metastasis, the infusion in vivo of tumour cells and the natural metastasis from a highly metastatic tumour.

16.4.2 Effect of n-6 polyunsaturated fatty acids

PUFA of the n-6 series have generally been shown to enhance metastatic spread of tumour cells. Thus a 23% maize oil diet has been shown to enhance the metastasis of a transplantable mammary tumour to the lung in intact and ovariectomized retired breeder rats (Katz and Boylan, 1987) although it had no effect in young virgin rats. Metastasis of a mammary tumour to the inguinal lymph node has also been shown to be significantly higher in animals fed on a 20% maize oil diet as compared with lower levels of fat (Kollmorgen et al., 1983). The level of linoleic acid in the diet required to influence significantly the spontaneous metastasis of a transplantable mammary tumour to the lungs is high (between 8 and 12% w/w) (Hubbard and Erickson, 1987b).

The survival of metastatic cells at distant sites appears to be increased in mice fed on diets containing 8–12% linoleic acid and the effect appears to be on host tissue (lungs) and not the tumour cells themselves, since the lodgement of tumour cells first cultured in the serum of mice fed on 8–12% linoleic acid and then injected into mice fed 1% linoleic acid was not affected (Hubbard and Erickson, 1989).

The suggested mechanism is that of immune suppression by the eicosanoids. This idea has been strengthened by the observation that an inhibitor of prostaglandin synthesis, indomethacin, inhibits the metastasis of a mouse mammary tumour enhanced by linoleic acid (Hubbard *et al.*, 1988). However, oleic acid, which is not a prostaglandin precursor, can also increase metastasis in mice fed on low linoleic acid diets, while having no effect on those fed on high linoleic acid diets (Buckman *et al.*, 1990). Metastasis appears to correlate more closely with the total unsaturated fatty acid in the diet rather than with the total PUFA or any individual fatty acid, suggesting that both MUFA and PUFA may stimulate mammary tumour metastasis. In contrast to these results in rodents, studies in cancer patients showed the n-6 PUFA content of membrane phospholipids to be lower in those tumours that gave rise to systemic metastases (Lanson *et al.*, 1990). The reason for this discrepancy is currently unknown.

16.4.3 Effect of n-3 polyunsaturated fatty acids

Diets rich in n-3 fatty acids have been shown not to reduce the number of pulmonary tumour foci from a syngeneic mammary tumour administered intravenously whereas there was inhibition of the primary tumour (Kort *et al.*, 1987). The total n-3 fatty acid content did not exceed 7% of total energy so the concentration may have been too low. In another study, a fish oil diet (at 4%) increased by four times the number of lung nodules from the Lewis lung carcinoma compared with the normal diet containing 5% mixed fats (Young and Young, 1989). Other studies have shown no significant differences between fish oil and maize oil on the spontaneous metastasis of a rat mammary tumour (Adams *et al.*, 1990). However, when the same tumour was injected intravenously, fish oil inhibited metastasis to a significantly greater extent than maize oil. The number of metastatic foci in the experimental model was twice that of the spontaneous model irrespective of the dietary treatment, which may explain some of the differences between these two experiments.

In contrast with these results, culturing of murine melanoma cells with pure EPA in ethanol caused a dose- and time-dependent decrease in invasiveness, collagenase IV production and a reduced ability to metastasize to the lung after intravenous injection (Reich *et al.*, 1989). In contrast arachidonic acid did not have any effect on the ability of the cells to metastasize. The difference between these results and the *in vivo* system was rationalised on the basis of the time of exposure to EPA required to reduced metastasis, since in the *in vivo* studies the host and not the tumour cells were exposed to the fat supplement.

In a study of metastasis of an experimental mammary tumour, animals fed diets containing *trans* fatty acids (Erickson *et al.*, 1984) showed fewer tumour cells in the liver and spleen than mice fed 20% *cis* fatty acid diets. Thus *trans* fatty acids appear to be less effective than *cis* fatty acids in facilitating dispersion and survival of tumour cells.

16.4.4 Conclusions

Cis-unsaturated fatty acids seem to play a role in the metastatic dissemination of tumour cells; fatty acids of the n-6 series seem to be most important (with linoleic acid playing a predominant role), but the mechanisms remain obscure.

16.5 ROLE OF POLYUNSATURATED FATTY ACIDS IN CANCER CACHEXIA

16.5.1 Introduction

Cachexia is defined as the total effects of the tumour on the host other than those resulting from mechanical interference with recognised structures. It is most dramatically seen as a loss of host body components, particularly muscle and adipose tissue, and this weight loss can eventually result in the death of the patient. Host weight loss cannot be effectively reversed by force-feeding or total parenteral nutrition (TPN). This suggests that the effect is a result of the metabolic effects of the tumour on the host. Although anorexia is invariably present in cancer patients, it tends to be a late effect and weight loss can occur in the absence of anorexia.

16.5.2 Effect of polyunsaturated fatty acids on the cachectic process

Studies using a murine colon adenocarcinoma which produces extensive loss of host body components without a reduction in food intake have shown a role for the n-3 fatty acids. Giving an isocaloric, isonitrogenous diet in which the carbohydrate calories were replaced by fat in the form of fish oil showed complete protection by the fish oil diet against the loss of body weight and a concomitant reduction in tumour size (Tisdale and Dhesi, 1990). The effect was associated with an increase in total body fat and muscle mass, which was maximal when the fish oil comprised 50% of the calories. The anti-cachectic effect of the fish oil exceeded the anti-tumour effect, but even the inhibition of tumour growth was greater than achieved with 5-fluorouracil, an agent used clinically to treat colonic cancer.

The anti-cachectic and anti-tumour effect of the fish oil diet has been attributed to the presence of EPA (20:5 n-3) which appears to be the only effective anti-cachectic agent *in vitro*. EPA inhibits the catabolic action of a tumour lipolytic factor by preventing cyclic AMP accumulation in fat cells (Tisdale and Beck, 1991). Other workers have also shown that giving rats a cod liver oil supplement, with a consequent increase in n-3 fatty acids in membrane phospholipids of cardiac cells, is also associated with a decrease in cyclic AMP, both in the basal state and in response to noradrenaline stimulation (Laustiola *et al.*, 1986). DHA, or other fatty acids of both the n-3 and n-6 series, were incapable of blocking the cyclic AMP response to lipolytic stimuli.

Although all fatty acids were cytotoxic to the tumour cells *in vitro*, only EPA was an effective anti-tumour agent when administered orally. This suggests that the anti-tumour effect of EPA, at least in this system, may derive from an indirect effect on the inhibition of the mobilisation of the EPA essential for tumour growth *in vivo*.

17

UNSATURATED FATTY ACIDS AND DIABETES

17.1 INTRODUCTION

Given the diversity of the many metabolic pathways which may be modulated by unsaturated fatty acids, it is not surprising that dietary modification involving polyunsaturated fatty acids (PUFA) has been the subject of a number of studies in people with diabetes. More work has been done on animal models of diabetes (Storlein, 1992), but this will not be covered here.

17.2 PATHOGENESIS

In the UK, there are two major types of diabetes mellitus: insulin-dependent diabetes mellitus (IDDM) and non-insulin dependent diabetes mellitus (NIDDM). They are both characterised by a raised concentration of glucose in the blood due to an absolute (IDDM) or relative (NIDDM) lack of the hormone insulin. In Western societies, NIDDM is about four times as common as IDDM. Together they affect some 2% of the population. In other countries such as Japan, IDDM is very much less common. NIDDM occurs more frequently among the obese. There is no evidence, however, to implicate any single dietary component, including unsaturated fatty acids, in the development of either of these conditions (Medalie *et al.*, 1975).

17.3 MANAGEMENT

The risk of coronary heart disease (CHD) is 2–3 times greater in diabetics than in non-diabetics and microvascular complications such as diabetic nephropathy and retinopathy are also a problem. In addition, the overweight common in individuals with NIDDM exacerbates many of the metabolic risk factors for CHD. Dietary management for diabetes, therefore, has two main aims:

(i)   to maintain blood glucose and lipid levels at as near normal levels as possible and so reduce the risk of both microvascular and macrovascular complications;

(ii) to achieve and maintain appropriate weight.

17.3.1 Lipid metabolism

Studies have been performed on diabetic subjects using PUFA of both the n-3 and n-6 series, as well as monounsaturated fatty acids. The latter appear to have similar qualitative effects on lipid metabolism in diabetic subjects to those in non-diabetic subjects (Reaven, 1988a) (see Chapter 11).

17.3.1.1   n-6 fatty acids

As in non-diabetic subjects, linoleic acid feeding consistently results in a reduction of total and low density lipoprotein (LDL) cholesterol concentrations (Heine and Schouten, 1989; Reaven, 1988) (see Chapter 12). Most, but not all, studies have also shown a reduction in high density lipoprotein (HDL) cholesterol, though to a lesser extent: the LDL/HDL ratio is therefore reduced (Sorisky and Robbins, 1989).

17.3.1.2   n-3 fatty acids

In view of the re-emergence of elevated serum triacylglycerol levels as a possible risk factor for CHD in diabetes as part of Reaven's syndrome (Reaven, 1988b), much interest has been devoted to the effects of n-3 fatty acids in reducing them. Usually, a mixture of eicosapentaenoic (EPA) and docosahexaenoic (DHA) acids has been used. A dose of around 3 g/day is associated with a reduction of elevated serum triacylglycerol concentrations in diabetic, as in non-diabetic subjects (Nestel *et al.*, 1984).

In contrast to fatty acids of the n-6 series, n-3 PUFA have been shown to have no consistent effects on cholesterol metabolism either in diabetic or in non-diabetic subjects (Friday *et al.*, 1989; Kasim *et al.*, 1988; Stacpoole *et al.*, 1989). One study on 18 IDDM subjects found no increase in

LDL cholesterol and an increase in HDL cholesterol on cod-liver oil supplementation (i.e. rich in n-3 fatty acids) compared with olive oil supplementation for eight weeks (Jensen *et al.*, 1989). Other studies in NIDDM and in normal and hyper-triglyceridaemic subjects have found that fish oil supplementation causes increases in plasma total cholesterol, LDL cholesterol and serum apoprotein-B (Glauber *et al.*, 1988; Stacpoole *et al.*, 1989).

17.3.2 Glucose and insulin metabolism

17.3.2.1 n-6 fatty acids

Specific data on the effects of n-6 PUFA on glucose or insulin metabolism are scanty. In a number of studies, an increase in the dietary P/S ratio was accompanied by a concomitant reduction in total fat, and an increase in complex carbohydrates. Such manoeuvres are not accompanied by worsening of diabetic control, though very high carbohydrate diets may elevate serum triacylglycerols (Heine and Schouten, 1989). One study demonstrated that carbohydrate tolerance and glycaemic control were unchanged in a diet containing about 39% energy as fat, whether the P/S ratio was 0.3 or raised to about 1.0 by linoleic acid enrichment (Heine *et al.*, 1989).

17.3.2.2 n-3 fatty acids

Preliminary work suggested that n-3 PUFA at a dose of 3 g/day increased insulin sensitivity in six NIDDM subjects (Popp-Snijders *et al.*, 1987). However, more comprehensive studies by other workers have failed to confirm this. Furthermore, some studies of dietary n-3 fatty acid supplementation using doses ranging from 1 to 10 g/day, have demonstrated a worsening of glycaemic control, as measured by fasting and meal-stimulated plasma glucose concentrations, and of glycosylated haemoglobin values. The effects are probably mediated by an increase in hepatic glucose output and decreased insulin secretion with no change in the peripheral glucose disposal rate (Friday *et al.*, 1989; Glauber *et al.*, 1988; Kasim *et al.*, 1988; Stacpoole *et al.*, 1989). A study in healthy non-diabetic subjects also found that fish oil supplementation causes a decrease in C-peptide excretion, a marker for insulin production (Stacpoole *et al.*, 1989).

These studies suggest that the use of fish oils in diabetic subjects should be viewed with caution, given their potentially adverse effects on glycaemic control.

17.3.3 Complications of diabetes

17.3.3.1 n-6 fatty acids

There are only few studies concerned with the effects of unsaturated fatty acids, mainly linoleic acid (18:2 n-6), on the development of specific diabetic complications affecting the nerves, eyes and kidneys (diabetic neuropathy, retinopathy and nephropathy). Two studies have shown that increasing linoleic acid intake causes a significant decrease in incidence of retinopathy in the treated group (Houtsmuller *et al.*, 1980; Howard-Williams *et al.*, 1985). However, in neither study was dietary information complete, nor was there rigorous assessment of diabetic control over the period of the study.

17.3.3.2 n-3 fatty acids

One well-conducted study using a double-blind cross-over design has demonstrated an effect of cod liver oil, but not olive oil, supplementation in decreasing transcapillary albumin leakage in IDDM (Jensen *et al.*, 1989). The importance of this finding in relation to clinical complications is not clear (Anonymous, 1990). Early proteinuria was decreased in diabetic rats fed on a diet enriched with fish oil, but the significance and mechanism of this finding are unknown (Sinha *et al.*, 1990).

Two studies have demonstrated that supplements of fish oil concentrates can reduce blood pressure in diabetic subjects (Jensen *et al.*, 1989; Kasim *et al.*, 1988). One study has demonstrated better arterial compliance in NIDDM and normal subjects who ate fish than in those who did not (Wahlqvist *et al.*, 1989). However, as there were no differences in biochemical parameters between the two groups, it is not clear that there was any influence of fatty acids.

18

UNSATURATED FATTY ACIDS AND SKIN DISEASES

18.1 NORMAL ROLES OF UNSATURATED FATTY ACIDS IN THE SKIN

18.1.1 Background

The two principal layers of the skin are the epidermis and the dermis (see Figure 18.1). The epidermis is a protective layer without blood vessels; it also has an immunoregulatory role. It has a deep layer of growing cells and a covering layer of dried dead cells that are constantly being shed and replaced from the growing layer. The dermis is made of tough, elastic connective tissue and has a rich network of blood vessels and nerves.

The skin is a highly active tissue for lipid synthesis with the epidermis being more active than the dermis (on a weight basis). The lipid synthesising activity of the dermis is mainly dependent on the production of sebum by the sebaceous glands.

18.1.2 Role of polyunsaturated fatty acids (PUFA) in the epidermis

18.1.2.1 Skin barrier function

The fatty acid content of the epidermis plays a major role in the integrity, water-retaining and barrier functions of the skin. Of the total fatty acids in normal rat skin, linoleic acid (18:2 n-6) accounts

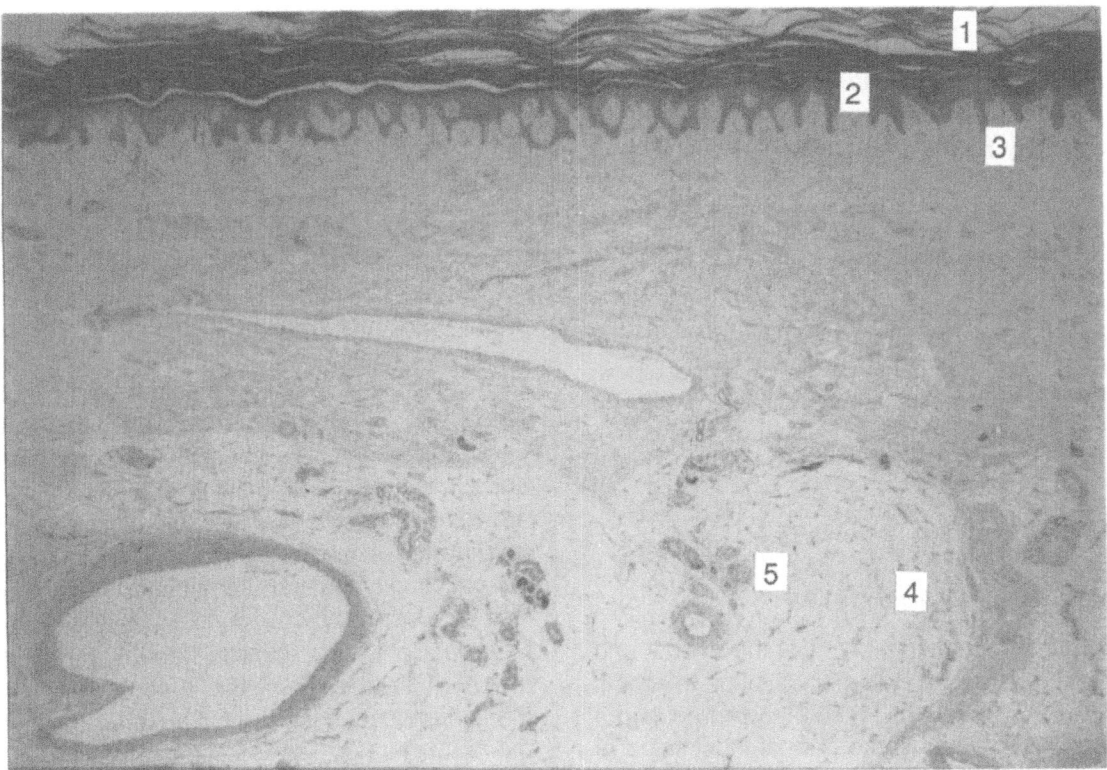

Figure 18.1 Transverse section through normal skin: 1 Stratum corneum; 2 Stratum malpighii; 3 Papillary dermis; 4 Subcutaneous fat; 5 Eccrine sweat gland.

for about 30% and only traces of alpha linolenic acid (18:3 n-3) are present (Ziboh, 1989). These fatty acids maintain the fluidity of the cell membrane since most of them are esterified in the form of membrane phospholipids. Arachidonic acid constitutes about 9% of total skin fatty acids but the epidermis does not synthesise arachidonic acid from precursors because it lacks the enzyme 6-desaturase. Thus epidermal arachidonic acid comes from circulating plasma lipids. Epidermal lipids, which play a role in water-proofing the skin, contain linoleic acid linked to some very long chain (C_{30}–C_{34}) fatty acids forming the acyl ceramides (see Chapter 7). These are found in the granular layer of the epidermis (Wertz and Downing, 1982).

18.1.2.2 Eicosanoids as mediators of inflammation

PUFA of the epidermis have recently received attention, mainly because these fatty acids are the precursors of the eicosanoids (see Chapter 8) which are important mediators of some major inflammatory skin diseases (Greaves, 1988).

Jessup et al. (1970) first showed that the skin is capable of synthesising prostaglandins following an appropriate challenge. The major prostaglandin of mammalian epidermis is PGE_2 (Van Dorp, 1971), and human skin homogenates were subsequently shown to be capable of transforming arachidonic acid via cyclo-oxygenase to prostaglandins (Greaves et al., 1972). More recently, it has been demonstrated that human epidermal cells are a rich source of the lipoxygenase products of arachidonic acid, including 12-hydroxyeicosatetraenoic acid (12-HETE) and leukotriene B_4 (Fincham et al., 1985). Both cyclo-oxygenase and lipoxygenase transformation products of arachidonic acid are important mediators of inflammation in a wide variety of skin disorders and this is undoubtedly a major function of essential fatty acids (EFA) in the skin.

Other non-eicosanoid hydroxy fatty acids have been identified in the human epidermis (Camp et al., 1983), e.g. 13-hydro-peroxyoctadecadienoic acid (13-HODE), and these are also known to be biologically active.

18.1.3 Role of PUFA in the dermis

18.1.3.1 Sebum production

Sebum is the main lipid product of the dermis and is formed in the sebaceous glands round the roots of the hairs. Downing et al. (1986) have proposed a role for EFA of sebum in the normal functioning of the hair follicle epithelial cells. Human sebum consists of triacylglycerols (57%), wax esters (26%) and squalene (10%) with small amounts of cholesterol and cholesteryl esters. Most of the fatty acids of these lipids are branched chain compounds and more than half are monounsaturated (Garton, 1985).

Active lipogenesis takes place in the sebaceous gland; there is no evidence yet of incorporation of circulating plasma lipids into differentiated sebaceous cells. In fact, the fatty acid composition of sebum wax esters and triacylglycerols is such that most of these acids are unlikely to be derived from plasma lipids (Downing, 1976).

Evidence for a role of sebum linoleic acid in the normal function of follicular epithelial cells has recently been reported (Downing et al., 1988). The amount of linoleic acid in sebum is inversely proportional to the rate of sebum production (Stewart et al., 1978). This is probably because of the diluting effect of endogenous lipid synthesis by sebaceous cells. Most of the linoleic acid is present in the cholesteryl ester fraction, with only small amounts in triacylglycerols and wax esters.

18.1.3.2 Eicosanoid production

The major eicosanoid of the human dermis is 15-hydroxy-eicosatetraenoic acid (15-HETE) and its major cellular source is the fibroblast (Kragbulle et al., 1986). Since 15-HETE is a powerful lipoxygenase inhibitor, a regulatory role for this hydroxy acid on eicosanoid biosynthesis by the epidermis has been proposed. The small blood vessels running through the dermis also represent an important source of eicosanoids in skin. The major product of cultured dermal microvascular endothelial cells is prostaglandin E (Bull et al., 1990).

18.2 ESSENTIAL FATTY ACID DEFICIENCY AND THE SKIN

18.2.1 Animal studies

EFA deficiency has been studied mainly in the rat (see Chapter 7). When fed for a prolonged period on a fat-free diet, rats developed a scaly skin (Burr and Burr, 1929) and showed extensive water loss from the skin. As well as developing scaliness (especially of the feet and tail), the skin showed haemorrhagic spots and cutaneous erythema. Hair loss occurred and the skin lost its elasticity. The EFA-deficient rats showed greatly reduced skin levels of linoleic acid and arachidonic acid, but there appeared to be a compensatory increase in monounsaturated acids and PUFA of the n-9 series. This enabled the fluidity of the cell membranes of the epidermis to be maintained to some degree. Epidermal acyl ceramides in EFA-deficient animals contained only 2% esterified linoleate compared to 38% in normal epidermis (Wertz et al., 1983). All these changes could be reversed by administration of PUFA of the n-6 series. Arachidonic acid proved less effective than its precursor, linoleic acid, in this respect (Prottey, 1977).

The ability of topically applied linoleic acid to reverse the changes associated with EFA deficiency in the rat in the presence of the prostaglandin synthetase inhibitor indomethacin (Elias et al., 1980) shows that linoleic acid deficiency must impair barrier function in the skin independent of its role as a precursor of prostaglandins.

18.2.2 Human studies

18.2.2.1 Causes of EFA deficiency in man

Infants fed on milk formulae which are deficient in linoleic acid develop a scaly dermatitis and have retarded growth and other systemic disturbances (Hansen et al., 1958).

There are now numerous reports that EFA deficiency can also occur in adults and children by surgical intervention, or through long-term parenteral feeding regimens.

Extensive bowel resection in adults is associated with severe EFA deficiency and patients have a dry scaly rash. The serum linoleic acid level falls to 2–3% of total fatty acids compared to 22% in controls. Application of topical linoleic acid for 2–12 weeks can correct the serum biochemical abnormality and the skin changes (Press et al., 1974a). As little as 2–3 mg linoleic acid per kg body weight is sufficient to correct any EFA deficiency, either systemically or in the skin, at least when it is delivered topically.

Total parenteral nutrition (TPN) can also cause cutaneous lesions of EFA deficiency although it takes at least 6–7 weeks for such lesions to appear (Riella et al., 1975). They are frequently flexural and can be corrected by parenteral administration of linoleic acid. The widespread use of TPN suggests that such cases are likely to become increasingly common.

18.2.2.2 Markers of EFA deficiency

Press et al. (1974b) showed that eicosatrienoic acid (20:3 n-9) levels rise in parallel with the fall in serum linoleic acid and that the presence in the serum of eicosatrienoic acid, or better still the trienoic/tetraenoic ratio in the serum, is a good biochemical index of EFA deficiency (see Chapter 7).

18.3 ACNE AND PSORIASIS

18.3.1 Background

Acne is a skin disorder in which the sebaceous glands become inflamed, the hair follicles become blocked and pustules form. The sebaceous secretion in acne has a lower proportion of acyl ceramides (6% versus 45% in the normal physiological state). These contain less linoleic acid than normal, yet their rate of synthesis is greatly increased. The differences could account for the abnormalities associated with acne.

Psoriasis is one of the commonest skin diseases in Britain, affecting about 1–2% of the population, and is often familial. It is a chronic skin disease in which scaly patches form on the elbows, forearms, knees, legs, scalp and other parts of the body. There is a reddening of the skin (erythema) and itching may occasionally occur in affected areas.

Elevated levels of arachidonic acid and its 5-lipoxygenase products, the leukotrienes of the 4-series, and the monohydroxy fatty acid 12-HETE have been found in psoriatic skin lesions (Plummer et al., 1978; Brain et al., 1984, 1985; Hammerstrom et al., 1975). The cellular source of 12-HETE in skin is likely to be the keratinocyte (Fincham et al., 1985), although the major source of leukotriene B_4 in affected skin is probably the neutrophil.

Both leukotriene B_4 and 12-HETE are powerful neutrophil and lymphocyte chemo-attractants and leukocyte accumulation is the earliest microscopic feature of a developing lesion of psoriasis. There is also some evidence to suggest that these lipoxygenase products may stimulate epidermal proliferation, although this remains to be confirmed.

18.3.2 Role of PUFA in the treatment of acne

Although EFA such as linoleic acid have been implicated in the pathogenesis of acne, there is little firm evidence from properly controlled clinical trials to show that administration of EFA improves acne.

18.3.3 The role of PUFA in the treatment of psoriasis

The pathogenic significance of the leukotrienes and 12-HETE has been demonstrated by the recent observation that topical application of a pharmaceutical preparation which is a selective 5-lipoxygenase inhibitor could cause differential suppression of leukotriene B_4, but not 12-HETE or arachidonic acid. This was followed closely by the

Table 18.1 EPA treatment of psoriasis

Study	Number of patients (controls)	Duration of trial	Dose of EPA	Clinical assessment	Biochemical indices
Ziboh et al. (1986)	13(0)	8 weeks	11–14 g/day EPA from fish oil supplements	8/13 were clinical responders[a]	Greater increase in ratio of EPA/DHA**, but not EPA/AA in epidermal lipids of clinical responders
Maurice et al. (1987)	10(0)	6 weeks	12 g/day as EPA capsules	8/10 patients were clinical responders[a]	Increase in the ratio of EPA/AA[ns] in platelets of clinical responders
				Decreased erythema in responders Decreased scaling in responders	Decrease in LTB_4 production in leukocytes (responders and non-responders)
Bittiner (1988)	14(14)	8–12 weeks	18 g/day as EPA capsules	Decreased[b] itching* Decreased[b] erythema*	–

ns = not significant; * = $p < 0.05$; ** = $p < 0.01$; [a] clinical responders = showing slight to moderate improvement compared with baseline; [b] compared with baseline and controls.

clinical clearance of the psoriatic skin (Black et al., 1991).

In view of the proposed role of 5-lipoxygenase in the pathogenesis of psoriasis, numerous attempts have recently been made to suppress the activity of this enzyme by dietary means. Several studies have been made using orally administered fish oil supplements rich in eicosapentaenoic acid (EPA) (20:5 n-3) and its metabolic product docosahexaenoic acid (DHA) (22:6 n-3) (see Table 18.1). EPA is preferentially transformed by lipoxygenase to the leukotrienes of the 5-series (see Chapter 8). These eicosanoids are much less biologically active than the leukotrienes of the 4-series (see Chapter 19). At the same time, there is a reduction in generation of leukotriene B_4 (Lee et al., 1984). It has therefore been proposed that increasing the amount of EPA in leukocytes could lead to an amelioration of the pathological changes in inflammation, including psoriasis (Lee et al., 1985) (Figure 18.2).

Ziboh et al. (1986) gave 13 psoriatic patients a fish oil-supplemented diet (11–14 g EPA daily) for 8 weeks. EPA and DHA were rapidly incorporated into the phospholipids of neutrophils and epidermal cells of these patients; the amounts incorporated increased with weeks of exposure to the fish oil supplement. Eight of the 13 patients demonstrated a small but favourable clinical response which correlated well with the biochemical changes: e.g. the ratios of EPA/DHA in the epidermal lipids.

In another trial, 8 out of 10 psoriatic patients receiving 12 g EPA daily for 6 weeks showed a modest improvement in their psoriasis, the principal effects being a slight lessening of erythema and scaling. There was a high degree of inhibition of leukotriene B_4 synthesis by peripheral blood mononuclear leukocytes although the effect of the diet on epidermal lipids was not determined (Maurice et al., 1987).

Bittiner et al. (1988) have conducted the only double blind randomised control trial of EPA in

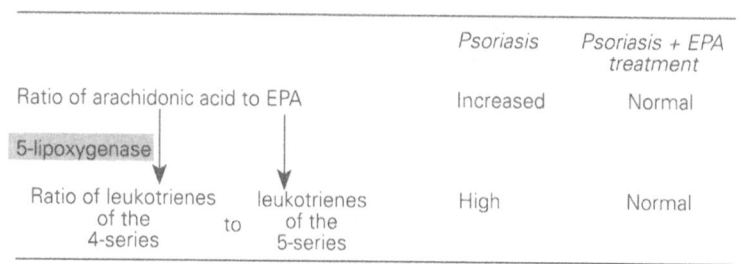

Figure 18.2 Rationale behind the use of fish oils for treating psoriasis.

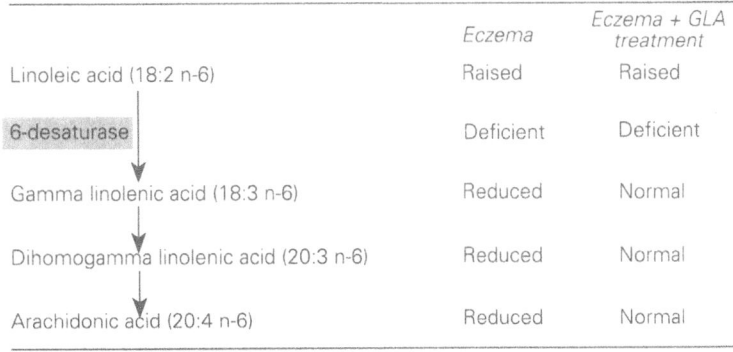

	Eczema	Eczema + GLA treatment
Linoleic acid (18:2 n-6)	Raised	Raised
6-desaturase	Deficient	Deficient
Gamma linolenic acid (18:3 n-6)	Reduced	Normal
Dihomogamma linolenic acid (20:3 n-6)	Reduced	Normal
Arachidonic acid (20:4 n-6)	Reduced	Normal

Figure 18.3   Rationale behind the use of evening primrose seed oil for treating atopic eczema.

psoriasis. Fourteen patients were given 18 g/day of EPA capsules whilst 14 controls received olive oil capsules for 8–12 weeks. The treated group showed small but statistically significant lessening of itching and erythema compared to the placebo group after 8 weeks.

In summary, it appears that biochemical modification of the lipid 'profile' can be achieved at least in blood leukocytes and possibly in epidermal lipids. However, the EPA-supplemented diet has only a modest therapeutic action of doubtful clinical value. The discrepancy between the biochemical effects of EPA and the modest therapeutic effects suggests that LTB_4 might not be the only mediator involved in the psoriatic lesion.

The extrapolation of the results from clinical trials using EPA capsules to the recommendation to eat more oily fish as a treatment for psoriasis must be given with caution. It is, however, encouraging that a dose of 18 g/day EPA could be provided by a realistic amount (150 g) of oily fish such as mackerel. Advice to psoriatic patients to eat plenty of oily fish, whilst not doing harm, might in some cases give some relief.

18.4   ATOPIC ECZEMA

18.4.1   Background

Eczema is a superficial inflammation of the skin, which affects both the dermis and the epidermis. Eczema causes itching with a red rash often accompanied by small blisters that 'weep' and become crusted. Atopic eczema is commonly found in children and is sometimes associated with a family history of allergy.

Current interest in the role of EFA in treating atopic eczema originates in an observation by Hansen et al. (1933) who showed that patients with eczema had elevated serum levels of linoleic acid (18:2 n-6) but reduced levels of its 6-desaturase

products, gamma linolenic acid (GLA) (18:3 n-6) and dihomogamma linolenic acid (DGLA) (20:3 n-6). Subsequent plasma analyses (Manku et al., 1984) supported these findings and extended them to show that plasma of atopic eczema patients also has reduced levels of arachidonic acid (20:4 n-6). The results are compatible with a proposed reduced activity of the 6-desaturase enzyme in atopic eczema (see Figure 18.3).

18.4.2   Role of PUFA, particularly evening primrose seed oil (EPSO) in the treatment of atopic eczema

In the 1980s there were several reports suggesting that oral replacement therapy using evening primrose seed oil (EPSO) (which contains 7–10% of GLA) effects clinical improvement in the skin of patients with atopic eczema (Wright and Burton, 1982; Lovell et al., 1981; Bordoni et al., 1987). The improvement was claimed to be correlated with increased plasma levels of DGLA (20:3 n-6) but the improvement reported by Wright and Burton (1982) between EPSO and placebo could have been due to differences in baseline scores, since actual scores were not compared.

On the other hand, a large single placebo-controlled trial failed to demonstrate any significant clinical benefit (Bamford et al., 1985). A general criticism of the EPSO trials is that concurrent usage of topical steroids and systemic antihistamines is not standardised between EPSO-treated and placebo-treated groups. Morse et al. (1989) have tried to reconcile these disparate results by conducting a meta-analysis of 10 placebo-controlled studies on the efficacy of GLA, in the form of capsules, in the treatment of atopic eczema (see Table 18.2). Four of the trials were parallel and six were crossover in design. In each trial, both doctors and patients assessed the severity of eczema by scoring measures of inflammation, dryness, scaliness, pruritus and overall skin

Table 18.2 Summary of ten trials of GLA treatment of psoriasis

Type of assessment	No of patients[a]	Improvement with GLA[b]	No of patients	Improvement with placebo	Difference between treatments
Patient, global	177	17.4%***	175	15.7%***	ns
Clinician, global	218	27.6%***	204	22.1%***	< 0.05
Itch	218	21.0%***	202	10.5%*	< 0.05

[a] One trial was on children, remainder on adults; [b] Doses ranged from 1 g to 6 g of 'Epogam'; * $p < 0.05$ compared with baseline; ** $p < 0.01$ compared with baseline; *** $p < 0.001$ compared with baseline; ns = not significant
(Adapted from Morse et al., 1989)

involvement. Individual 'symptoms scores' were combined to give a single global 'severity score' and 'itch' was assessed separately. The magnitude of the response was expressed in percentage terms by relating the change in response to either GLA or placebo as a percentage of the starting severity score.

Table 18.2 concentrates on the differences between the last assessment point of all the trials and the first assessment. There was a significant improvement from baseline in patients receiving either GLA or placebo for all three parameters, although the improvement in 'itch' when placebo was given was smaller than the rest. The improvement in patients treated with GLA was significantly greater than in those treated with placebo when assessed by doctors and for 'itch' but not, unfortunately, for the patient's own global assessment of improvement.

This meta-analysis highlights the marked placebo effect that characterises this type of dermatological study, but it does give some evidence that GLA and EPSO might have a modest beneficial effect on mild atopic eczema. However, claims that this treatment 'substantially improves the clinical symptoms of atopic eczema after only 4 weeks of treatment' (Beagi et al., 1988) seem over-optimistic when set against current published evidence. Unfortunately, there is no readily available dietary source of GLA as it is only found in a few seed oils.

19

UNSATURATED FATTY ACIDS AND IMMUNE DISORDERS

19.1 INTRODUCTION

Many patients become interested in the dietary management of chronic diseases with a suspected immunological basis because they have proved difficult to manage by other means. This approach has the emotional advantage of allowing the patient to participate in the management of their complaint. The attraction of alternative medicine and the fear of drug toxicity have given further impetus to this interest. The efficacy of supplementing the diet with unsaturated fatty acids is one of the many dietary issues which has been debated. Given the scope for unscrupulous, misguided exploitation of such patients, it is particularly important to examine the issues as objectively as possible.

Investigators have used n-3 PUFA such as docosahexaenoic acid (22:6 n-3) (DHA) and eicosapentaenoic acid (20:5 n-3) (EPA) and n-6 PUFA found in sunflower seed oil and evening primrose seed oil. Monounsaturated fatty acids have not been tested.

Trials have either tested just one form of unsaturated fatty acid or have compared both n-3 and n-6 PUFA in patients with immune disorders. In contrast with many forms of alternative or homeopathic medicine, investigators have tested a valid idea about the effects of unsaturated fatty acids on immune-mediated inflammation and have sought to correlate biochemical with therapeutic effects. In general, both forms of unsaturated fatty acid treatment have achieved the desired effects in terms of altering the lipid composition of cell membranes with a reduction in pro-inflammatory activity. Where there has been clinical benefit, it is still difficult to determine which form of unsaturated fatty acid supplement is preferable. In many diseases, biochemical modulation has not been accompanied by clinical improvement.

19.2 THE IMMUNE SYSTEM (see Figure 19.1)

19.2.1 Background

The immune system is responsible for host defence

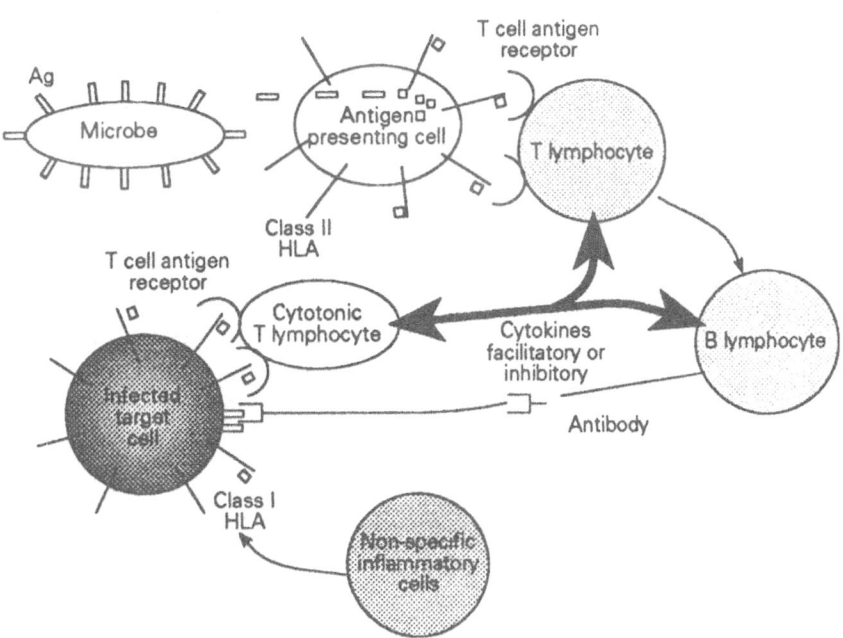

Figure 19.1 The immune system.

against infective agents. It also protects against the potentially injurious effects of allergens absorbed through the skin, respiratory system and gut.

The hallmark of immune reactions is specificity, so that molecules are synthesised which recognise the structures peculiar to micro-organisms and foreign substances. The generation of specific immune responses allows the acquisition of immunological 'memory' by which a more rapid response is mounted when the inciting antigens are next encountered.

Not all immune responses are specific but many are mediated by cells and their products which are able to engage micro-organisms and allergens in an entirely uninstructed manner. The components of the inflammatory response are non-specific although they are commonly activated by immunologically specific mechanisms. In most situations, such as infectious diseases, the first line of defence is non-specific. A specific immune response is eventually generated which allows other protective cells and molecules to be directed to the threatened site. Immunological 'memory' blunts the severity of subsequent challenges and may be completely protective.

A complex hierarchy of cells and their products makes up the immune response. Specific immune responses are mediated by different families of lymphocytes. The brief account which follows greatly over-simplifies the complexities of the immune system but excellent introductory texts are available (e.g. Roitt, 1988). It is crucially important to appreciate the variety of mechanisms involved and their interdependence.

19.2.2 T lymphocytes

19.2.2.1 Antigen recognising T lymphocytes

T lymphocytes express surface receptors which recognise the unique composition and structure of an antigen. Antigen-presenting cells are found in the blood, lymphatic system, and other tissues. They express cell membrane surface proteins of the histocompatibility system termed 'class II' molecules which contain a groove occupied by antigens. This arrangement makes it easier for T lymphocytes to encounter antigens concentrated in particular sites and in a form which their receptors can readily engage.

19.2.2.2 Cytotoxic T lymphocytes

T lymphocytes can also kill cells bearing foreign antigens thereby aiding their removal. Cytotoxic T lymphocytes also express cell membrane proteins which recognise foreign antigens. These antigens occupy the grooves of a different class of surface

histocompatibility antigen expressed by target cells termed 'class I' molecules. This form of immunity is particularly important because it allows protective molecules to be transported by mobile cells to sites which large macromolecules cannot reach. It corresponds to classical 'cell-mediated (delayed) hypersensitivity'.

19.2.3 B lymphocytes

B lymphocytes synthesise antibodies which combine with foreign antigens in the intra- and extra-vascular compartments. There are several classes and sub-classes of antibody which differ functionally according to where they are synthesised and in the avidity with which they combine with different antigens. Antibodies largely, but not exclusively, mediate what has classically been termed 'immediate hypersensitivity'. B cell function is largely dependent on signals from T lymphocytes which have encountered foreign antigens.

19.2.4 Non-specific inflammatory cells

T and B lymphocytes are the main instruments of antigen-specific immune responses. Their interaction with antigens recruits non-specific populations of inflammatory white cells (mainly polymorphonuclear leukocytes and eosinophils), phagocytic cells (mainly macrophages), and non-specific 'killer' cells. These cells not only assist in the elimination of foreign antigens, but also contribute to tissue damage in allergic diseases.

19.2.5 Immune mediators

Several protein 'families' also contribute to specific immune responses and the associated inflammatory reaction. These are mainly secreted by the participating cells, such as lymphocytes and macrophages, but also by other cell types including hepatocytes and vascular endothelial cells:

(i) *Leukotrienes* are synthesised predominantly by granulocytes and mediate many of the inflammatory phenomena characteristic of immediate hypersensitivity reactions.
(ii) *The complement system* comprises a number of serum proteins which function as a triggered cascade. Activated complement components assist in the clearance of antigens bound to antibody and also in attracting cells non-specifically to sites of immune-mediated inflammation.
(iii) *Cytokines* are synthesised by activated cells

of the immune system and enhance the proliferation and differentiation of other cells in this system in response to immune stimulation. Cytokines affect many other systems. Their effects are dose-dependent and modified by interactions with other cytokines.

19.2.6 Immunoregulation

There are also elaborate regulatory mechanisms for limiting the extent of conventional immune reactions to foreign antigens and for preventing immune responses to self-antigens (autoimmunity). The main components are T lymphocytes with 'suppressor' activity, antibodies which inhibit the proliferation of B cells responding to the initial immune stimulus, and cytokines at appropriate sites and in the correct concentration.

19.3 IMMUNE-MEDIATED DISEASES

The great diversity of immunologically mediated disorders arises from two main factors:

(i) disease through antigen persistence;
(ii) disease induced by the immune system itself.

19.3.1 Disease through antigen persistence

Defects in the immune response can arise when the response is inadequate to eliminate the infectious agent completely and the associated inflammation can be injurious. Immuno deficiency may be overt, as in a disorder such as common variable hypogammaglobulinaemia, or subtle, as in some disorders of granulocyte function. In addition, many infectious agents are able to evade immune elimination by a variety of strategies.

19.3.2 Disease induced by the immune system itself

The common element in most primary immunopathological disorders is defective regulation of the immune response itself.

In atopic disorders, the defect appears to reside in the manner in which commonly encountered allergens are presented to T cells by antigen-presenting cells. In autoimmune disorders, self-antigens provoke an immune response by seemingly analogous mechanisms. Genetic factors are important in determining susceptibility to atopic and autoimmune disorders, most obviously by determining the structure of class I and II histocom-

patibility antigens and the repertoire of antigen receptors on T lymphocytes.

However, many immunopathological disorders are clearly multifactorial and both immunodeficiency and defective regulation operate synergistically. Thus in systemic lupus erythematosus (SLE/lupus), inherited deficiency of the C2 molecule in the complement cascade confers susceptibility to the disease, yet autoimmunity is a major contributing factor.

There are also many chronic degenerative diseases in which abnormal immune mechanisms are encountered, yet their precise role is unclear. These reactions may be of primary aetiological importance, a major cause of tissue damage, or entirely an epiphenomenon. It is, for example, highly probable that joint inflammation in patients with rheumatoid arthritis results from both T and B cell activation but it is still far from certain that this activity initiates the disease. Similarly, perivascular immune-mediated inflammation is readily detectable in the brain and spinal cord of patients with multiple sclerosis yet it is still not known whether or not this process accounts for the characteristic demyelination. Autoimmune reactions are observed in many persistent infections, particularly those caused by tropical parasites, but in this situation they are usually harmless. The complexities of immune disorders are well described in many authoritative texts (e.g. Lachmann and Peters, 1982).

There are obvious implications for those wishing to manipulate the immune response for therapeutic advantage. Non-selective perturbation is likely to have unpredictable consequences. Indeed, even when the target for manipulation has been well defined, the interactions in the immune system ensure that the outcome may be unexpected. Furthermore, the train of pathological events in most chronic diseases with an immune component remains unclear. Thus pragmatic observation is still more important than theoretical speculation in most attempts to control such disorders by dietary manipulation.

19.4 THE THEORETICAL BASIS FOR DIETARY MANIPULATION IN PATIENTS WITH IMMUNE DISORDERS

19.4.1 Biochemical considerations (see Figure 19.2)

19.4.1.1 Potential effects of n-3 unsaturated fatty acids

Manipulating the fatty acid composition of the diet demonstrably affects the pro-inflammatory activities

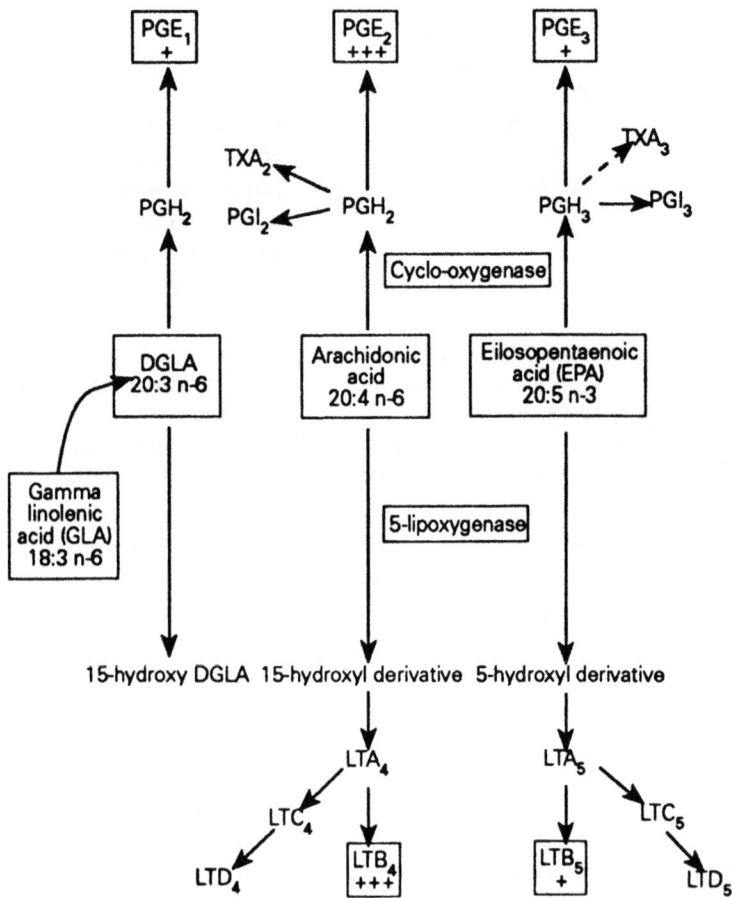

Figure 19.2 Biochemical basis for the potential effect of n-3 and n-6 unsaturated fatty acids on inflammation: $+++$ strongly pro-inflammatory; $+$ weakly pro-inflammatory; $\rightarrow$ less efficient reactions.

of many cell types involved in immune response and inflammatory reactions. The relevant cell types include lymphocytes, macrophages, granulocytes (often referred to as leukocytes), and non-specific 'killer' cells.

Cell activation leads to the activation of phospholipase A_2 (PLA_2) resulting in the selective release of free fatty acids. Arachidonic acid, dihomogamma linolenic acid (DGLA) and eicosapentaenoic acid (EPA) are readily released by PLA_2 action but docosahexaenoic acid (DHA) is not released so readily (Sperling, 1989).

Several pro-inflammatory products are produced by the oxidative metabolism of arachidonic acid (20:4 n-6) by the cyclo-oxygenase and 5-lipoxygenase pathways (see Chapter 8). Prostaglandins and thromboxanes are the active metabolites derived by the cyclo-oxygenase pathway; the leukotrienes, lipoxins, and mono- and di-hydroxy fatty acids are derived by the 5-lipoxygenase pathway. Leukotriene B_4 (LTB_4) is a chemoattractant thereby promoting the arrival of neutrophils in inflammatory lesions. Its ability to increase neutrophil adhesion to vascular endothelial cells is a key step in this process. It stimulates the secretion of

lysosomal enzymes and superoxide radicals by neutrophils. Other 5-lipoxygenase products of arachidonic acid have similar effects (Sperling, 1989).

The cyclo-oxygenase pathway of arachidonic acid metabolism leads first to the formation of the intermediate prostaglandins PGG_2 and PGH_2 followed by the generation of bioactive prostaglandins and thromboxane (see Chapter 8). EPA inhibits cyclo-oxygenase in intact cells while both EPA and DHA inhibit this enzyme in a cell-free system.

Both arachidonic acid and EPA are metabolised initially to 5-hydroxy derivatives by the 5-lipoxygenase pathway. These are further metabolised to leukotriene A_4 (LTA_4) and leukotriene A_5 (LTA_5) respectively. Hydrolysis of LTA_4 by a leukotriene A epoxide hydrolase yields leukotriene B_4 (LTB_4) with potent pro-inflammatory activities. The action of the same enzyme on LTA_5 yields LTB_5 which has markedly less pro-inflammatory activity than LTB_4.

Increased intake of n-3 fatty acids from diets rich in fish oils can therefore competitively inhibit the formation of 2-series prostaglandins and the 4-series leukotrienes from arachidonic acid (i.e. prostacyclin I_2, PGE_2, thromboxane A_2 and LTB_4).

Increased consumption of n-3 PUFA, such as EPA and DHA, favours the production of prostaglandins of the 3 series and leukotrienes of the 5 series. The overall effect is to reduce the inflammatory response and the efficiency of granulocyte mobilisation activation.

19.4.1.2 Potential effects of n-6 unsaturated fatty acids

Similar anti-inflammatory effects could be expected with plant seed oils, notably those extracted from seeds of borage and evening primrose. These contain relatively high proportions (18% and 9%) of gamma linolenic acid (18:3 n-6) which is converted in the body to DGLA (20:3 n-6). The metabolism of DGLA yields prostaglandin E_1 which differs structurally from prostaglandin E_2 by the absence of a double bond in the 5-position. Prostaglandin E_1 is pro-inflammatory but also immunosuppressive in some experimental systems. However, these effects are dose-dependent and the outcome is largely determined by the conditions of the assay system. Since DGLA competes with arachidonic acid for oxidative enzymes, it also reduces the production of cyclo-oxygenase products derived from arachidonic acid. Moreover, DGLA is converted to 15-hydroxy DGLA by the action of 15-lipoxygenase. It is not a substrate for 5-lipoxygenase like arachidonic acid and the inflammatory leukotriene LTB_4 is not produced.

19.4.1.3 General

The overall effect of dietary manipulation is to alter the lipid content of cell membranes in a rather non-specific manner; there is no evidence that different inflammatory and immune cell populations are selectively affected. Furthermore, while the secretion of inflammatory mediators has received particular attention, other membrane changes have received less attention. These include changes in membrane fluidity, other enzyme activity, and receptor function, all of which might be expected to influence immunological functions such as antigen presentation and mediator secretion.

19.4.2 Effects on inflammation

19.4.2.1 Animal studies

Early animal studies gave some hint that dietary manipulation of the fatty acid composition of the cell membrane of inflammatory cells influences their reactivity. Some of these experiments involved the behaviour of neutrophils and other cells *in vitro*. Others were based on the reactions of

animals *in vivo* to non-specific inflammatory stimuli. In general, selective enrichment with metabolites of n-3 series fatty acids depresses inflammation. Local concentrations of PGE_2 and LTB_4 are reduced and leukocyte effector functions such as chemotaxis and the release of lysosomal enzyme proved especially vulnerable. Similar effects have been noted with the n-6 fatty acids gamma linolenic acid and DGLA (Tate *et al.*, 1989).

19.4.2.2 Studies in man

Diets rich in EPA reduced LTB_4 production (Prescott, 1984) and inflammatory reactions (Payan *et al.*, 1986) in normal individuals. Daily dietary supplements of 18 g of fish oil increased the EPA content of blood neutrophil and monocyte cell membranes (Lee *et al.*, 1985). This regimen significantly reduced the chemotactic response of these neutrophils to LTB_4 and their adhesion to vascular endothelial cells. Labelling studies indicated that fish oil supplements inhibited arachidonic acid metabolism by suppressing phospholipase A_2 and 5-lipoxygenase activity. Fish oil supplements also suppressed the generation of platelet activating factor (PAF) by blood monocytes from normal human donors (Sperling *et al.*, 1987a). This dietary manoeuvre also markedly suppressed the synthesis of the cytokines interleukin-1 and tumour necrosis factor (TNF) (Endres *et al.*, 1989) which are important mediators of immune reactions. In many chronic inflammatory disorders maintained by immune reactions there is a cascade of cytokine production resulting in fresh activation of lymphocytes and non-specific inflammatory cell populations. Thus inhibition of interleukin-1 and TNF production is not only anti-inflammatory but also offers the possibility of breaking the sequence of self-perpetuating pro-inflammatory events.

Similar changes in the fatty acid profile of human blood monocytes have been shown after the daily administration of 1.1 g GLA in the form of borage seed oil (Pullman-Mooar *et al.*, 1990). After 12 weeks on this diet, there was a significant reduction in prostaglandin E_2, LTB_4 and leukotriene C_4 produced by stimulated monocytes. This reduction was noted in association with a marked increase in the ratios of the metabolite DGLA (20:3 n-6) to both arachidonic acid and stearic acid in these cells. Thus DGLA appears to block the generation of oxygenation products from arachidonic acid.

19.4.3 Effects on specific immune responses

Immune responses mediated by both T and B lymphocytes are blunted by administering diets rich in n-3 and n-6 fatty acids. These effects have been shown in experiments performed *in vitro* involving lymphocyte stimulation with mitogens and responses given *in vivo* to specific antigens (reviewed by Buchanan *et al.*, 1991). However, such experiments assess the immunosuppressive effects of dietary manipulation at the time of primary antigenic challenge when both cell-mediated and humoral immune responses are most vulnerable to manipulations of this kind. There is little evidence that established responses are similarly affected. This is an important distinction since immune responses relevant to human disease are analogous to persistent immune responses. Nor is there any evidence that manipulating the dietary fatty acid content influences vital immune functions such as immunological memory and the induction of immunological tolerance. In most chronic immune disorders, the real goal is to re-establish tolerance to autoantigens in order to avoid the need for continued immunosuppression.

Nevertheless, there is good evidence that dietary manipulation influences human lymphocyte responses. An increased dietary intake of n-3 PUFA reduced the production *in vitro* of the key regulatory cytokine interleukin-1 by human blood monocytes (Endres *et al.*, 1987). Similarly, DGLA suppressed the reactivity of human peripheral blood lymphocytes measured *in vitro* by mitogen-induced production of interleukin-2 and the growth of these cells in response to this cytokine (Santoli *et al.*, 1990). Interestingly, these effects appeared to be independent of PGE production.

19.5 ANIMAL MODELS OF IMMUNE-MEDIATED DISEASE

There are two kinds of animal model of immune disease which can give information of possible relevance to human disease:

(i) models induced by immunisation designed to initiate immune-mediated inflammation, often in association with an auto-immune response to defined self-antigens;
(ii) spontaneous models of human disease, usually arising on a genetic basis.

19.5.1 Diseases induced by immunisation

Diets rich in unsaturated fatty acids have an unpredictable effect in these situations; they can either improve or augment the disease.

19.5.1.1 Improvement of the disease by
unsaturated fatty acids

Rats or mice immunised with a particular mycobac-

terial oligopeptide develop chronic relapsing arthritis with synovial immunopathological features resembling rheumatoid arthritis. The inciting oligopeptide has been sequenced and the disease is mediated by T lymphocytes specifically immune to this sequence. The disease can be transferred to histocompatible recipient animals by T lymphocyte lines expressing receptors for the relevant oligopeptide. The severity of the disease can be reduced by a diet enriched with EPA and GLA (Leslie *et al.*, 1985).

19.5.1.2 Augmentation of the disease by
unsaturated fatty acids

In contrast to the findings in adjuvant arthritis, dietary fish oil augmented the severity of arthritis in rats immunised with type II collagen (Prickett *et al.*, 1984): a manoeuvre which also induced a chronic relapsing polyarthritis.

19.5.2 Spontaneous immune disorders

Several inbred mouse strains develop spontaneous autoimmune diseases on the basis of a complicated polygenic pattern of inheritance. The principal features are B lymphocyte proliferation, vasculitis and nephritis. There are associated defects in immunity including reduced capacity to synthesise the cytokines interleukin-2 and tumour necrosis factor (TNF). The aetiology and pathogenesis of these disorders are controversial, but current attention is focused on evidence for defective tolerisation of T lymphocytes to self-antigens. These strains of mice (especially NZB × NZW and MRL lpr/lpr) have been extensively used as models for testing novel immunotherapeutic strategies including dietary manipulation.

Overall, the results in these models have been inconsistent. Several experimental diets have proved successful in reducing the severity of disease features such as nephritis and vasculitis; these include calorie reduction, limited intake of essential fatty acids, and increased dietary EPA (reviewed by Denman, 1987). However, amelioration of disease has not been the invariable outcome and the results have often depended on the precise composition of the modified diet. Thus menhaden oil, but not EPA capsules, decreased the severity of the nephritis characteristic of MRL mice, even though both preparations contained the same amount of EPA. Survival was unaffected (Westberg *et al.*, 1989).

19.5.3 Relevance to human immune disorders

It is not clear that any reported improvement in the

animal experiments resulted specifically from the altered fatty acid composition of the diet rather than from reduced calorie intake or other factors. Furthermore, it is not possible to extrapolate directly from these animal models to human immune disorders, such as lupus. For the most part, it is possible to start dietary manipulation at a much earlier stage in the disease than would be feasible in the analogous human disorders. The mechanisms of the resulting suppression of inflammatory and autoimmune responses have not yet been analysed in sufficient detail to confirm that modifications in arachidonic acid metabolism and membrane lipid content account for any reduction in disease activity.

19.6 ROLE OF UNSATURATED FATTY ACIDS IN TREATING CLINICAL IMMUNE DISORDERS

19.6.1 General considerations

Hypersensitivity to dietary constituents has been invoked in the aetiology and pathogenesis of almost all unexplained diseases. The precise nature of the alleged hypersensitivity has often been poorly defined. Different investigators have attributed any claimed improvement to a combination of food intolerance, non-immunological hypersensitivity, and true allergy resulting from immunological mechanisms. Many studies have been flawed by inadequate clinical definition of the patients, poorly designed trial protocols, and an

abundance of mystique at the expense of scientific analysis. Fortunately, the situation is changing and more recent trials have yielded valuable information (Brostoff and Challacombe, 1987). Several trials also have focused on the issue of dietary fatty acid composition.

Features such as disturbed immunity, autoimmune manifestations, and associated inflammation have been studied in great detail and so there is a broad background of information against which trials of dietary manipulation can now be conduced. Moreover, the morbidity of these disorders encourages patient compliance in such trials.

Virtually all chronic diseases with a major immunological component have been the subject of such studies. The results have been discouraging so far for multiple sclerosis and asthma, but have been mildly encouraging for the group of diseases termed 'inflammatory connective tissue diseases' which includes systemic lupus erythematosus (SLE) and rheumatoid arthritis.

19.6.2 Treatment of multiple sclerosis

A possible link between diet and multiple sclerosis has been pursued for over 40 years. This link has been suggested by epidemiological studies supported by claims that there is a lipid abnormality in patients with multiple sclerosis (MS). In particular, low levels of linoleic acid have been reported in the serum, platelets and red cells of MS patients. However, these and related observations have proved controversial and have not been confirmed by all investigators. It is still possible that there is

Table 19.1 Effect of unsaturated fatty acid on multiple sclerosis – summary of clinical trials

Study	Design	Biochemical changes	Clinical outcome
Millar et al. (1973)	Linoleic acid 17.2 g/day given to 36 patients for 2 years; Oleic acid 7.6 g/day as placebo	35% increase in serum linoleate level (as % of total fatty acids)	Relapses blunted, disease progression unaltered
Paty et al. (1978)	Linoleic acid 17.0 g/day given to 38 patients for 30 months	44% increase in serum linoleate levels	No effect on relapses, progression unaltered
Dworkin et al. (1984)	Linoleic acid 17–23 g/day given to 87 patients for 2½ years; Oleic acid as placebo	Not measured	Reduced severity and duration of relapses; reduced disability (especially if minimal disease)
Bates et al. (1989a and 1989b)	1.7 g/day EPA and 1.1 g/day DHA given to 155 patients for 2 years; Olive oil as placebo	Significant increase in serum and adipose tissue concentration of EPA and DHA as % of total lipids	Slight benefit (ns) in relapse rate and disease progression

ns = not significant

an abnormality of n-6 unsaturated fatty acids in MS but, even if this is demonstrated, it is still not clear whether the abnormality reflects poor diet, a metabolic abnormality or damage induced by a persistent immune disorder (Bates, 1989; Swank and Dugan 1990).

Dietary trials have been conducted in multiple sclerosis over a period of some 20 years with the object of correcting the alleged lipid abnormalities by increasing the intake of PUFA and reducing the dietary content of saturated fatty acids. In addition to formal trials of n-3 and n-6 unsaturated fatty acids, many patients supplement their diet in this manner either of their own accord or with medical approval. Controlled trials (Table 19.1) have shown only a mild reduction in the frequency of relapses or disease progression.

Nevertheless, it is now clear that dietary lipids influence many processes which may be relevant to the pathogenesis of MS. Moreover, changes in dietary habits over the past 40 years make it difficult to test the original hypotheses about the epidemiology of MS in terms of lipid intake. Future trials designed to manipulate lipid intake may be more informative as neuroradiological methods for monitoring central nervous system changes have greatly improved.

19.6.3   Treatment of bronchial asthma

Lipid-derived mediators contribute to the inflammation and hyper-responsiveness of the respiratory passages characteristic of bronchial asthma (Ritter and Taylor, 1988). Controlled studies of unsaturated fatty acid supplements (Table 19.2)

Table 19.2   Effect of unsaturated fatty acids on bronchial asthma – summary of clinical trials

Study	Design	Biochemical changes	Clinical outcome
Parm et al. (1988)	3.2 g/day EPA + 2.2 g/day DHA given to 12 patients for 10 weeks	Reduced neutrophil leukotriene B_4 and B_5 generation	No clinical improvement in cough, wheeze or peak flow
Picado et al. (1988)	Cross-over trial. 3 g mixed n-3 fatty acid given to 10 patients Lactose as placebo	Serum thromboxane B_2 reduced	No clinical improvement in peak flow or symptom score

showed that such measures can achieve marked increase in the EPA content of neutrophils with a corresponding reduction in LTB_4 and LTB_5 production and neutrophil chemotaxis. However, these changes did not, unfortunately, result in clinical benefit to the subject.

19.6.4   Treatment of SLE ('lupus')

SLE may affect connective tissues anywhere in the body. There are skin changes and the kidneys and other organs can also be involved.

Table 19.3   Effect of unsaturated fatty acids on SLE (lupus) – summary of clinical trials

Study	Design	Biochemical changes	Clinical outcome
Clark et al. (1989)	Fish oil (6 or 18 g daily) given to 12 patients for 10 weeks	Platelet arachidonic acid level reduced and EPA increased. Neutrophil leukotriene B_4 release decreased	No obvious improvement in nephritis
Westberg and Tarkowski (1990)	EPA capsules (0.2 g/kg/day) given to 17 patients for 12 months in double blind, crossover trial. Olive oil as placebo	Not studied	8/17 patients improved in treated group; 2/17 improved in control group; Improvement short-lived

Table 19.3 summarises trials which have investigated the effect of unsaturated fatty acids on SLE. The findings are equivocal. Any improvement with fish oil therapy is short-lived.

19.6.5   Treatment of rheumatoid arthritis

The main long-term problem in rheumatoid arthritis is progressive erosion of articular cartilage and underlying bone leading to irreversible joint destruction.

Table 19.4 Effect of unsaturated fatty acids on rheumatoid arthritis – summary of clinical trials

Study	Design	Biochemical changes	Clinical outcome
Sperling et al. (1987b)	Fish oil (20 g/day) given to 12 patients for 6 weeks	As expected, decreased AA:EPA and decreased LTB$_4$ in neutrophils	Reduction in severity of some, but not all, measurements of disease activity
Belch et al. (1988)	18 month trial: (i) GLA (540 mg/day) given to 16 patients (ii) EPA (240 mg/day) + GLA (450 mg/day) given to 15 patients	As expected, reduced eicosanoid concentration in plasma and red cell membranes	Patients in both treated groups improved; No improvement in control group given liquid paraffin
Cleland et al. (1988)	3.2 g/day EPA + DHA given to 23 patients for 12 weeks	As expected, decreased LTB$_4$ in stimulated granulocytes	Improvement achieved
Jantti et al. (1989)	Evening primrose oil (20 ml/day 9% GLA) or olive oil given to 9 patients for 12 weeks	Linoleic acid, DGLA and arachidonic acid increased but oleic acid and EPA decreased	No improvement
Kremer et al. (1990)	2 different doses of fish oil (27 mg/kg EPA + 18 mg/kg DHA or 54 mg/kg EPA + 36 mg/kg DHA) given to 20 patients for 24 weeks. Olive oil (6.8 g/day) given to 12 patients for 24 weeks	As expected, decrease in interleukin-1 and LTB$_4$ production in macrophages and neutrophils	Subjective and objective improvement in joint inflammation, particularly in higher dose group
Van der Tempel et al. (1990)	Double-blind, placebo-controlled cross over trial. EPA (2.04 g/day) + DHA (1.32 g/day) given to 16 patients for 12 weeks	As expected, increase in membrane EPA and DHA. Decrease in LTB$_4$; Increase in LTB$_5$	Modest clinical improvement Reduction in morning stiffness and joint swelling

In contrast with the equivocal findings in SLE, a more consistent response to dietary manipulation has been observed in some studies in rheumatoid arthritis (Belch, 1990). Several trials giving supplements of unsaturated fatty acid have been conducted in this disease, many with great clinical, biochemical, and statistical sophistication. Table 19.4 gives selected trials in chronological order and shows that the anticipated anti-inflammatory effects are accompanied by a varying reduction of active synovitis in some trials. However, the duration of follow-up was relatively short and the effects on cartilage and bone erosion have not yet been recorded. There are still many problems to be solved before dietary supplements of this nature can be recommended as standard treatment in immune disorders, even in a relatively well-studied disease such as rheumatoid arthritis.

19.6.6 Conclusions from clinical trials

19.6.6.1 Efficacy

Chronic diseases run a chronic course marked by unpredictable remissions and relapses. While diets enriched with PUFA blunt the inflammatory features of the disease, their long-term efficacy has not yet been established. These comments apply equally to such diverse conditions as the progressive demyelination in multiple sclerosis and the renal impairment in SLE.

19.6.6.2 Relevant mechanisms

Investigations have centred on the biochemical effects of dietary supplements which are of perceived theoretical importance. However, while clinical improvement may accompany these events, it does not necessarily follow that the altered lipid composition of cell membranes contributing to immune-mediated inflammation accounts for this improvement. The complexity of the immune system and the universal effects on cell membranes produced by these manipulations means that several immuno-modulatory effects could be responsible (Kremer et al., 1990).

Unexpected alternative explanations must also be kept in mind, although a recent study excluded the possibility that alpha-tocopherol added to fish oil preparations might account for their clinical benefit (Tulleken et al., 1990).

19.6.6.3 Side effects

Any obvious side-effects of dietary supplementation are minor and have usually taken the form of gastro-intestinal intolerance.

19.6.6.4   Long-term consequences

The nutritional consequences of dietary supplementation in the longer term have not yet been evaluated and must take account of the aberrations imposed by the diseases being treated. There is also justifiable concern that the outlook of those patients with chronic illnesses which attracts them to dietary experiments may encourage them to manipulate their diet in other respects. This attitude may lead to malnutrition and poor compliance with orthodox therapeutic recommendations.

20

SUPPLEMENTARY SOURCES OF
UNSATURATED FATTY ACIDS

Supplementary sources of unsaturated fatty acids can be broadly divided into two categories:

(i) those which are sold as medicines and have a product licence;
(ii) those which are sold over the counter as food supplements.

20.1 UNSATURATED FATTY ACIDS SOLD AS MEDICINES

20.1.1 Availability and uses

Evening primrose seed oil (Epogam) has been licenced as a medicine under the Medicines Act 1968 with indications for the treatment of atopic eczema and cyclical mastalgia. MaxEPA, a fish oil concentrate, has a medicines licence with indications for triacylglycerol lowering in patients at risk from pancreatitis and coronary heart disease. Cod liver oil has a medicines licence as a traditional remedy for the relief of joint pains and stiffness.

Medicines, unlike foods, are granted a licence after a full review of their safety, efficacy and formulation. Cod liver oil is obtainable over the counter, whereas MaxEPA and Epogam are only available on prescription.

20.1.2 Unsaturated fatty acid content

Epogam is made from a variety of evening primrose seed oils and contains 8.5–9.0% gamma linolenic acid (18:3 n-6). The recommended daily dosage is 6–8 g of oil which provides about 600 mg of gamma linolenic acid (Horrobin, 1990).

The recommended daily dosage of MaxEPA as a hypotriglyceridaemic agent is 10 g, which provides 1.8 g EPA and 1.2 g DHA and 10 IU of vitamin E (10 mg dl alpha tocopherol acetate).

Cod liver oil is sold as 'Pure cod liver oil' or as 'Cod liver oil BP'. The latter is fortified with retinol and vitamin D. The recommended dosage of pure cod liver oil for adults is 10 ml which provides 0.83 g EPA, 0.74 g DHA, 1200 retinol equivalents and 10 μg of vitamin D. The recommended dosage of cod liver oil BP is 5 ml. Purity criteria for cod liver oil are laid down in *Martindale* and the *European Pharmocopeia*. Most brands of cod liver oil on sale in the UK are fortified with additional vitamin E, usually 1 mg/g oil.

20.2 UNSATURATED FATTY ACIDS SOLD AS FOOD SUPPLEMENTS

20.2.1 Availability and uses

Various mixtures of vegetable oils and fish oils are offered for sale, usually encapsulated in soft gelatine shells. In most cases these consist of the plant or fish oil triacylglycerols, but some products are fatty acid concentrates derived from edible oils. The compositional standards for medicines do not apply to 'health food' supplements. These supplements are covered by food laws and are not classified as medicines even though they sometimes make implied health claims.

20.2.2 Unsaturated fatty acid content

Evening primrose seed oil, blackcurrant seed oil and borage seed oil are sold as sources of supplementary gamma linolenic acid. Evening primrose seed oil is also sold in combination with fish oil. One brand of evening primrose seed oil, Efamol, is identical to the Epogam material and has the same formulation criteria. Numerous claims are made for high potency gamma linolenic acid (GLA) products but a review of the label claims reveals that they often contain relatively small amounts of GLA.

Most fish oil capsules on sale usually contain 12–18% EPA and 10–12% DHA (percentage of the total fatty acids) (see Table 20.1). There are some high potency products containing approximately 50–60% EPA and DHA. These are all made from fish flesh oils rather than fish liver oils and are low in vitamins A and D. Recommended intakes usually provide less than 1 g of EPA and DHA per day. Fish oil products are usually fortified with additional alpha tocopherol (vitamin E) at a concentration of 2 mg/g of oil and may contain synthetic antioxidants such as dodecylgallate, because encapsulated oils are prone to oxidation if not stored under dry conditions.

Table 20.1 EPA and DHA acids in selected fish oils (g per 100 g)

Fish oils	EPA (20:5 n-3)	DHA (22:6 n-6)
Cod liver oil	9.0	9.5
Herring oil	7.1	4.3
Menhaden oil	12.7	7.9
MaxEPA™	17.8	11.6
Salmon oil	8.8	11.1
Mackerel oil	7.3	12.6
Anchovy oil	19.6	9.3
Sardine oil	11.0	13.0

Consumer misconceptions about food additives have resulted in supplements that do not always contain synthetic antioxidants; alpha tocopherol acetate which is added to some oils is a poor antioxidant 'ex vivo'. There is a need to ensure that such preparations are monitored for peroxidation products.

No independent review of materials on sale has been published in the UK. In the United States, Chee et al. (1990) reported the composition of fish oil supplements and showed that they provided between 140 and 260 mg EPA and 90–170 mg DHA, 0.6–2.2 mg vitamin E, and 1.3–298 μg retinyl esters per gram of oil. The major fish oil manufacturers routinely check all batches of oil for heavy metal and organic contaminants such as polychlorinated biphenyls in their oils but there are no clear guidelines for the formulation of these products.

20.3 FUTURE POTENTIAL FOR UNSATURATED FATTY ACID SUPPLEMENTS FROM BIOTECHNOLOGY

It is now possible to use single cell organisms to produce polyunsaturated oils (Kyle, 1992). Most microorganisms do not produce large amounts of triacylglycerol in their biomass, but a few yeasts and fungi can be efficient oil producers. Transgenic modification of yeast can be used to modify the composition of the oil produced. Fungal sources include the organisms *Mucor javanicus* and *Mortierella isabellina* which produce oils containing significant amounts (18%) of gamma linolenic acid. *Mortierella alpina* can produce arachidonic acid as well as GLA.

Various microalgae species are also prolific oil producers and can produce EPA and DHA. One strain of single cell microalgae can produce DHA almost as its sole polyunsaturated fatty acid; it is present at 30–35%. Some of these microalgae are related to *Gonyaulux catenalis* which is implicated in paralytic shellfish poisoning but the strains of microalgae used for oil production do not produce the poison saxitoxin.

20.4 CONTROLS ON SUPPLEMENTS

None of the single cell oil products is currently being offered for sale in the United Kingdom and it is likely they will require government approval as novel foods before being marketed.

Food supplements in the UK are often marketed in packagings and with advertising material that make implied health claims containing statements that may be misleading according to the joint MAFF and Department of Health's Report on Dietary Supplements and Health Foods (MAFF, 1991).

21

LABELLING OF FOODS IN RELATION TO UNSATURATED FATTY ACIDS

21.1 NUTRITION LABELLING

For individuals to be encouraged to reduce their intakes of fat and saturated fatty acids, it is necessary that foods carry information about their nutrient composition. Such information must be provided in a form that is both easily understood and useful.

21.1.1 Food Labelling Regulations 1984

The Food Labelling Regulations (Statutory Instrument, 1984) stated that a list of ingredients (in descending order by weight at the manufacturing stage) must be included on product labels. However, the provision of data on nutritional content was not compulsory, unless certain claims appear on the labelling or in advertising. Many manufacturers voluntarily included nutrient composition on product labels, but a number of different formats were used, making comparisons between products difficult.

There were regulations regarding particular claims made on food advertisements and labels. For example, when a food is characterised by the presence of a particular ingredient, special emphasis on the presence of that ingredient is not permitted, unless accompanied by a declaration of the minimum percentage of that ingredient in the food determined at the time of its use in the preparation of the food. Similarly, when a food is characterised by the low content of a particular ingredient, special emphasis on the low content of that ingredient is not permitted, unless accompanied by a declaration of the maximum percentage of that ingredient in the food, determined at the time of its use in the preparation of the food.

There were specific controls on claims relating to the polyunsaturated fatty acids (PUFA) and cholesterol contents of foods. For any PUFA claim to be made, the food had to satisfy the following criteria:

(i) It must contain at least 35% fat by weight.
(ii) At least 45% of the fatty acids must be PUFA

and not more than 25% may be saturated.
(iii) The claim must be accompanied by the words 'low in saturates' or 'low in saturated fatty acids'.
(iv) The label must include a declaration (expressed in g/100 g or ml/100 ml) stating the amount of fat or oil, the amount of PUFA which are *cis*, *cis* methylene-interrupted PUFA and the amount of saturated fatty acids. Each part of the declaration must be given equal prominence.

For a claim to be made relating to the presence or absence of cholesterol in a food, the following criteria had to be satisfied:

(i) The food must contain no more than 0.005% cholesterol.
(ii) A claim for PUFA must also be made and all the conditions for this satisfied.
(iii) The claim relating to cholesterol must not be in larger letters than, nor precede, the PUFA claim.

Where a PUFA or cholesterol claim was made, any suggestion, whether expressed or implied, that the food is beneficial to human health because it is high in PUFA or low was cholesterol was not permitted.

21.1.2 MAFF guidelines on nutrition labelling

Guidelines on nutrition labelling were produced by the Ministry of Agriculture, Fisheries and Food (MAFF, 1988) as a temporary measure before the introduction of the EC directive, in an attempt to encourage the provision of information in a standard form so that consumers could compare products more easily.

Three formats were recommended:

Category I Energy, protein, carbohydrate, fat
Category II Energy, protein, carbohydrate, fat

with a breakdown to show saturates.

Category III Energy, protein, carbohydrate with a breakdown to show sugars, fat with a breakdown to show saturates, sodium, fibre.

Optional additions to category III were: starch, *trans* fatty acids and PUFA. Monounsaturated fatty acids could only be added if PUFA were shown.

Energy had to be expressed in kilojoules and kilocalories and other nutrients in grams, the amounts being presented per 100 g or 100 ml. The information had to appear together in a table with the decimal points lined up, or, if necessary, in a linear form.

21.1.3   EC Directive on Nutrition Labelling for Foodstuffs

Under the EC Directive on Nutrition Labelling for Foodstuffs adopted by the EC in 1990, nutrition labelling was still not compulsory, unless a nutrition claim, such as 'low fat' or 'reduced energy', was made. Where nutrition information was provided, only two formats were permitted. These were as follows:

Group I   Energy, protein, carbohydrate and fat.
Group II Energy, protein, carbohydrate, sugars, fat, saturates, fibre and sodium.

These two groups are essentially the same as category I and category III of the MAFF guidelines. One or more of starch, polyols, monounsaturates, polyunsaturates and cholesterol could also be stated, subject to the provisions and formats specified.

Nutrient values had to be given per 100 g or per 100 ml. The information could also be given per serving as quantified on the label or per portion, provided that the number of portions contained in the package was stated. Data had to be presented together in one place in tabular form, with the numbers aligned if space permits. If this was not possible, the information must be presented in a linear form.

The Directive also contained provisions for the declaration of any of 18 vitamins and minerals which are present in significant amounts (vitamins A, D, E, C, thiamin, riboflavin, niacin, B_6, folate, B_{12}, biotin, pantothenic acid, calcium, phosphorus, iron, magnesium, zinc and iodine). A 'significant amount' was defined as the provision of 15% of the directive-specified recommended daily allowance per 100 g or 100 ml, or per package if this contained only a single portion. However, it is not intended to implement the vitamin and mineral provisions into UK law as they were only agreed on

a provisional basis and are at present being reviewed.

The Group II format will be introduced in 1995 and will have to be used whenever claims are made for any of the nutrients: sugars, saturates, fibre or sodium. No progress has been made on a second directive which would enable the introduction of compulsory labelling of selected nutrients if required for health reasons. A decision on this is likely to be made once experience has been gained with the voluntary system.

21.1.4   FAC proposals for regulating additional nutritional claims

Claims that products are, for example, 'low fat' or 'very low fat' have undoubtedly been a source of confusion to consumers, with products of very different fat contents carrying the same claim on the label. For this reason, MAFF's Food Advisory Committee (FAC) (MAFF, 1989) recommended that such claims should be controlled through legislation rather than voluntary guidelines.

FAC recommendations included the following:

(i)   A 'reduced fat' claim should only be permitted if the product contains at least 25% less fat than a comparable product for which no claim is made.

(ii) A 'low fat' claim should only be permitted if the product contains 5 g or less fat per 100 g and per normal serving; a 'low in saturates' claim should only be permitted if the product contains 3 g or less saturated fatty acids per 100 g and per normal serving.

(iii) A 'free from fat' claim should only be permitted if the product contains 0.15 g or less fat per 100 g or per 100 ml; a 'saturates free' claim should only be permitted if total saturate content of the food is 0.1 g or less.

These, and other proposals for other nutrients, have been formally notified to the EC Commission. The proposals will be considered as part of an EC Directive on Claims and will only be incorporated into UK Law once they are issued in that form.

21.2   CONSUMER UNDERSTANDING OF NUTRITION LABELLING

A study of consumer attitudes to, and understanding of, nutrition labelling was conducted to assess the most appropriate ways of presenting the information so that it can be understood and used by consumers (Consumers' Association, 1985). Subjects were 820 women and men who claimed substantially to influence the choice of food consumed by their household. The survey showed that although approximately 90% of the respondents had heard of the terms 'calories', 'energy' and 'fat', the proportions who recognised the terms 'kilocalories', 'kilojoules' and 'trans fatty acids' were very much lower (31%, 27% and 18% respectively). People thought it was important for fat and calories to appear on food labels, but gave a higher priority to protein and vitamins.

Six label formats were presented to subjects and they were asked to identify pieces of information on them and, in some cases, to carry out a simple calculation. From the responses, particular characteristics of the various formats were identified as helpful or confusing. Labels which abbreviated the term kilocalories to kcal performed poorly, presumably because the word 'calories' did not appear on the label and those with the energy value at the head of the nutrient list proved more 'off-putting' than those with the energy value at the foot of the list. There was evidence that the term 'unsaturated fatty acids' caused some confusion, with respondents looking at this when they should have been looking at total fat. Specifying values as one might speak: X g of fat, appeared to perform better than: fat . . . . X g. Formats with a bar chart or which described the content for a nutrient as 'high', 'medium' or 'low' tended to be preferred by the subjects. Labels on which information was given numerically in g/100 g and also graphically as percentage of energy caused confusion; unfamiliarity with the concept of 'percentage of energy' probably contributed to this.

A recent survey (MAFF, 1990) was conducted throughout England and Wales of 1028 adults who were mainly responsible for shopping for food for their household. When respondents were shown a prompt list of the types of information that appear on food labels and asked what they particularly looked for, 44% stated that they looked for nutrition information. About 72% of the people agreed that nutrition information made it easier to choose food for a healthy diet, while only 8% admitted finding such information confusing. The main reason for the confusion was a lack of understanding of the information presented.

21.3 TASK FORCE RECOMMENDATIONS FOR LABELLING OF FOODS IN RELATION TO UNSATURATED FATTY ACIDS

21.3.1 Provision of nutrition information

This Task Force would encourage all manufacturers to follow the Group II format as specified in the EC Directive on Nutrition Labelling, i.e. the labelling of the 8 core nutrients: energy, protein, fat, carbohydrate, saturates, sugars, sodium and fibre in a standard format.

This Task Force would also encourage manufacturers to give further information about the unsaturated fatty acid content of their products. Values for unsaturated fatty acids should be given as mg/100 g food. The n-6/n-3 ratio of the food might be helpful once these terms become better established through education campaigns (see Chapter 22). The use of the general non-specific terms 'vegetable oils' and 'fish oils' should be discouraged.

21.3.2 Publicity and education about nutrition information

If individuals are to alter their diets in line with published guidelines, more general publicity about food labelling is needed to improve consumer understanding of the information provided. If information is to be useful, publicity about labelling must be backed up by effective nutrition education programmes. The food label cannot be expected to provide an education in nutrition.

Furthermore, although the labelling of products with their nutrient content as g per 100 g will help consumers to obtain a lower fat intake in absolute terms, consumers must understand that this may not necessarily result in a lower percentage of energy obtained from fat if their total energy intake is reduced concurrently.

22

RECOMMENDATIONS FOR INTAKES OF UNSATURATED FATTY ACIDS

22.1 EXISTING RECOMMENDATIONS FOR UNSATURATED FATTY ACIDS

22.1.1 UK recommendations

22.1.1.1 Adults

Over a period of almost 20 years there have been a series of reports from professional and governmental committees advocating changes in the British diet in an attempt to reduce the incidence of various diseases, particularly coronary heart disease (CHD). In the main, dietary recommendations were made because of the known positive association between plasma cholesterol concentration and risk of CHD, and because plasma cholesterol can be reduced to a certain extent by diet (see Chapter 11). Recommendations were also made to prevent and/or treat obesity since it is associated with several risk factors for CHD.

The various reports show a consensus of opinion in recommending a reduction in the consumption of both total fat and saturated fatty acids. Nevertheless, particular levels of intakes are not specified in all the reports. Reductions in total fat to 30% of total energy (Royal College of Physicians, 1983; National Advisory Committee on Nutrition Education (NACNE), 1983) or 35% of food energy (DHSS, 1984) and in saturated fatty acids to 10% of total energy (NACNE, 1983) or 15% of food energy (DHSS, 1984) have been advocated. All these dietary recommendations assume that people are in energy balance. The NACNE recommendations were intended as means for population intakes, whereas the 1984 Panel of the Committee on Medical Aspects of Food Policy (COMA) intended their guidelines as maxima for individuals.

In some of the reports, people who have been shown by clinical and laboratory investigation to have an increased risk of CHD were advised to follow the recommendations more strictly (Royal College of Physicians, 1976), or more stringent recommendations were made for them: a reduction in total fat intake to 30% of food energy, a reduction in saturated fatty acids to 10% of food energy and a reduction in dietary cholesterol to 100 mg/1000 kcal (DHSS, 1984).

In none of the reports was an increase in the intake of polyunsaturated fatty acids (PUFA) advocated as a specific measure in itself. However, in almost all of them, it is recognised that a reduction in intake of saturated fatty acids sufficient to produce a significant fall in plasma cholesterol level may be unacceptable without a degree of substitution by PUFA. To facilitate achievement of the recommendation for saturated fatty acids, COMA recommended that the ratio of PUFA to saturated fatty acids (P/S) may be increased to approximately 0.45 (DHSS, 1984). NACNE (1983) stated that an increase up to a value of 1.0 'is likely to be beneficial', but commented that the highest ratio in free-living populations is 0.6–0.8 and that a P/S ratio of 2.0 or more 'is not conducive to a reduction in total mortality rates even though the rate of CHD may decline'.

Most of the reports advocated compensating for the reduction in total fat intake by increasing the consumption of 'fibre'-rich carbohydrates (e.g. bread, cereals, fruit and vegetables) in order to maintain energy intake.

The latest COMA Panel (Department of Health, 1991) saw no physiological requirement for any fatty acid other than the essential fatty acids (EFA) and recommended that for infants, children and adults, linoleic acid (18:2 n-6) should provide at least 1% of total energy and that alpha linolenic acid (18:3 n-3) should provide at least 0.2% of total energy. These recommendations were based on the prevention of clinical signs of EFA deficiency.

The 1991 COMA Panel based its recommendations for fat intakes on the summation of the dietary reference values (DRVs) for individual classes of fatty acids (see Table 22.1). These DRVs were mainly based on the relationship between fatty acid intake and plasma cholesterol levels. If the DRVs could be achieved, the prevalence of CHD and obesity should decrease. A desirable level of plasma cholesterol is considered to be less than

5.2 mmol/l, whereas the average level in the UK is 5.8 mmol/l. Based on Keys (1980) equation, the Panel calculated that a reduction in SFA intake to 10% of total energy (particularly if this reduction was confined to SFA of 14 or 16 carbon atoms) would achieve a desirable reduction in plasma cholesterol for the population.

Table 22.1 Dietary reference values for fat for adults

	Individual minimum	Population average	Individual maximum
Saturated fatty acids		10 (11)	
Cis-polyunsaturated fatty acids		6 (6.5)	10
n-3	0.2		
n-6	1.0		
Cis-monounsaturated fatty acids		12 (13)	
Trans fatty acids		2 (2)	
Total fatty acids		30 (32.5)	
Total fat		33 (35)	

Figures refer to percentages of daily total energy intake (percentage of food energy)
(Source: Department of Health, 1991)

The properties of *trans* fatty acids were considered to be different both from their *cis* counterparts and from saturated fatty acids and current evidence could not be used to relate them conclusively to long-term health concerns. Because of this concern, the Panel recommended that they should not increase further than the currently estimated intake of 2% total energy.

The COMA Panel considered that the evidence for monounsaturated fatty acids (MUFA) on plasma cholesterol levels was insufficient and recommended that intakes of MUFA should continue to provide 12% of total energy.

Cis-polyunsaturated fatty acids should also continue to provide an average of 6% of total energy and be derived from a mixture of n-6 and n-3 PUFAs. The COMA Panel recommended that intake by individuals should not exceed 10% of total energy, because the existing evidence justified a cautious approach. No recommendations were given for P/S ratios.

22.1.1.2   Infants

Recommendations have also been made regarding the fatty acid content of infant formulae (DHSS, 1980) since humans have a specific requirement for linoleic acid (18:2 n-6). Breast milk is regarded as the ideal food for the young infant. However, as its fatty acid composition resembles that of maternal dietary fat if energy intake is adequate, there is *considerable* variation both within and between

individuals and in different populations (see Tables 9.1 and 9.2). Among different UK populations, for example, the average linoleic acid content of human milk has been reported to vary from 3.2% to 12.4% of total energy (see Table 22.3). Within the UK, it has been recommended that, in infant milk formulae, linoleic acid and alpha linolenic acid should together provide at least 1% of the total energy (DHSS, 1980; DHSS, 1989). The latest report from COMA (DH, 1991) has raised these requirements slightly to recommend that a minimum of 1.2% energy be provided by linoleic acid and alpha linolenic acid together for infants, children and adults.

22.1.2   Recommendations from other countries and organisations

There are recommendations to reduce total fat intake by scientific and medical committees in many Western countries including Ireland, France, The Netherlands, Finland, Sweden, Denmark, Norway, Australia, USA and Canada. An increase in PUFA consumption has been recommended in a number of countries including The Netherlands, Finland, Sweden, Norway, Germany, USA, Australia and New Zealand (Turner and Gray, 1982; Brussaard and Hautvast, 1983; James, 1988). Where the recommendations include the P/S ratio, most recommended values are within the range 0.5–1.0.

The World Health Organisation (WHO) published guidelines on nutrient intake in 1982. Dietary recommendations were made with the aim of achieving the stated optimum mean population plasma total cholesterol level of 5.2 mmol/l (200 mg/dl). A reduction in saturated fatty acids to 10% of total energy and in dietary cholesterol to 300 mg/day was advocated, with some of the saturated fatty acids being replaced by monounsaturated fatty acids and PUFA. They recognised that PUFA may have beneficial effects other than lowering plasma cholesterol, but stated that the available evidence could not justify the giving of general advice to increase the intake of any particular fatty acid; 10% of energy from PUFA was recommended as an appropriate upper limit. The lowering of total fat intake to an average of between 20 and 30% of total energy was advocated for populations in whom the level of physical activity is low and obesity is common. An increase in complex carbohydrates was also recommended to emulate the diets characteristic of populations with little CHD.

More recent recommendations from WHO (1990) placed lower and upper limits on population nutrient goals which indicated the range of values for which nutrient goals might be seen for different

Table 22.2 Existing recommendations for average population intakes of unsaturated fatty acids (as % total energy)

Source	Total fat	All MUFA	All PUFA	n-6 PUFA (usually 18:2 n-6)	n-3 PUFA (usually 18:3 n-3)	EPA+DHA	P/S	P/S/M[a]	n-6/n-3
NATO Advance Workshop (1989)	24–28%	12–14%	6–7%	4.8%	1.0	0.27	1/1	1/1/2	4
Nordic Recommendations (PNUN, 1989)	20–30% individual	–	–	3–10% individual	0.5% individual	–	–	–	–
NRC (1989)	< 30% individual	10–16% by difference	7% (10% maximum)	1–2%	–	–	1/1	1/1/1	4–10
WHO (1990)	15–30%	0–27% by difference	3–7%	–	–	–	–	–	–
Health and Welfare Canada (1990)	30% (SFA <10%)	16.5% by difference	3.5% minimum	3% minimum	0.5% minimum	–	> 0.35	–	4–10
COMA (DH 1991)	33%	12%	6%	1% minimum	0.2% minimum	–	–	–	–

[a] Ratio of polyunsaturated to saturated to monounsaturated fatty acids

populations. For saturated fatty acids, 10% of total energy was recommended as the upper limit with 0% as the lower limit, since there is no specific requirement for saturated fatty acids. For PUFA, the recommended lower limit was 3% of total energy and the upper limit was 7% of total energy.

One of the problems about dietary recommendations for unsaturated fatty acids is the many ways in which these can be expressed. Table 22.2 attempts to summarise some of the existing recommendations from a variety of Reports published in the last three years and demonstrates how difficult it is to compare like with like.

In some instances, one average population nutrient goal is given (e.g. DH, 1991) whilst others suggest lower and upper limits between which average population intakes could be set (e.g. WHO, 1990). A few recommendations are still given as maximum intakes for individuals. These are always higher than the corresponding recommendations for average population intakes: e.g. 10% for an individual maximum for PUFA, corresponding to an average PUFA intake of 7% for the population (NRC, 1989).

22.1.3 What further recommendations are needed in the UK?

The UK guidelines set by COMA (DHSS, 1984; DH, 1991) have been made primarily in an attempt to reduce the mean plasma total cholesterol concentration, and to reduce mean LDL cholesterol concentration. However, relatively little attention was paid to other factors influencing risk of CHD such as HDL cholesterol, triacylglycerols, fibrinogen, clotting factor VII, viscosity and platelet aggregation. Indeed, haemostatic factors have been found to be more strongly predictive of CHD than plasma cholesterol (see Chapter 13). Furthermore, there is increasing evidence that n-3 PUFA, such as eicosapentaenoic acid (EPA) (20:5 n-3), which have little or no effect on plasma cholesterol concentration, may have beneficial effects on a number of other CHD risk factors and, more importantly, may reduce risk of death from thrombosis. This Task Force therefore decided to propose recommendations for unsaturated fatty acids which were more extensive than the dietary reference values given by the 1991 COMA Panel because they were based on this rather different philosophy. The Task Force also attempted to define safe ranges of intakes for individuals as well as setting average intakes for populations to achieve.

22.2 RATIONALE FOR CALCULATING AND EXPRESSING TASK FORCE RECOMMENDATIONS

22.2.1 Biochemical, physiological and clinical requirements

With nutrients that are essential such as EFA, a lower limit for the safe range can be set which is based on minimum requirements for individuals. Intakes below this will almost certainly be inadequate.

The problem with the concept of requirements is that the criteria used to define the requirements can be clinical or biochemical and physiological. For EFA, biochemical and physiological criteria invariably give higher requirement levels than clinical criteria. Although clinical deficiencies are rarely observed at very low levels of intakes, biochemical measurements of the fatty acid metabolites in plasma lipids often show abnormalities implying that the very low intakes of the essential fatty acids have been limiting, i.e. below the biochemical and physiological requirement for the nutrient. Another major problem is that many definitions of requirements are of an essentially arbitrary nature and not based on either biochemical or clinical criteria.

22.2.2 The toxicological approach

The acceptable daily intake (ADI) is the amount of a substance which can be safely consumed every day during the entire lifetime of an individual. ADI are usually expressed per kg body weight of the consumer. This approach, which is commonly used in toxicology, is not one which has been extensively used in setting nutritional guidelines, but it is useful to set an upper limit beyond which intake becomes undesirable. This Task Force has used this approach in making recommendations for what appear to be upper limits of the safe range for intakes of unsaturated fatty acids. The limits are not based on evidence of clinical toxicity but are prudent guidelines based on:
 (i) known biochemical changes that are produced by high doses.
 (ii) concern for the lack of experience at present with the safety or risk of higher amounts.

22.2.3 The value of expressing recommendations on the basis of energy intakes

Whether requirements are based on biochemical or clinical criteria, there are large individual variations in nutrient requirements. Individual requirements are, however, invariably strongly related to energy intakes. This makes a strong argument for expressing intakes as a proportion of the dietary energy intake and this Task Force has proposed recommendations on this basis.

One extra complication is that some committees use total energy intake as their basis for recommendations while others use food energy intake (i.e. excluding alcohol) as their basis (e.g. DHSS, 1984). This Task Force has based its recommendations on total energy intake although it agreed with reservations expressed elsewhere (DH, 1991) that individuals should not try to achieve the recommendations by increasing total energy intake without increasing total energy expenditure, or by increasing alcohol intake.

22.2.4 The value of expressing recommendations on the basis of absolute weight

When recommendations have to be converted into practical dietary advice, it is helpful to have them expressed on a weight basis as well as a percentage of energy basis. Then people can choose to eat specific items using the information that they might find on the label of the product or from food tables. This Task Force therefore expressed its recommendations for unsaturated fatty acids on an absolute weight basis as well as on a percentage of energy basis.

22.2.5 Average population targets or individual guidelines?

Many reports have been ambiguous about the intended application of the goals or guidelines presented. In the 1984 COMA Report, the recommendations were principally concerned with individual dietary habits in contrast to the more usual approach which merely advocates a desirable mean intake of a nutrient for whole populations. Thus the 1984 COMA Report recommended 35% food energy for total fat (15% for saturated fatty acids) as an acceptable maximum for individuals. The 1991 COMA Report and the 1990 WHO Report, however, were explicit in emphasising that the population nutrient goals referred to population averages. Thus the WHO Report proposed that the average population intake for total fat (as a percentage of total energy) should be set between 15% and 30%. The upper WHO limit of 30% is therefore below the 1991 COMA recommendation for an average population intake of 33% of total energy from fat.

This Task Force also gave its recommendations in terms of average intakes for populations. However, in terms of practical recommendations for individuals, it is important to know the safe ranges of intakes around the population averages. This Task Force therefore recommended an upper limit of the safe range for unsaturated fatty acids using an ADI approach, i.e. a limit which it is sensible for individuals not to exceed. The Task Force recommended a lower limit of the safe range on the basis of minimal requirements.

22.3 TASK FORCE RECOMMENDATIONS FOR UNSATURATED FATTY ACIDS

22.3.1 Requirements and Task Force recommendations for n-6 unsaturated fatty acids

22.3.1.1 Infants (see Table 22.7)

The normal minimum requirement for linoleic acid (18:2 n-6) in the young infant appears to be 0.5% to 1.0% of total energy, so long as alpha linolenic acid (18:3 n-3) supplies a further 0.5% of total energy and has a sparing effect on linoleic acid (Naismith et al., 1978). This requirement is based on weight gain and absence of EFA-specific skin lesions (see Chapter 9).

Energy intakes of infants are high in relation to body weight (approximately 110 kcal/kg body weight) so that a requirement for linoleic acid of 1% on an energy basis would work out at about 100 mg linoleic acid per kg body weight.

However, this minimum requirement of at least 0.5–1% is a clinical requirement and might not reflect a comprehensive biochemical and physiological requirement since babies fed on formulae containing this amount of linoleic acid, show

elevated proportions of eicosatrienoic acid (20:3 n-9) and a low proportion of arachidonic acid (20:4 n-6) (Crawford et al., 1978) indicating preferential metabolism of the n-9 MUFA in the presence of limiting amounts of n-6 PUFA (see Chapter 7).

In the absence of definitive data on the biochemical and physiological requirements for infants, recommendations are best estimated by assuming that human milk is the ideal food for babies and by basing the safe range of recommendations on the linoleic acid content found in human milk.

Table 22.3 shows that the average linoleic acid content of breast milk of UK omnivores is about 4% of total energy. Using this assumption and others, Table 22.4 shows that 100 ml of human milk contains 282 mg linoleic acid. An infant weighs about 6 kg at 6 months of age and consumes about 800 ml of milk a day. Thus total linoleic acid intake will be 282 × 8 = 2260 mg linoleic acid or 377 mg/ kg body weight.

Table 22.4 also shows the amounts of other n-6 PUFA present in human milk, together with their potencies relative to linoleic acid. A case could be argued that requirements for n-6 PUFA in infancy could be based on a system of linoleic acid equivalents, once further research can support the use of the figures for relative potency. In this case, infant requirements for linoleic acid equivalents would be 347 mg/100 ml human milk or 462 mg/kg body weight.

The energy from linoleic acid in human milk is not always 4% of the total energy content. It can vary from 3% to 12% of the total energy content depending on the mother's own intake of linoleic acid (see Table 22.3). The amounts of the other n-6 PUFA do not vary. The range of requirements of infants, and hence subsequent recommendations for n-6 fatty acids, should therefore be based on this range and expressed as a safe range of intakes. The recommendations proposed are 280 mg linoleic acid/kg body weight as the lower limit and 1130 mg linoleic acid/kg body weight as the upper limit of the safe range corresponding to the upper and lower limits of linoleic acid in UK breast milk.

The Task Force recommended that the infant

Table 22.3 Summary of fatty acid composition of UK human milk (as % total energy)

	Linoleic acid 18:2 n-6	Alpha linolenic acid 18:3 n-3	Arachidonic acid 20:4 n-6	DHA 22:6 n-3
All UK omnivores	4.4 (range 3.2–5.5)	0.39	0.18	0.21
All UK vegetarians	12.4	0.87	0.25	0.10

For references to individual studies, see Table 9.1. Summary as % total energy provided by Dr T. A. B. Sanders

Table 22.4 n-6 fatty acids in human milk

	Amounts (mg/100 ml)	Potency relative to linoleic acid	Linoleic acid equivalents (mg/100 ml)
Linoleic acid (18:2 n-6)	282	1.0	282
Gamma linolenic acid (18:3 n-6)	4	1.5	6
Dihomogamma linolenic acid (20:3 n-6)	9	1.5	14
Arachidonic acid (20:4 n-6)	15	3.0	45
		Total =	347

Fat content of human milk is assumed to be 4.1 g/100 ml
Linoleic acid content of human milk is assumed to be 4% of total energy and 7200 mg/100 g fatty acids
Fatty acids (mg) = 0.956 × fat (mg)
Linoleic acid content of human milk = 0.072 × 4.1 × 0.956 × 1000 (mg/100 ml)
The energy content of human milk is 69 kcal/100 ml

formulae provided for non-breast fed babies should provide sufficient linoleic acid at least for these minimum recommendations to be met for all infants. Table 22.3 shows that human milk contains small amounts of preformed arachidonic acid (15 mg/100 ml or 0.2% of total energy). The ability of the infant to convert linoleic acid (18:2 n-6) to arachidonic acid is not known. This Task Force therefore recommended that infant formulae should contain preformed arachidonic acid in an attempt to replicate the fatty acid profile of human milk.

22.3.1.2 Children (see Table 22.7)

As requirements for nutrients are usually greater in infancy than at other times of life, the recommendations for infants which are based on the linoleic acid content of breast milk (i.e. 3–12% energy intake) are assumed to be more than adequate for older children. Unfortunately there are few data to confirm this assumption.

22.3.1.3 Adults (see Table 22.5)

In the absence of any better guide to biochemical requirements, the recommendations for healthy adults are again broadly based on the linoleic acid content of breast milk.

This Task Force proposed an average population intake of 6% of total energy from all n-6 unsaturated fatty acids and a safe range of intakes between 3 and 10% of total energy. The lower limit of the safe range is proposed as 3% of total energy because 1–2% of the energy as linoleic acid is needed to correct skin symptoms. The skin EFA component is predominantly linoleic acid but the inner membrane lipids of most internal organs contain arachidonic acid as the major n-6 PUFA. Therefore, to allow for the low efficiency of conversion of linoleic acid, or genetic or hormonal differences in metabolism, the Task Force recommended a minimum requirement of 3% of total energy.

The upper limit of the safe range for n-6 fatty acids is recommended as 10% because intakes higher than this (12 to 15%) have been shown to lower HDL cholesterol (see Chapter 11) and to increase the risk of gallstones in man (Sturdevant et al., 1973). No specific dietary requirements for the derivatives of linoleic acid such as GLA can be given because there is no good basis for making them.

Although there is an argument for proposing much lower values for the safe range of intakes (say between 1 and 4% of total energy) on the basis of the threshold level for tumour promotion in animals (see Chapter 16), the Task Force thought that this low intake of linoleic acid would not help in

Table 22.5 Task Force recommendations for dietary intakes of unsaturated fatty acids for healthy adults

Fatty acid	Average population intake (as % total energy)	SAFE RANGES as % of total energy	g/day MEN	g/day WOMEN
Linoleic acid (18:2 n-6)	6	3 to 10	8 to 26	6 to 20
Alpha linolenic acid (18:3 n-3)	1	0.5 to 2.5	1 to 6	1 to 5
EPA (20:5 n-3) + DHA (22:6 n-3)	0.5	0 to 2.0	0 to 5	0 to 4
All PUFA	7.5	3.5 to 14.5	9 to 38	7 to 29
MUFA	12	0 to 20	0 to 51	0 to 39

All recommendations for safe ranges expressed as intakes for individuals
Reference man assumed to require 2550 kcal energy/day
Reference woman assumed to require 1940 kcal energy/day (DH, 1991)

the ultimate objective of reducing deaths from coronary heart disease (CHD). The Task Force concluded that the hoped-for benefits in preventing CHD would outweigh the theoretical possibility of an increased cancer risk. Furthermore, the evidence from prospective studies has shown no increase in total deaths in men who consumed linoleic acid at levels above 4% of total energy (see Chapter 10).

This safe range of 3 to 10% on the basis of energy intake can be expressed on a weight basis assuming that the average adult man has an energy intake of 2550 kcal/day and that the average adult woman has an energy intake of 1940 kcal/day, which are the estimated average requirements (EAR) for energy given by the 1991 COMA panel (DH, 1991).

These safe ranges of intake could also be expressed on a body weight basis assuming reference man to weigh 75 kg and reference woman to weigh 60 kg. In this case, they are 0.1 g/kg body weight at the lower limit and 0.33 g/kg body weight at the upper limit. Thus the range of requirements for infants expressed on a body weight basis are much greater than those for adults (see Section 22.3.1.1).

Adults on *weight-reducing diets* need not consume any linoleic acid as adequate amounts will be released from their fat stores which typically contain 8% of their energy as linoleic acid (Wood *et al.*, 1987). If half of their energy requirements are met by the mobilisation of their body fat, then linoleic acid would provide 4% of the total energy available for use.

22.3.1.4 Pregnant and lactating women (see Table 22.7)

Essential fatty acid (EFA) requirements during pregnancy and lactation have been reviewed and discussed in Chapter 9. Evidence was presented to show specific quantitative requirements for fetal growth and to show that mothers with low EFA intakes tended to produce low birthweight babies. Thus there is a case for advocating additional requirements (e.g. about 2% extra energy from EFA as advocated by the FAO/WHO (1978) Consultation).

However, there is also evidence to show that metabolic adaptation occurs during pregnancy leading to increased efficiency of energy utilisation but it is not known whether this involves decreased linoleic acid oxidation or increased deposition in fat stores. Mobilisation from these fat stores during the first few months of lactation will contribute to the supply of n-6 fatty acids. Proponents expressing this line of thought would argue that there do not need to be recommendations for specifically increased EFA intakes during pregnancy.

Taking all the evidence into account, the Task Force concluded that it would not give specifically increased recommendations for n-6 PUFA intake for pregnant or lactating women, but that it would advocate the *counselling* of pregnant and lactating women, with particular advice to ensure that their diets contained adequate n-6 fatty acids. Counselling should also be given in the form of *pre-conceptual advice* to women who are planning to become pregnant. This advice should result in higher average intakes of n-6 PUFA by pre-pregnant, pregnant and lactating women.

22.3.2 Requirements and Task Force recommendations for n-3 fatty acids

22.3.2.1 Infants (see Table 22.7)

The requirement of infants for alpha linolenic acid was not doubted by the Task Force. Approximately 0.5% of the energy intake as alpha linolenic acid is adequate to support the normal growth of rat pups (Pudelkowicz, 1968) and a requirement for alpha linolenic acid for cognitive and visual development has often been demonstrated (see Chapter 9). Human milk provides about 0.4% of its energy as alpha linolenic acid (see Table 22.3) and this corresponds to a daily intake of about 40 mg alpha linolenic acid per kilogram body weight in a breast fed infant.

Human milk also provides about 0.2% of its energy as docosahexaenoic acid (DHA) (22:6 n-3) which corresponds to a daily intake of about 20 mg DHA per kilogram body weight in breast fed infants. Chapter 9 summarises the experimental evidence that would support a case for ensuring that *all* infants receive adequate amounts of DHA in their diet on the assumption that the ability of infants to elongate and desaturate dietary alpha linolenic acid is minimal.

Against this is the argument that no apparent developmental problems have ever been encountered in infants raised exclusively on cows' milk formulae which contain only traces of DHA although it is impossible to know how many of these never achieved their full intellectual or visual potentials.

Taking all the evidence into account, this Task Force thought it would be prudent for DHA to be present in all infant formulae at the same level as it is present in human milk (i.e. 0.2% total energy).

It has been suggested that the *pre-term infant* may require 11 mg DHA per kilogram body weight (Liu *et al.*, 1986) but this Task Force recommended that pre-term formulae should contain DHA at the higher level of 20 mg DHA per kilogram body weight, equal to that provided by milk.

22.3.2.2 Adults (see Table 22.5)

This Task Force recommended that the average population intake of alpha linolenic acid (18:3 n-3) should be 1.0% of total energy.

The minimum requirements for alpha linolenic acid as a proportion of energy are assumed to be less than those for infants so that a limit of 0.5% energy intake seems a reasonable recommendation to propose for a lower limit of safe intake for individuals. On the basis of toxicity studies, 2.5% of energy intake is proposed as the upper limit of the safe range.

As far as long chain n-3 PUFA are concerned, an argument can be made for not setting a lower limit for EPA and DHA for adults. In spite of the important metabolic role played by EPA and DHA in, for instance, inhibiting thrombogenesis (see Chapter 13), there is no scientific evidence that preformed EPA and DHA are required because the majority of the world's vegetarian population manage to exist quite well without either. It could be argued that vegetarians can only manage without EPA and DHA because they have correspondingly lower arachidonic acid intakes than omnivores. Their eicosanoid profile, therefore, would resemble that of omnivores eating meat and fish who have higher intakes of all three long chain fatty acids. Alternatively, it might be that some type of metabolic adaptations have taken place within the vegetarian population so that they are able to synthesise EPA and DHA from alpha linolenic acid more efficiently than omnivores. Taking all the evidence into account, this Task Force decided not to recommend a lower limit for EPA and DHA, but to suggest that research is required on the metabolism of fatty acids by vegetarians.

Upper limits for the sum of EPA and DHA are proposed at 2% of total energy intake. This is equivalent to the amounts of long chain n-3 fatty acids that would be consumed from a regular intake of fatty fish. Intakes above this upper limit of the safe range are regarded as pharmacological doses and may adversely modify a number of cell functions.

This Task Force therefore recommended that the safe range of total n-3 PUFA intakes for individuals is between 0.5% and 4.5% of total energy intake. (Total n-3 PUFA includes intakes from alpha linolenic acid as well as from EPA and DHA.)

22.3.2.3 Pregnant and lactating women

Using the same philosophy as that applied to recommendations for n-6 PUFA (see Section 22.3.1.4), the Task Force decided against making a specific set of recommendations for n-3 PUFA for pregnant and lactating women but again decided it would be prudent to *counsel* pregnant and lactating women to ensure that their diets contained adequate amounts of n-3 PUFA. Again, *counselling* should also be directed at women who plan to become pregnant.

22.3.3 The P/S ratio

P/S ratio is a frequently used term, yet its definition and biological significance are not entirely clear. P is most often calculated as the total intake of PUFA and S as the total intake of saturated fatty acids. *Trans* fatty acids are sometimes included in the S component of the ratio, as for example in the 1984 COMA recommendations. More recently, they have not been included in the S component as in the MAFF guidelines on nutrition labelling (MAFF, 1988) and the EC Directive on nutrition labelling (1990).

If the ratio is of importance in relation to plasma total and LDL cholesterol concentrations then it might be argued that the S component of the ratio should be the sum of those saturated fatty acids which raise cholesterol levels and the P component should be simply 18:2 n-6 or the total n-6 PUFA. The possible role of *cis*-MUFA in lowering LDL cholesterol is another reason for regarding the classical P/S ratio as redundant. Finally, the effects of PUFA differ according to whether they are n-6 or n-3. This Task Force Report has made it abundantly clear that there are real biochemical and physiological differences between fatty acids of the two families.

For these reasons, the Task Force recommended that the P/S ratio should no longer be used.

22.3.4 Other ratios

The Task Force considered the possibility of recommending a ratio based on total unsaturated fatty acids to saturated fatty acids (U/S). Although this would allow for the role of MUFA, the Task Force decided that the other disadvantages of the P/S ratio would not be overcome.

Very recently, Ulbricht and Southgate (1991) have proposed that the P/S ratio, as a measure of the propensity of the diet to influence the incidence of CHD, should be replaced by indices of atherogenicity and thrombogenicity.

The derivation of these indices is based on similar thinking to that summarised in Section 22.3.3 and on much of the evidence presented in this Report, certainly as far as CHD is concerned.

Suggestions of new indices such as these are bound to lead to heated debate about their ability to predict CHD. It is pertinent to note Ulbricht and

Southgate's conclusion that 'the incidence of CHD cannot be related to any single attribute of the diet, ... it is determined in a multidimensional way in which one of the dimensions is time'.

It is also important to note that the indices of atherogenicity and thrombogenicity have been devised solely on the basis of the relationship between fatty acids and heart disease. Although they are most certainly more accurate predictors for CHD than the old P/S ratio, they are not suitable, and were never intended, for use in a practical sense and they do not necessarily relate to the relationship between the fatty acid content of the diet and other diseases such as skin diseases and cancer.

22.3.5 The n-6/n-3 PUFA ratio

The n-6/n-3 PUFA ratio could be defined as the ratio of *all* n-6 PUFA to *all* n-3 PUFA or, more simply, as the ratio of linoleic acid (18:2 n-6) to alpha linolenic acid (18:3 n-3). The long chain metabolites of both PUFA families are known to have greater potency than their C18 parent fatty acids; this complicates and distorts a ratio based on all fatty acids. This Task Force recommended that until more is known about their relative potencies and rates of conversion, the simpler ratio of linoleic acid to alpha linolenic acid should be used.

The metabolism of linoleic acid and alpha linolenic acid are not independent of each other, since these two fatty acids compete for the same enzyme system (6-desaturase). However, relatively little is known about the ideal dietary balance between these two families of PUFA. In experimental animals, an increase in the ratio of linoleic acid to alpha linolenic acid in the diet has been found to result in an increase in docosapentaenoic acid (DPA) (22:5 n-6) and a decrease in DHA (22:6 n-3) in brain ethanolamine phosphoglycerides.

22.3.5.1 Infants

The mean n-6/n-3 ratio for UK human milk is about 11:1. Manufactured infant formulae, until recently, had ratios of approximately 10:1 to 15:1. There are now some infant formulae with ratios nearer to 5:1.

A ratio of 5:1 or less has some justification in terms of cell membrane lipid composition but there is no evidence, as yet, that the brain lipids of human infants would be adversely altered by providing a milk with a higher n-6/n-3 PUFA ratio, although the results of some animal experiments suggest that this cannot be ruled out. Further research is needed to ascertain the optimum dietary n-6/n-3 ratio but, in the meantime, the Task

Force felt that the n-6/n-3 ratio in infant formulae should approximate that of human breast milk as long as the formulae also provide adequate amounts of DHA and arachidonic acid.

22.3.5.2 Adults

An FAO/WHO Committee (1978) recommended an n-6/n-3 ratio of 5:1 and at least two countries have made recommendations along these lines: Canada, between 4 and 10:1 and Sweden 5:1. The ratio of the mean intakes of n-6 and n-3 PUFA in the diets of British adults is 7:1 (Gregory *et al.*, 1990).

The average adult population intakes for linoleic and alpha linolenic acid recommended by this Task Force correspond to an n-6/n-3 ratio of 6:1.

22.3.6 Requirements and Task Force recommendations for monounsaturated fatty acids (MUFA)

MUFA include palmitoleic (16:1 n-9), oleic (18:1 n-9), erucic (22:1 n-9), cetoleic (22:1 n-11) and nervonic (24:1 n-9) acids.

22.3.6.1 Oleic acid

There is increasing evidence that oleic acid like n-6 PUFA, reduces both plasma total cholesterol and LDL cholesterol, when it replaces saturated fatty acids in the diet. The effects of oleic acid on platelets, other haemostatic factors and on mortality are unknown, however and further study is required before any firm or specific recommendations can be made concerning the intake of oleic acid.

This Task Force recommended that the average population intake of oleic acid should be 12% which reflects the current average intake. As MUFA can be synthesised *de novo*, there can be no lower limit for the safe range of intakes.

There is a case for suggesting that as long as the diet provides appropriate energy and EFA, MUFA intakes do not need an upper limit. The average intakes of MUFA in some Mediterranean populations are 20% of total energy.

However, the recommendation for the upper limit of oleic acid intake to be 20% of total energy is consistent with the fact that the oleic acid accounts for 20% total energy in human milk. It also keeps the total fat content in accordance with the principles of a balanced diet.

22.3.6.2 Long chain MUFA

Certain C20–22 MUFA can cause transient myocardial lipidosis when fed to experimental animals in excess of 5% of the total energy intake.

Current EC legislation restricts the amount of erucic acid (22:1 n-9) in edible oils to less than 5% by weight. This does not apply to other C20–22 MUFA. A 200 g portion of herring provides about 10 g of C20–22 MUFAs which is 5% of total energy. There is no evidence that the regular consumption of fatty fish containing these monounsaturated fatty acids is harmful to man. The Task Force thought it prudent, however, to set an upper limit of 5% energy of these long chain monounsaturated fatty acids.

22.3.7 Recommendations for antioxidants

If population intakes of unsaturated fatty acids continue to increase this will also increase requirements for antioxidants such as vitamin E, selenium, vitamin C and beta-carotene. The 1984 COMA Panel concluded that current intakes of these nutrients should be adequate for a diet resulting from their recommendations (P/S of up to 0.45) and therefore made no specific recommendations about the dietary intake of antioxidants. The 1991 COMA Panel (DH, 1991) set the reference nutrient intake (RNI) for vitamin C at 40 mg/day and for selenium at 75 µg/day (men) and 60µg/day (women). They set the minimum intake of vitamin E at 4 mg/day (men) and 3 mg/day (women).

The requirement for vitamin E is linked to the intake of linoleic acid. As dietary sources of linoleic acid and vitamin E are often similar, this requirement can be simultaneously met. Industrial processing of fats and oils and the deep frying of these oils can lead to some losses of vitamin E. These are not usually of nutritional significance, providing the vitamin E/linoleic acid ratio is well above adequate in the initial product. The effect of repeated deep frying using the same oil is not known.

The extent to which n-6 unsaturated fatty acids other than linoleic acid and n-3 fatty acids increase the requirement for vitamin E is uncertain.

This Task Force agreed with others (DH, 1991) that 0.4 mg vitamin E/g linoleic acid should be adequate (i.e. a safe range of 3.2 to 10.4 mg per day for men and 2.5 to 8 mg per day for women).

22.4 SUMMARY OF TASK FORCE RECOMMENDATIONS FOR HEALTHY ADULTS AND COMPARISON WITH COMA (1991) RECOMMENDATIONS

The 1991 COMA Panel proposed dietary reference values for EFA 'only on the basis of prevention of EFA deficiency'. This Task Force decided that its extensive overview of the role of *all* unsaturated fatty acids in many different aspects of health and disease justified a different and somewhat less cautionary approach and has produced recommendations that *extend and enlarge* on those given by COMA. Table 22.6 presents a side-by-side comparison of recommendations for easy reference and the following points should be noted:

(i) The recommendations for MUFA are similar.
(ii) The recommendations for average population intake of n-6 PUFA as linoleic acid (18:2 n-6) are similar but the Task Force recommends that individual minimum requirements are higher.
(iii) The Task Force recommendation for intakes

Table 22.6 Comparison of recommendations for unsaturated fatty acids made by the Task Force and the 1991 COMA Panel for healthy adults

	1991 COMA (% total energy)	1992 BNF Task Force (% total energy)
Average population intake of MUFA	12.0	12.0
Minimum individual intake of MUFA	Not given	0
Maximum individual intake of MUFA	Not given	20.0
Average population intake of n-6 PUFA as 18:2 n-6	6.0	6.0
Average population intake of n-3 PUFA as 18:3 n-3	Not given	1.0
Average population intake of n-3 PUFA as EPA and DHA	Not given	0.5
Average population intake of total PUFA	6.0	7.5
Minimum individual intake of n-6 PUFA as 18:2 n-6	1	3
Maximum individual intake of n-6 PUFA	10	10
Minimum individual intake of n-3 PUFA as 18:3 n-3	0.2	0.5
Maximum individual intake of n-3 PUFA as 18:3 n-3	Not given	2.5
Minimum individual intake of n-3 PUFA as EPA and DHA	Not given	0
Maximum individual intake of n-3 PUFA as EPA and DHA	Not given	2.0
Minimum individual intake of *all* PUFA	1.2	3.5
Maximum individual intake of *all* PUFA	10.0	14.5

Table 22.7 Summary of Task Force recommendations for dietary intakes of PUFA in special groups

Women			
Pre-pregnant, pregnant and lactating	No specific increased recommendation for PUFA but *counselling* to ensure that recommendations for average population intakes are met, i.e. average total PUFA intakes of at least 7.5% to be achieved.		
Infants and children			
Pre-term *and* full-term infants[a] not on human milk	Formulae should contain quantities of linoleic acid, arachidonic acid, alpha linolenic acid and DHA to replicate the amounts found in human milk:		
	Linoleic acid (18:2 n-6)	4% of total energy	(377 mg/kg body weight with safe range of 280 to 1130 mg/kg body weight)
	Arachidonic acid (20:4 n-6)	0.2% of total energy	(20 mg/kg body weight)
	Alpha linolenic acid (18:3 n-3)	0.4% of total energy	(40 mg/kg body weight)
	DHA (22:6 n-3)	0.2% of total energy	(20 mg/kg body weight)
Children on mixed diets	The polyunsaturated fatty acid composition of the diet should replicate that of human milk until more information is forthcoming		

[a] The recommendations given on a body weight basis assume a 6 kg infant drinking 800 ml milk per day

of n-3 PUFA both as alpha linolenic acid (18:3 n-3) and as the long chain n-3 fatty acids, EPA and DHA, are much more extensive with more explicit recommendations on intakes for average populations and for safe ranges for individuals.

(iv) The Task Force recommendations for total PUFA intake (7.5% of energy) are higher than the COMA Panel recommendations (6% of energy) mainly because of the extra importance attached to the n-3 PUFA. The difference in the maximum intake of *all* PUFA (14.5% Task Force versus 10% COMA) is again attributable to the contribution from n-3 PUFA.

22.5 SUMMARY OF TASK FORCE RECOMMENDATIONS FOR SPECIAL GROUPS

Table 22.7 summarises the Task Force recommendations for pre-pregnant, pregnant and lactating women, for pre-term and full-term infants and for children on mixed diets.

Although no specific increase in requirements was recommended for these women, the Task Force was *very concerned* that these physiological states represent times of particular importance for *ensuring* that individuals meet the average recommendations made for non-pregnant women and that they better them if possible.

The Task Force was also particularly concerned that the unsaturated fatty acid intakes of formula-fed infants matched that of breast-fed infants. This is of *vital* importance in the special formulae given to pre-term infants because of the essentiality of

arachidonic acid and DHA for brain development but the Task Force considered it prudent to adopt the same principle for all infant formulae.

The Task Force was concerned about the lack of data on which to base recommendations for growing children. It advised a cautionary approach based on the composition of human milk until further information is available.

22.6 TASK FORCE RECOMMENDATIONS FOR DIETARY CHANGES FOR ADULTS

Comparing the Task Force recommendations made for unsaturated fatty acids with the data given on intakes of unsaturated fatty acids in Chapter 3, the major change proposed for the British population is an increase in n-3 PUFA intake. Unfortunately, the Adult Survey (see Chapter 3) only gave data on total n-3 PUFA intake and did not give data on intakes of long chain n-3 PUFA. It can be assumed, though, that intakes of EPA (20:5 n-3) and DHA (22:6 n-3) should be increased as well as those of alpha linolenic acid (18:3 n-3).

The richest sources of long chain n-3 PUFA are found in fatty fish such as herring, mackerel, sardine and salmon (see Chapter 2) and increased consumption of these long chain n-3 PUFA is the best practical advice for the majority of the British population. People who do not eat fish, such as vegetarians and vegans, are advised to ensure adequate intakes of alpha linolenic acid by increasing consumption of green leafy vegetables such as lettuce, broccoli and spinach and to rely on metabolic processes to form EPA and DHA.

The whole population, however, must be advised that increasing n-3 PUFA intake will only be beneficial if it is done in the context of a balanced diet. In particular, a good antioxidant status must be ensured by eating plenty of fruit and vegetables.

Further advice about dietary changes can be found in publications such as 'Eight guidelines for a healthy diet' produced by the Ministry of Agriculture, Fisheries and Food (1990) and 'Enjoy healthy eating' by the Health Education Authority (1991).

23

CONCLUSIONS AND SUGGESTIONS FOR FUTURE PROGRESS

CHAPTER 1   CHEMISTRY OF UNSATURATED FATTY ACIDS

Conclusions

- Unsaturated fatty acids are predominantly straight chain monocarboxylic acids. They can vary according to their number of carbon atoms, their number of double bonds, the position of those double bonds within the carbon chain and the geometrical configuration of the double bond. Monounsaturated fatty acids (MUFA) have one double bond and polyunsaturated fatty acids (PUFA) more than one double bond.
- The physical properties of the fats found in food are largely influenced by the nature of their constituent fatty acids, their chain lengths, the position and geometry of their double bonds and the position of the fatty acids in the triacylglycerol molecule.

Suggestions

- Many systems of nomenclature are available for fatty acids. This Task Force recommends the adoption of the n-3/n-6/n-9 nomenclature and the *cis/trans* nomenclature to denote respectively the positions of double bonds relative to the terminal methyl group and their geometry.
- The shorthand chemical notation (e.g. 18:3 n-3 for alpha linolenic acid and 20:4 n-6 for arachidonic acid) is particularly useful for showing metabolic relationships between fatty acids and should be used alongside the trivial name whenever possible to emphasise the importance of chemical structure in determining function.

CHAPTERS 2 AND 3   SOURCES OF UNSATURATED FATTY ACIDS IN THE DIET AND INTAKES IN THE POPULATION

Conclusions

- A wide range of foods provide significant quantities of fat in the average British diet. In many cases, the fat is present naturally (e.g. meat, milk, cheese) and yet there is considerable scope for modifications and possible reductions. In other cases, when the fat is added during preparation of the product, there is also scope for changes in the end-product.
- In general, foods containing vegetable oils are relatively rich sources of PUFA whereas foods containing animal fats are relatively rich sources of

saturated fatty acids (SFA) and MUFA. There are some notable exceptions to this general rule and the vast majority of fat-containing foods contain fatty acids of all three classes.

- The average intake of fat in the UK has fallen from 121 g/person/day in 1969 through 97 g/person/day in 1984 to 86 g/person/day in 1990. The average proportion of food energy from fat has remained unchanged at about 42% which exceeds any dietary recommendations.
- The average consumption of SFA as a percentage of food energy has declined from 18.5% in 1959 through 20.3% in 1975 to 16.6% in 1990. These consumption figures are higher than any dietary recommendations for SFA.
- The average consumption of PUFA as a percentage of food energy has increased from 3.2% in 1959 to 6.7% in 1990.
- The average ratio of polyunsaturated to saturated fatty acids in the diet (the P/S ratio) has increased from 0.17 in 1959 through 0.22 in 1979 to 0.40 in 1990. This change in P/S ratio is attributable largely to the increased consumption of margarine at the expense of butter and wider use of vegetable oils in food preparation. The 1984 COMA Report proposed that the P/S ratio should rise to about 0.45, but the 1991 COMA Report did not suggest a desirable P/S ratio.
- Linoleic acid intake tends to be higher among vegans and Asian Indians than in the general population. In contrast, vegans and Hindus consume diets which are probably very low in long chain (C20–22) PUFA of the n-3 series. Fatty acid intakes have been confirmed by measurement of fatty acids in plasma phospholipids.

Suggestions

- Further information is required on intakes of individual fatty acids by different population groups particularly when the intakes may be related to health or physiological indicators of risk such as plasma cholesterol.
- Research is needed to reduce the cost of dietary surveys so that larger numbers of individuals may be studied. The reliability of survey data needs to be improved by overcoming problems related to under-estimates and low response rates.
- More data are required on fatty acid composition of foods with particular regard to the longer chain (C20 to C24) unsaturated fatty acids. Values for these fatty acids should be expressed as mg/100 g food. Surveys should not have to rely exclusively on published fatty acid data.
- More data is needed on the oxidation, degradation and polymerisation products of oils present in foodstuffs and those which may be produced during their storage and processing. The chemical and biological consequences of these effects should be studied.

CHAPTERS 4 AND 5 DIGESTION AND ABSORPTION OF UNSATURATED FATTY ACIDS

Conclusions

- For healthy adult subjects consuming a varied diet, there are minor variations in the rate at which different fats are digested due to their intramolecular structure.
- Unsaturated fatty acids may be absorbed and metabolised more rapidly than saturated fatty acids of equivalent chain length. A high proportion of the

former are probably delivered directly to the liver via the portal venous route.

- In the case of the long chain n-3 polyunsaturated fatty acids, eicosapentaenoic acid (EPA) and docosahexaenoic acid (DHA), there is evidence that they are absorbed more readily as free fatty acids than when they are ingested in ester combination as triacylglycerols or ethyl esters.

Suggestions

- More research is needed on the influence of triacylglycerols rich in unsaturated fatty acids, on gastric emptying and the possible effects on satiety and gastro-intestinal disorders.
- It is possible that differences in the rate of uptake of saturated versus unsaturated fatty acids may have metabolic consequences and warrant further investigation. What is the role of the portal venous route?
- The more rapid absorption of EPA and DHA as their free fatty acids has implications for their metabolism, and hence the formulation of unsaturated fatty acid supplements merits further attention.
- More research is needed on the absorption of the products of lipid oxidation and peroxidation.

CHAPTER 6   TRANSPORT OF UNSATURATED FATTY ACIDS

Conclusions

- There are parallel exogenous and endogenous pathways by which lipids from the diet and lipids synthesised within the body are transported in the blood.
- Triacylglycerols are formed in the intestinal mucosa from the products of lipid digestion and are transported in the form of chylomicrons.
- There is a continuum of lipoprotein particles, each containing a characteristic apoprotein, whose composition, size range and physico-chemical properties are influenced in part by the amounts and types of unsaturated fatty acids in the diet. Very low density lipoproteins (VLDL) are involved in the transport of endogenously synthesised triacylglycerols to peripheral tissues. Low density lipoproteins (LDL) are the end-product of VLDL metabolism and are responsible for delivery of cholesterol to the cells. High density lipoproteins (HDL) apoproteins and cholesterol from VLDL and are involved in the transport of cholesterol to the liver. HDL may have other important metabolic roles.
- Regulation of uptake of lipids into cells and tissues from circulating lipoproteins depends on the activities of specific cell surface lipoprotein receptors and enzymes, especially lipoprotein lipase (LPL). These in turn are regulated by the concentrations both of circulating lipoproteins themselves and of various hormones which, again, are influenced in part by the nature and amount of fat in the diet.

Suggestions

- The immediate fate of dietary lipids after a meal needs more investigation using stable isotope tracer techniques. The mechanism for the occasional occurrence of two postprandial lipaemic peaks and its significance are obscure. The short-term regulation of tissue LPL by dietary substrates and hormones needs to be investigated further. Chylomicron remnants formed during post-prandial lipoprotein metabolism may be atherogenic and require special attention.

- The regulation of HDL metabolism, its postulated role in reverse cholesterol transport and the interaction of this lipoprotein particle with other lipoproteins need further study.

CHAPTER 7   METABOLISM OF UNSATURATED FATTY ACIDS

Conclusions

- The high fat content of the UK diet largely suppresses the biosynthesis of fatty acids *de novo* in human tissues, but the detailed mechanisms involved are poorly understood. Two fatty acids, linoleic acid (18:2 n-6) and alpha linolenic acid (18:3 n-3), cannot be synthesised and must be supplied in the diet from plant sources which contain the enzymes to make them. These fatty acids, or their derivatives, are known as the essential fatty acids (EFA).
- Low level endogenous fatty acid synthesis does occur, in particular, retailoring of dietary fatty acids to suit the body's requirements; e.g. the conversion of dietary EFA into longer chain, more highly unsaturated fatty acids and the reverse process, retroconversion, in which long chain PUFA are partly degraded.
- The regulation of desaturation and elongation of EFA is poorly understood. Man's potential to form the longer chain n-3 derivatives is not adequately known and this uncertainty is reflected in prudent dietary recommendations for these fatty acids. Progress should become more rapid now that the different desaturases are being isolated and purified and their genes cloned.
- The channelling of fatty acids and their metabolites into different metabolic pathways is highly complex. The relative quantitative importance of the pathways of esterification, oxidation, desaturation, elongation and retroconversion and the conversion of fatty acids into eicosanoids depends on competition between the pathways for substrates. The competitiveness of a pathway depends, in turn, on nutritional and hormonal status which influences the amounts and activities of enzymes in the pathways.

Suggestions

- The mechanisms by which dietary fatty acids, especially polyunsaturated fatty acids, suppress the biosynthesis *de novo* of saturated fatty acids in different tissues need to be investigated.
- The regulation of fatty acid desaturation, elongation and the conversion of long chain PUFA into eicosanoids should be further studied. Better characterisation of the desaturation enzymes is needed, particularly with respect to 4-desaturation. Is there any evidence for genetic variability of fatty acid desaturation? Further definition is required of the biochemical controls for interactions between the n-6 and n-3 PUFA families in the desaturation and elongation pathways.
- Factors governing the location of fatty acyl groups in certain phosphoglycerides require further investigation.
- Studies are needed of the mechanisms controlling the balance between microsomal fatty acid elongation and peroxisomal fatty acid shortening.
- Information about all, or some, of the above will allow recommendations for the optimal amounts of different unsaturated fatty acids in the diet to be made on a much firmer scientific basis.

CHAPTER 8   FUNCTIONS OF UNSATURATED FATTY ACIDS

Conclusions

- Unsaturated fatty acids contribute to the provision of energy. There is some evidence that essential fatty acids (EFA) may be conserved in the body at the expense of the non-essential fatty acids.
- Unsaturated fatty acids have important structural roles in maintaining the fluidity, permeability and conformation of membranes. For example, the myelin sheath of nerves and the rods of the retina have specialised functions and highly specific lipid compositions.
- Unsaturated fatty acids in membranes also play an important role in metabolic control via the inositol lipid cycle which is involved in cell responses to a range of hormones, neurotransmitters and local growth factors. The dominant species of phosphatidyl inositol (PtdIns) in the plasma membrane contains arachidonic acid (20:4 n-6) as well as stearic acid (18:0).
- Unsaturated fatty acids in membranes also play a critical role in metabolic control as precursors of the eicosanoids, particularly the prostanoids and the leukotrienes which are formed via the cyclo-oxygenase and lipoxygenase pathways. The spectrum of eicosanoids produced can be influenced by the unsaturated fatty acid composition of the diet.

Suggestions

- The balance between the structural and metabolic roles of unsaturated fatty acids and their oxidation to yield energy should be studied. Is there any preferential 'sparing' of unsaturated fatty acids as energy providers?
- The effect of variations in the diet on the fatty acid composition of PtdIns in different cell membranes needs to be investigated. Could the replacement of arachidonic acid by other fatty acids have implications for the role of the derived second messenger, diacylglycerol? What are the effects of dietary PUFA on cell growth factors?
- Further definition is required of the structural/functional roles of PUFA in the phospholipids associated with G protein systems.
- The control of eicosanoid formation via phospholipase and via intracellular calcium needs further investigation. The interrelationship between eicosanoid metabolism and the inositol lipid cycle should be studied to explore its physiological significance e.g. in the amplification of cell signals.
- More needs to be known about the conversions of PUFA catalysed by cytochromes P_{450} and by lipoxygenases other than 5-lipoxygenase, including the physiological functions of these products.
- The possible interactions of dietary PUFA in the production of eicosanoids from cyclo-oxygenase, the lipoxygenases and cytochromes P_{450} should be investigated.

CHAPTER 9   UNSATURATED FATTY ACIDS AND EARLY DEVELOPMENT

Conclusions

- The supply of the long chain PUFA, arachidonic acid and DHA, to the fetus is crucial for normal development. Neural development, retinal function, cognitive and learning ability and the vascular system could all be adversely affected if the supply is inadequate.
- The most active period of cell division and organogenesis is in the first weeks

of pregnancy. It is important that the essential fatty acid (EFA) intake during the preconception period satisfies the quantitative and qualitative need for a fat store to guarantee energy and EFA provision in preparation for the early part of pregnancy when appetite may be affected.

- The premature and low birthweight infant has an increased risk of neurodevelopmental handicap. This may be due to a deficiency of long chain EFA whilst the infant is *in utero*. The premature infant is specifically denied the high input of long chain EFA selectively provided by the placenta during the fetal brain's rapid period of growth in the last trimester of pregnancy.
- In human milk, longer chain derivatives of linoleic and alpha linolenic acid (i.e. arachidonic acid and DHA) account for about 1% of the total fatty acids. The provision of the equivalent amount of linoleic acid and alpha linolenic acid in infant milk formulae is probably not an adequate substitute. Infant milk formulae which provide the main nutrient intake of infants should supply DHA, and possibly arachidonic acid, in the same concentration as human milk.

Suggestions

- Food tables do not permit, or encourage, proper assessment of essential fatty acid (EFA) intakes as there are insufficient facts. Part of the problem is due to expressing the intakes as g/100 g of food. This means that small but significant amounts could be neglected. Polyunsaturated fatty acids, particularly the long chain PUFA, should therefore be expressed as mg/100 g of food, as are the essential amino acids.
- Most nutritional recommendations for pregnancy are given for the second half of pregnancy which may be too late for the optimal development of the fetus. Recommendations should be extended to include nutrient intakes, particularly EFA intakes, during early pregnancy and the preconception period.
- Infant milk formulae should approach the fatty acid composition of human milk. Infant formulae should contain n-3 fatty acids; the balance of n-6/n-3 fatty acids should be similar to that found in human milk, provided adequate DHA is present.
- The provision of long chain EFA in formulae intended for *preterm infants* should be considered a priority. The appropriate amounts of long chain EFA in these formulae should be established.
- The relationship between maternal and fetal EFA status and developmental disorders such as retardation of mental development, retinopathy of prematurity, periventricular haemorrhage and respiratory disorders in the neonate should be investigated.
- The relationship between maternal and fetal EFA status and the development of the cardiovascular, cerebrovascular and immune systems should be investigated. The role of EFA status in the relationship between low birthweight and placental weight and subsequent risk of hypertension and heart disease in adult life should also be examined.

CHAPTER 10 UNSATURATED FATTY ACIDS AND CORONARY HEART DISEASE

Conclusions

- There is little direct evidence of an association between consumption of monounsaturated fatty acids and CHD risk.
- Partial replacement of saturated fatty acids by n-6 PUFA may reduce the risk of a CHD event, but has not been shown to reduce risk of death from all causes.

- There is evidence that consumption of fish, particularly fatty fish rich in long chain PUFA of the n-3 series, is effective in the secondary prevention of CHD death.

Suggestions

- There needs to be further evaluation of the effect of monounsaturated fatty acids on CHD incidence and total mortality by means of randomised controlled trials.
- Trials must be planned which evaluate the effects of unsaturated fatty acids on CHD incidence and CHD risk factors other than plasma lipids.
- Further randomised controlled long-term trials are needed to investigate the effect of fatty fish on CHD incidence and total mortality.

CHAPTER 11   UNSATURATED FATTY ACIDS AND PLASMA LIPIDS

Conclusions

- Between-population studies relating the dietary intake of fatty acids to plasma lipids have demonstrated a consistent association with saturated fatty acids, but not as clearly with unsaturated fatty acids. Within-population studies can only demonstrate an association between unsaturated fatty acid intake and plasma lipids when there is a wide range of fatty acid intakes. Intervention studies in the free-living population are most likely to demonstrate that increasing unsaturated fatty acid intake can reduce plasma cholesterol levels if they are carried out on people with high levels of cholesterol and if good dietary compliance is ensured.
- Experimental studies have shown that when unsaturated fatty acids replace C12-C16 saturated fatty acids in the diet, they lead to a reduction of total and LDL cholesterol in the plasma. More consistent effects are found with n-6 PUFA than with n-3 PUFA. Intakes of linoleic acid greater than 12% of dietary energy also lead to a reduction in HDL cholesterol concentrations and are therefore not advised. Oleic acid does not appear to affect HDL cholesterol concentrations.
- Single gene defects may result in missing or poorly functioning cell receptors for apoproteins, but differences between individuals in lipoprotein metabolism within the 'normal' range may be governed by poorly understood polygenic patterns of inheritance. Modification of dietary fat composition, i.e. the substitution of unsaturated fatty acids for saturated fatty acids, plays an important role in the clinical management of the hyperlipoproteinaemias.
- Dietary EPA and DHA lead to a marked reduction in plasma triacylglycerol and VLDL cholesterol concentrations if they are consumed in amounts greater than 2–3 g/day. HDL cholesterol concentrations are increased but may be decreased by much higher intakes of n-3 PUFA (more than 10 g/day).

Suggestions

- Further studies are needed to study the influence of unsaturated fatty acids on lipoprotein synthesis and catabolism using stable isotope techniques.
- Studies are needed to identify those genetic variations that respond to changes in dietary fat composition, particularly with regard to apoprotein polymorphisms and lipoprotein receptor status.
- Studies on the influence of dietary unsaturated fatty acids on sterol metabolism are warranted.

CHAPTER 12 UNSATURATED FATTY ACIDS AND ATHEROSCLEROSIS

Conclusions

- A reduction in plasma total and LDL cholesterol concentrations and an increase in plasma HDL cholesterol and apoprotein A1 lead to decreased atherosclerosis in animals and in man.
- The normal cell receptor for the apoprotein of LDL cannot interact with apoproteins that have been biochemically modified by the products of lipid peroxidation within the lipid moiety of the particle. Modified LDL may be taken up by 'scavenger receptors' on macrophages which may have pathological consequences for CHD such as arterial injury, foam cell formation, platelet aggregation and vasoconstriction.
- Experimental atherosclerosis is most strongly influenced by the saturated fatty acid content of the diet. Substitution of saturated fatty acids by MUFA and n-6 PUFA will usually reduce experimental atherosclerosis.
- Dietary EPA and DHA inhibit the development of atherosclerosis in dogs, pigs and primates by mechanisms independent of plasma cholesterol concentrations. It is not yet certain whether n-3 PUFA can inhibit atherosclerosis in man but fish oils may be of some benefit in preventing restenosis following angioplasty.

Suggestions

- Attention needs to be focused on the influence of dietary unsaturated fatty acids on the oxidative modification of LDL.
- Further studies are needed on the potential of EPA and DHA for reducing fibrous plaque formation and intimal thickening and in preventing restenosis.

CHAPTER 13 UNSATURATED FATTY ACIDS, THROMBOGENESIS AND FIBRINOLYSIS

Conclusions

- Changes in dietary unsaturated fatty acids cause changes in the fatty acid composition of phospholipids of platelet membranes and changes in the subsequent eicosanoid spectrum.
- Dietary induced enrichment of platelet membranes with n-3 PUFA reduces the responsiveness *ex vivo* of platelets to the aggregating agents collagen, ADP and thrombin. The shift in thromboxane metabolism towards TXA_3 formation at the expense of the more strongly pro-thrombotic TXA_2, formed from n-6 PUFA, might be one of the factors responsible for the reduced tendency of platelets to aggregate. Large oral doses of n-3 PUFA also appear to prolong the bleeding time and reduce platelet adhesiveness.
- The techniques for measuring platelet responsiveness *ex vivo* are inadequate in several ways and do not necessarily correlate with, or contain information about, the likelihood of thrombosis *in vivo*.
- Although two of the factors involved in the coagulation pathway (factor VII activity and fibrinogen concentration) are potent coronary risk factors, there is very little evidence that they can be influenced significantly by the amounts of unsaturated fatty acids in the diet.
- There is very little evidence for the involvement of dietary unsaturated fatty acids in the fibrinolytic pathways.

Suggestions

- The use of optical densitometry to measure platelet response *ex vivo* is probably too insensitive to measure small dietary effects on platelets which might represent biologically important changes in platelet behaviour. Improved methods should be developed which can take account of any changes in platelet size, number and turnover and changes in plasma fibrinogen which could, in turn, influence platelet aggregability.
- Experiments should be designed to test whether dietary unsaturated fatty acids in acceptable quantities in a mixed diet can promote anti-aggregatory responses in platelets and endothelial cells which have the potential to reduce the risk of thrombosis.
- The use of pure fatty acids, singly or in combinations, and at various dosages, rather than ill-defined fish oils or seed oils, might resolve some of the conflicting findings with respect to effects on platelet aggregability, adhesiveness and bleeding time. There is a need to quantify dose–response relationships.
- Functional assays are needed to assess the activity of the coagulation and fibrinolytic pathways *in vivo*. Markers of the prothrombinase complex activity and of plasmin activity might be more informative than the concentrations of individual components of these pathways.

CHAPTER 14 UNSATURATED FATTY ACIDS AND BLOOD PRESSURE

Conclusions

- Fish oils induce a small reduction in blood pressure provided they are given in daily doses supplying about 3 g or more of the n-3 PUFA. The reduction is more evident in those with mild to moderate hypertension than in normotensive people.
- The mechanism for the reduction in blood pressure is unknown. One possible mechanism may be a restoration of the normal release of endothelium-derived relaxing factor (EDRF) in endothelial cells which have been damaged by atherosclerosis or injured by exposure to high blood pressure.
- PUFA of the n-6 series, in amounts normally consumed, do not have any demonstrable effect on blood pressure.

Suggestions

- The possibility that supplementing the diet with fish oil will enhance the effect that sodium restriction has on lowering blood pressure merits further study. So do the interactions of unsaturated fatty acids with all other physiological factors which influence blood presssure.
- The effect of fish oils in reducing blood pressure in people of different age groups needs to be studied.
- Further work is needed to elucidate the mechanism(s) by which n-3 PUFA reduce blood pressure.

CHAPTER 15 UNSATURATED FATTY ACIDS AND CARDIAC ARRHYTHMIAS

Conclusions

- While animal studies are suggestive, human studies have not yet firmly established a protective role of either n-6 or n-3 unsaturated fatty acids against cardiac arrhythmias.
- There is better evidence for a pro-arrhthymic role of saturated fatty acids in man than there is for an anti-arrhythmic role of unsaturated fatty acids.

Suggestions

- Larger studies are needed of the risk of serious arrhythmias in people who have survived heart attacks, or who have clinical evidence of CHD, in relation to the fat composition of their habitual diet. Studies must control for confounding variables such as age, smoking habit, extent of ischaemic damage, potassium levels, preceding drug therapy and coincident diseases.
- Continued work with animal models which are the most appropriate to the human situation will help to elucidate mechanisms of action of the fatty acids. Animal studies should employ realistic levels of fish oil supplementation, be large enough to demonstrate small effects with conviction, and conform to recognised guidelines for the study of arrhythmias in ischaemia, infarction and reperfusion.

CHAPTER 16 UNSATURATED FATTY ACIDS AND CANCER

Conclusions

- The epidemiological studies relating PUFA intake to human cancer are inconclusive. Evidence is scanty.
- The evidence from animal models suggests that PUFA of the n-6 series are a necessary extra pre-requisite for the promotion and maintenance of tumour growth. PUFA of the n-3 series inhibit tumour growth.
- The discrepancy between the animal and human studies could arise if n-6 PUFA show their promotional effects in animals up to a certain threshold level (4% of total energy). Once this threshold has been exceeded, as it would be in most human diets, then the total amount of fat and the enhancing effect of total energy would become more predictive than the amount of n-6 PUFA in the diet.
- The anti-proliferative effects of certain unsaturated fatty acids such as EPA and GLA result from a pharmacological effect of the fatty acids, rather than a nutritional one, since the effective doses are so high (25 to 50% total energy).

Suggestions

- More information about the threshold value for n-6 PUFA intake in humans is needed. Clinical studies should be initiated to study the effects of modification of the intake of n-6 PUFA in a group of women at high risk of developing cancer. An attempt should be made to reduce the intake of n-6 PUFA and/or increase the intake of n-3 PUFA, while at the same time reducing both the total fat and caloric intake.
- The mechanism(s) of modification of tumour growth by PUFA needs to be determined. Studies should be performed with pure fatty acids, rather than

mixtures of oils. The effect of the individual fatty acids could then be determined, as well as possible interactions between them.
- Clinical studies should be performed to evaulate the pharmacological effectiveness of certain PUFA as antitumour agents: e.g. EPA and gamma linolenic acid (GLA) tested as pure substances. The tumours most suitable for clinical evaluation would appear to be carcinomas of the breast and of the internal organs which are usually unresponsive to chemotherapy. The limited degree of structural stability imposed on these PUFA by the *cis* double bonds and the hydrophobic nature of these molecules might allow important interactions with enzymes and receptors.
- The possible role of PUFA as intracellular mediators and modifiers of the action of known oncogene products should be evaluated. Do PUFA act directly in cellular metabolism in addition to their role as precursors of eicosanoids?

CHAPTER 17   UNSATURATED FATTY ACIDS AND DIABETES

Conclusions

- The qualitative effects of unsaturated fatty acids on lipid and lipoprotein metabolism appear similar in diabetics and non-diabetic subjects. Fatty acids of the n-6 series lower total and LDL cholesterol; fatty acids of the n-3 series have less consistent effects.
- Unsaturated fatty acids of the n-6 series have no consistent effects on carbohydrate metabolism in diabetes, but the long chain n-3 fatty acids tend to cause impaired glycaemic control in non-insulin dependent diabetic subjects.
- Preliminary data suggest a possible beneficial effect of dietary enrichment with unsaturated n-3 fatty acids on some microvascular complications of diabetes.

Suggestions

- Can an effect of n-3 unsaturated fatty acids on microvascular complications be confirmed? What differences exist between effects on different tissues? What are the underlying mechanisms for these effects?
- More research on the relative benefits of fish oil supplementation for diabetics is needed. The possibility that fish oil may reduce microvascular diabetic complications will have to be viewed in the context of total diabetic therapy and possible concomitant effects on glucose and lipid metabolism.

CHAPTER 18   UNSATURATED FATTY ACIDS AND SKIN DISEASES

Conclusions

- Human skin is functionally highly dependent on unsaturated fatty acids which not only contribute to the integrity and barrier function of the skin but also act as a source of mediators that are important in inflammatory skin diseases. Sebum also contains essential fatty acids which may play a role in hair growth.
- Supplementation with fish oil or with EPA causes a modest clinical benefit in psoriasis, probably via effects on eicosanoid biosynthesis.

- Evening primrose seed oil, which contains GLA, produces a symptomatic improvement for atopic eczema but does not change the underlying disease state.

Suggestions

- New information on the role of immunoregulatory and proinflammatory cytokines suggests that these agents, which are formed in the skin, play a key role in psoriasis and other inflammatory disorders. However, their interactions with other eicosanoids need to be evaluated.
- Dryness and itching of the skin in the elderly is an increasing problem and the status of epidermal unsaturated fatty acids in this condition needs to be evaluated.
- Although EPA supplementation is only moderately effective in psoriasis, its ability to potentiate other treatments for psoriasis should be examined.

CHAPTER 19 UNSATURATED FATTY ACIDS AND IMMUNE DISORDERS

Conclusions

- The immune system is immensely complex and it has many interactions with non-specific inflammation. While many immunopathological mechanisms have been clarified, there is still much uncertainty about their role in acute and chronic disorders of unknown aetiology. This is also true of those situations in which immune mechanisms are clearly operating.
- The increased consumption of n-3 fatty acids such as EPA and DHA favours the production of prostaglandins of the 3 series and leukotrienes of the 5 series. The overall effect is to reduce the efficiency of granulocyte mobilisation.
- Certain PUFA supplements have become popular for the treatment of rheumatoid arthritis, multiple sclerosis, asthma and lupus. The proven clinical benefit of such measures is modest and the long-term benefits and disadvantages need further evaluation. The observed changes in cell membrane lipid composition and leukotriene spectrum may not necessarily account for any clinical improvement.

Suggestions

- The biochemical and functional consequences of dietary manipulation can be investigated in relatively well-defined systems *in vitro* complemented by studies *in vivo* in experimental models of immune-mediated inflammation. More information is needed on the effects on receptor expression and response to immunoregulatory signals such as cytokines, and lymphocyte activation.
- Future clinical studies of immune disorders should be meticulously designed long-term trials with suitable controls, which will permit accurate assessment of the biochemical and functional effects of dietary PUFA including GLA and DGLA.
- In treating immune disorders, dietary manipulation is just one of the therapeutic options. Future trials of dietary supplementation should keep convenience and risk in mind and try to identify those patients who could most benefit because they are uneasy about drug treatment.

CHAPTER 20 SUPPLEMENTARY SOURCES OF UNSATURATED FATTY ACIDS

Conclusions

- Certain unsaturated fatty acids are recognised as medicines. These supplements, with the exception of cod-liver oil, have clear indications and quality criteria. They are generally used under medical supervision.
- Supplements of unsaturated fatty acids available over the counter in pharmacies or 'health food' shops are regarded as foods, even though many make implied health claims. As yet, no criteria for their formulation exist.

Suggestions

- In view of the potential for contamination of fish oils by offshore pollutants and by oxidised lipids, agreed purity criteria should be established.
- All fish liver oil preparations should carry a declaration of their retinol and vitamin D contents, with appropriate warnings to pregnant women if they exceed 750 retinol equivalents in a portion.
- Novel sources of oils should undergo careful review before being allowed on the market.

CHAPTER 21 LABELLING OF FOODS IN RELATION TO UNSATURATED FATTY ACIDS

Conclusions

- The EC Directive on Nutrition Labelling lays down rules for declaring total fat and saturates and has replaced the MAFF guidelines. However, labelling is only voluntary at present unless some type of claim is made. This Task Force would encourage manufacturers to give as much information as possible in the standard format including information on unsaturated fatty acids. Values should be given as mg/100 g food (especially for the long chain PUFA) and the n-6/n-3 ratio of the food might be helpful. The general terms 'fish oils', 'vegetable oils' and polyunsaturated fats should be discouraged.
- Claims and 'special emphasis' declarations concerning the polyunsaturated fatty acid and cholesterol contents of foods are controlled by the Food Labelling Regulations 1984. Additional nutritional claims such as 'reduced fat' and 'low fat' are the subject of UK recommendations, but they are unlikely to be incorporated into UK law until they form part of an EC Directive.
- Consumer surveys have shown that whereas the food label is the correct place for information about nutrients, it cannot be expected to provide an education in nutrition.

Suggestions

- There is an urgent need for EC legislation about claims and 'special emphasis' declarations so that there is a standardised system of control in all EC countries, such as that proposed for the declaration of nutrition information.
- An assessment is needed of the effect of current nutrition labelling practices on food choices and on nutrient intakes.
- Independent nutrition education campaigns need encouragement and support to allow the information on labels to be understood and used effectively.

CHAPTER 22 RECOMMENDATIONS FOR INTAKES OF UNSATURATED FATTY ACIDS

Conclusions

- Task Force recommendations are given for average population intakes of unsaturated fatty acids: n-6 PUFA (6%), alpha linolenic acid (1.0%) and EPA + DHA (0.5%). The recommendations are more extensive than those given by the COMA Panel in their 1991 Report on Dietary Reference Values. They call for slightly higher population intakes of polyunsaturated fatty acids, particularly those of the n-3 series.
- Safe ranges of intakes are given as percentages of energy intake and on an absolute weight basis. The recommendations for the lower limit of the safe ranges are based on the unsaturated fatty acid content of human milk. The recommendations for the upper limit of the safe range have been derived using a toxicological approach.
- The Task Force considers the use of the polyunsaturated fatty acid to saturated fatty acid ratio (P/S ratio) to be obsolete because of new evidence about the varied role of the different saturated fatty acids and of the monounsaturated fatty acids. The U/S ratio (unsaturated fatty acids to saturated fatty acids) is no more useful than the P/S ratio.
- The Task Force is particularly concerned that pre-pregnant, pregnant and lactating women should heed these recommendations and advises that special counselling about EFA intake during these times is warranted.
- The Task Force recommends that formulae for pre-term and full term infants replicates the EFA content of human milk.
- The general adult British population can best achieve these recommendations by continuing to substitute foods rich in unsaturated fatty acids for those rich in saturated fatty acids and by increasing the proportion of n-3 PUFA in particular. At the same time, intakes of complex carbohydrates and antioxidant nutrients should be increased.
- In practical terms, this nutritional message translates into 'Eat more fatty fish such as herring, mackerel, sardine and salmon, and eat more fruit, vegetables, and wholegrain cereals'.

Suggestions

- As human milk contains n-6 fatty acids other than linoleic acid which have higher relative potencies, the possibility of basing requirements for infants on linoleic acid equivalents should be considered.
- Although the requirement of the pre-term infant for long chain n-6 and n-3 unsaturated fatty acids is established, there is controversy over the requirement of full-term infants for these fatty acids. Further research is required to ascertain whether the elongase and desaturase enzymes can function adequately in the neonate. Meanwhile, it is prudent to advise the addition of these fatty acids (DHA and, possibly, arachidonic acid) to infant milk formulae.
- The physiology and metabolism of vegans must be examined to explore how they can seemingly manage without long chain n-3 fatty acids in their diet. Do they make special metabolic adaptations which are unnecessary for people eating mixed diets or do they require less long chain n-3 PUFA because of lower intakes of preformed arachidonic acid?
- Further work is required on the possible adverse effects of long chain MUFA.
- More information is required on the proportions of antioxidants that should accompany unsaturated fatty acids in the diet. Meanwhile, manufacturers and processors should be encouraged to monitor and minimise the loss of natural antioxidants.

24

GENERAL CONCLUSIONS AND RECOMMENDATIONS

- Fats and oils contain many different fatty acids, both saturated and unsaturated, which can have various physiological effects. This Task Force recommends that research workers define much more carefully which fatty acids they are considering. General terms such as fish oils, vegetable oils or polyunsaturated fats should be avoided. Ideally, the effects of each of the different types of unsaturated fatty acids should be investigated independently so that the effects of monounsaturated fatty acids (MUFA), n-6 polyunsaturated fatty acids (n-6 PUFA) and n-3 polyunsaturated fatty acids (n-3 PUFA) can be documented separately. This is somewhat unrealistic because they do not occur separately in foods. Specifying the fatty acid composition of oils, is however, more realistic.
- Although the body is able to synthesise most saturated fatty acids (SFA) and MUFA from carbohydrates, the British diet provides preformed versions of the majority of fatty acids. Two polyunsaturated fatty acids, linoleic and alpha linolenic acid, cannot be synthesised by the body and these must be provided from plant sources in the diet. These acids and their derivatives are generally known as the essential fatty acids (EFA). It is possible that under certain circumstances, such as in the premature infant, the rate of synthesis of other fatty acids which are derivatives of EFA (such as arachidonic acid, docosahexaenoic acid (DHA) and eicosapentaenoic acid (EPA)) may not meet the body's requirements and that such fatty acids could be said to be 'conditionally' essential.

  The source of EPA and DHA in body tissues of vegetarians is a relatively unexplored area. There are no apparent pre-formed plant sources of EPA and DHA; fatty fish normally provide the major dietary source of these long chain n-3 PUFA. Metabolic adaptation might take place in long-term vegetarians; this possibility should be investigated.
- PUFA of the n-6 and the n-3 families are metabolised by the same enzyme systems and thus compete with each other. It is possible that there may also be competition between PUFA within the same family which would mean that the number of competitive interactions could be even greater. The effect of supplementing the diet with a fatty acid is difficult to predict with certainty because it might act as a precursor, a product or an intermediate in a chain of reactions.
- PUFA of both families also compete between themselves for the enzyme systems which convert them into the metabolically-active eicosanoids. Thus, dihomogamma linolenic acid (DGLA), arachidonic acid and EPA all compete for the enzymes cyclo-oxygenase and lipoxygenase which convert them to the prostaglandins and other eicosanoids with differing physiological activities. Again, the balance of the eicosanoids produced in the tissues is difficult to predict and caution is advised in the interpretation of results.
- Different unsaturated fatty acids have different effects on plasma cholesterol and lipoproteins. When saturated fatty acids (SFA) are replaced by MUFA, plasma LDL cholesterol concentrations are reduced, but plasma HDL

cholesterol concentrations are unaffected. Although n-6 PUFA can also reduce plasma LDL cholesterol levels, they should not be used to replace SFA completely. When n-6 PUFA provide more than about 12% of dietary energy, they can lead to an undesirable reduction in plasma HDL concentration. n-3 PUFA have no consistent effects on LDL and HDL cholesterol but they do consistently lower VLDL.

- It is important that experimental studies in both animals and man are not extrapolated uncritically to the clinical situation. Conditions *in vitro* do not necessarily reflect those *in vivo*. Likewise, the results of well-controlled intervention studies cannot always be assumed to predict the results in free-living subjects. Caution must be applied in interpreting studies which could involve possible interactions between fatty acids of the same family, or of different families, and with other constituents of the diet such as antioxidants. Gene–nutrient interactions leading to inter-individual variation in the effect of diet must also be considered.

- Altering the unsaturated fatty acid content of the diet to manipulate the balance of eicosanoids in cell membranes has been attempted as a treatment for various pathological conditions ranging from eczema and rheumatoid arthritis to coronary heart disease, with differing degrees of success: e.g. diets rich in certain n-6 PUFA can change the eicosanoid balance in cells of the immune system and diets rich in certain n-3 PUFA can change the eicosanoid balance in platelet membranes. Many studies have been hampered by the use of mixtures of fatty acids which have not been well defined.

  Currently, there is reasonable evidence that increasing dietary levels of unsaturated fatty acids can reduce the incidence and deaths from coronary heart disease via effects on blood pressure, atherosclerosis and thrombogenesis, and may reduce the severity of several skin conditions and immune diseases, such as rheumatoid arthritis. The results achieved depend upon the balance of MUFA, n-6 PUFA and n-3 PUFA and many other factors.

- The developing fetal brain is critically dependent on certain unsaturated fatty acids, namely DHA and arachidonic acid. It is therefore particularly important that women who intend to become pregnant, as well as those who are pregnant, should have adequate dietary intakes of essential fatty acids and their derivatives.

- It is just as important that the lactating mother has adequate supplies of EFA and their derivatives in her diet to ensure that these fatty acids are available to the breast fed infant. Manufacturers of infant formulae should ensure that their products contain amounts of linoleic acid, alpha linolenic acid, arachidonic acid and DHA which replicate those found in human milk. This is particularly important in formulae intended for premature and low birthweight babies. The premature and low birthweight infant has an increased risk of neurodevelopmental handicap which may be associated with deficiency of essential fatty acids, or their metabolic derivatives, when the infant is *in utero*.

- Taking account of current intakes, the advice of the Task Force is that average intakes of n-6 PUFA have increased sufficiently in recent years, and do not need to be increased further. It is prudent however to increase intakes of n-3 PUFA, particularly long chain PUFA. Although there is no clear evidence to suggest that an absence of long chain n-3 PUFA in the adult diet is 'unsafe', there are several independent pieces of evidence for long chain n-3 PUFA having a protective role. Increasing the amount of PUFA will also increase the requirement for anti-oxidants such as vitamin E and beta-carotene.

- More comprehensive nutritional labelling of products will help consumers to identify sources of fatty acids. There is an urgent need for food labelling legislation relating to claims concerning the fat and unsaturated fatty acid content of foods. This Task Force recommends that manufacturers should be encouraged to give full information about unsaturated fatty acids. Values for polyunsaturated fatty acids should be given as mg/100 g food and the n-6/n-3 ratio in the food would be helpful. The meaning of this ratio must be explained to consumers through independent nutrition education campaigners.

- This nutritional advice can be translated into practical terms by recommending that the general population increases intakes of fatty fish such as herring, sardine, mackerel and salmon. At the same time, it is prudent to recommend increased intakes of fruit, vegetables and wholegrain cereals to raise antioxidant status.

GLOSSARY

Acetylcholine chemical transmitter released upon stimulation of cholinergic nerves

Adenylate cyclase enzyme which catalyses the synthesis of cyclic adenosine monophosphate from adenosine triphosphate

Adjuvant arthritis arthritis provoked by a substance which enhances an immunological reaction

Adrenoceptor of alpha and beta type, which combine with catecholamines to evoke a cellular response. Alpha receptors evoke a constrictive response, and beta receptors a relaxation response. Noradrenaline acts almost exclusively on alpha receptors: adrenaline has both alpha and beta receptor actions

Amphiphilic (literally liking both) – a compound which is soluble in both aqueous and non-aqueous solvents

Aneurysm localised dilation of the wall of an artery, most commonly that of the aorta

Angioplasty the surgical joining of blood vessels

Angiotensin angiotensin I is generated from angiotensinogen by the enzyme renin. Angiotensin I is converted to angiotensin II by a hydrolysing enzyme. It is a very potent vasoconstrictor which also stimulates the release of aldosterone from the adrenal cortex

Antibody a protein (an immunoglobulin), synthesised by B lymphocytes, which combines with antigens, rendering them harmless

Antigen a substance, usually a protein, which the body recognises as foreign, against which it produces an antibody

Apical membrane the membrane at the apex

Arginine vasopressin a neuropeptide, synthesised in nuclei of the hypothalamus, and secreted into the systemic circulation. Vasopressin causes vasoconstriction, hepatic glycogenolysis, secretion of adreno-corticotrophic hormone, and exerts an antidiuretic effect on the renal tubules

Arrhythmia any form of irregular electrical activity in the heart muscle. It can be intermittent or continuous

Arthritis inflammation of the joints characterised by swelling, redness, pain and loss of function. It can be caused by a large number of diseases

-ase suffix signifying the compound in question is an enzyme

Atherosclerosis disease of the arteries, including those supplying the heart muscle, in which fatty, fibrous plaques develop on the inner walls of the artery. This may eventually disrupt blood flow, especially if a thrombus forms

Bradykinin nonapeptide released from its precursor, high molecular weight kininogen, by the action of plasma kallikrein. It is a potent vasodilator, a bronchoconstrictor in asthmatic subjects, and constrictor of gut and uterine muscle. It also appears to play a role in eicosanoid formation and pain sensation

Balance study method used to determine the amount of a nutrient retained in the body, i.e. the difference between intake and excretion measured over a number of days

Cancer cachexia syndrome of weight loss, electrolyte, water and metabolic abnormalities often accompanying cancer

Carcinogen an agent capable of inducing tumour formation

Chylomicron the largest of the circulating lipoprotein particles, normally present in plasma only in the post-prandial state. Responsible for transport of triacylglycerol of dietary origin from the intestine to peripheral tissues. Its basic structure is similar to the very low density lipoprotein, but its surface protein composition is distinctive

Coagulation the process by which blood is converted from a liquid to a solid state, i.e. blood clotting

Cohort study a longitudinal follow-up study of a defined group of subjects

Collagen structural protein of connective tissue and basement membrane, organised into microfibrils. Eight types are described, seven of which are found in the blood vessel wall

Cysternae thin walled dilated vessels carrying blood and other fluids

Cystoskeleton the structural framework of the cell

Cytokines compounds, synthesized by activated cells of the immune system, which enhance the proliferation and differentiation of other cells in the immune system in response to immune stimulation. They also affect many other systems and their effects are dose-dependent and modified by interaction with other cytokines

Cytoplasm the jelly-like substance that surrounds the nucleus of a cell

Dense tubular system a discrete system of membranous channels running through the cytoplasm of the platelet

Dermis the deeper zone of the skin comprising mainly connective tissue, blood vessels, nerves, and sebaceous and sweat glands

Docosa- prefix meaning 22

Eicosa- prefix meaning 20

Eicosanoid derivative of the 20 carbon fatty acids, dihomogamma linolenic acid, arachidonic acid and eicosapentaenoic acid

Enterocyte the predominant cell-type of the intestinal mucosa, specialised for the transport of nutrient molecules

Epidemiology the study of disease in relation to its occurrence in different population groups

Epidermis the outer layer of the skin comprising living epidermal cells and dead corneocytes

Erythema redness of skin

Esterification the chemical reaction that occurs between an organic acid group (–COOH) and an alcohol group (–OH) resulting in the elimination of water. The two original molecules are joined by a –OOC– linkage

Fibrinogen a circulating glycoprotein synthesised mainly in the liver, and composed of three pairs of subunits. Two of the subunits are cleaved by thrombin during fibrin formation

Fibrinolysis the process by which blood clots are dissolved and removed from the circulation

Fibroblast the predominant cell type in connective tissue, responsible for the production of collagen, proteoglycans and fibronectin

Fibronectin an adhesive protein produced by smooth muscle cells, fibroblasts and endothelial cells, and thought to play a role in binding endothelial cells to the subendothelium

Glycogen a highly branched storage polysaccharide of glucose found in many animal cells including muscle cells and hepatocytes

Glycolipid a lipid with a carbohydrate chain attached

Glycoprotein a molecule consisting of protein linked to carbohydrate by covalent bonds

Guanylate cyclase an enzyme which catalyses the hydrolysis of guanosine triphosphate to cyclic guanosine monophosphate: i.e. GTP → cGAMP

Hexa- prefix meaning six

Histocompatibility the matching of tissue components, mainly specific glycoprotein antigens in cell membranes

Hydrolysis the breakdown of long chain like molecules to smaller units with the addition of the constituents of water

Hydrophilic (literally 'liking water') – a compound that is soluble in water

Hydrophobic (literally 'fearing water') – a compound that is insoluble in water

Hypercholesterolaemia a high plasma total cholesterol concentration, usually due to a high concentration of circulating low-density lipoprotein (LDL) particles. It is currently said to be mild-to-moderate when the cholesterol concentration is between 5.2 and 6.2 mmol/l, and severe at higher cholesterol levels

Incidence the number of new cases arising in a defined population over a period of time

Inflammation the immediate defensive reaction to tissue injury. Local blood vessels dilate, increasing blood flow to the area and causing redness and swelling

Interleukin-I a cytokine consisting of two distinct molecules which are both recognised by the interleukin-I receptor. It is produced by macrophages and a wide variety of cells when stimulated, and is thought to be the cytokine mainly responsible for inflammatory reactions induced by pathogenic organisms and the autoimmune processes

Inuit the people commonly known as Eskimos

Ischaemia an inadequate flow of blood to a tissue or part of a tissue caused by constriction or blockage of the vessel supplying it

Juxta-glomerular apparatus a collection of highly granular cells within the wall of the afferant renal arteriole feeding the capillaries of the glomerulus (see also Renin)

K_m a parameter describing the rate of an enzyme controlled reaction. It is equal to the substrate concentration at which the initial reaction velocity is half the maximum possible velocity ($\frac{1}{2} V_{max}$)

Leukocytes white blood cells; cells involved in the protection against foreign substances and in antibody production

Leukotrienes substances synthesised predominantly by granulocytes which mediate many of the inflammatory phenomena characteristic of immediate hypersensitivity reactions

Lipo- prefix meaning lipid or fat

Lipolysis the breakdown of a triglyceride or phospholipid into its constituent fatty acids and alcohol backbone

Liquid-crystalline state a property of certain long chain molecules which, as their temperature is raised, pass through a 'liquid-crystalline' state before becoming clear liquids at higher temperatures

Lymphocyte there are two forms of lymphocyte, T and B. T lymphocytes express surface receptors which recognise the unique composition and structure of antigens and then engage them. Cytotoxic T lymphocytes then kill cells bearing foreign antigens. B lymphocytes synthesise antibodies in response to signals from T lymphocytes which have encountered antigens

Metastasis the dissemination of malignant cells from the primary tumour to other parts of the body to form secondary growths

Microvilli finger-like protrusions of the enterocyte membrane, through which nutrients are absorbed from the intestinal lumen

Mitochondria organelles within cells which are the site of energy production

Mucosa the tissues lining the absorptive surfaces of the intestine, containing enterocytes, blood vessels and other specialised cells

Myocardial infarction the series of processes immediately following occlusion of a coronary artery, culminating in necrosis of the heart muscle in the zone supplied by the obstructed vessel and causing the heart to beat irregularly

Myocardium the muscle of the heart

Neoplasm new and abnormal growth; it can be either benign or malignant

Noradrenaline one of the catecholamines, and responsible for the humoral transmission of the impulse from sympathetic nerves to the target cell (see Adrenoceptor)

Penta- prefix meaning five

Phospholipid a fat-soluble substance containing phosphorus

Phosphoglyceride a particular type of phospholipid consisting of a glycerol backbone to which are attached two fatty acid side-chains and a phosphorylated nitrogenous base such as choline or ethanolamine. They are constituents of biological membranes

Photoreceptor the outer segments of the rods or cones of the retina containing layers of membrane lined with a light-sensitive compound whose decomposition on exposure to light triggers nerve signals to the brain

Placebo a dummy treatment with which the active treatment is compared

Plasma the fluid in which blood cells are suspended

Plasmalogen a phospholipid with a structure similar to lecithin and found in brain and muscle cells

Platelet-derived growth factor a polypeptide produced by platelets, but also synthesised by other cells including endothelial cells and activated macrophages. Its actions are directed primarily at connective tissue cells and smooth muscle cells

Polyarthritis inflammation of several or many joints causing swelling, redness, warmth, tenderness and loss of function

Polygenic a characteristic which is controlled by a number of genes, each of which has only a slight effect. The expression is the result of their combined interaction

Prevalence the number of cases existing either at a particular time or over a stated period, in a defined population

Prostaglandin one of a group of hormone-like substances produced from the long chain polyunsaturated fatty acids

Pruritis itching

Renin a peptide secreted by the cells of the juxtaglomerular apparatus of the kidney, which has enzymic properties catalysing the conversion of angiotensinogen to angiotensin I

Re-perfusion the re-establishment of blood flow after blood supply had been stopped

Re-stenosis re-narrowing of a blood vessel after it has been opened

Rheumatoid arthritis an inflammatory disease of the joints

Sarcolemma the sheath enveloping a muscle fibre

Sarcoplasmic reticulum the interfibrillar material of muscle cells

Sebaceous glands glands in the skin producing an oily substance called sebum. They open on to hair follicles

Serosal the side of the cell facing the blood

Serotonin also called 5-hydroxytryptamine. A compound found in the granules of the platelet and in the brain. It is a potent vasoconstrictor

Scavenger receptors receptors for oxidised LDL particles on macrophages. Oxidised LDL is not recognised by the normal LDL receptors on cells and is removed from the plasma by the scavenger receptors on macrophages

Serum the fluid that separates from clotted blood. It is essentially similar to plasma but lacks the substances involved in blood coagulation, e.g. fibrinogen

Smooth muscle cell a cell in the media of the arterial wall, responsible for both vascular tone and integrity

Systemic treatment taken internally, either by mouth or by injection

Tetra- prefix meaning four

Thrombin an enzyme which catalyses the conversion of fibrinogen to the insoluble fibrin polymer in clot or thrombus. Thrombin is also a platelet agonist and has actions on the vascular endothelium

Thromboxane compounds produced by platelets which cause vasoconstriction and platelet aggregation

Topical treatment treatment applied to the skin

Total parenteral nutrition a form of feeding in which solutions of nutrients are given directly into the blood supply

Transition metals metals, such as iron, which are capable of existing in two different ionised states. Iron can form the Fe^{2+} or the Fe^{3+} ion and can change from one to the other by the removal or acceptance of an electron

Tri- prefix meaning three

Triacylglycerol a glycerol molecule to which three fatty acid chains are attached. It forms the basic constituent of fat. Formerly known as triglycerides

Triene/tetraene ratio the ratio of 20:3 n-9 to 20:4 n-6 fatty acids. During essential fatty acid deficiency, the proportion of 20:3 n-9 increases. A value of 0.4 is taken to indicate the onset of EFA deficiency

Triglyceride see Triacylglycerol

Tumour promoter agent acting after a carcinogen and increasing tumour promotion

Vascular endothelium the monolayer of spe-cialised cells which lines the lumenal surface of the vessel and normally presents an anticoagulant surface to the circulating blood

Very-low density lipoproteins particles secreted into the plasma by the liver, containing a triacyl-glycerol-rich core surrounded by a hydrophilic shell consisting of a monolayer of phospholipid, non-esterified cholesterol and specific proteins (apolipoproteins)

Vinyl a double bond between two carbon atoms

V_{max} a parameter of an enzyme controlled reaction describing the maximum velocity achieved when the enzyme is saturated with reactant

REFERENCES

Abbey, M., Clifton, P., Belling, B. and Nestel, P. J. (1990) Effect of fish oil on lipoproteins, lecithin:cholesterol acyltransferase, and lipid transfer protein activity in humans. *Arteriosclerosis*, **10**, 85–94.

Abbot, W. G., Swinburn, B., Ruotolo, G., Hara, H., Patti, L., Harper, I., Grundy, S. M. and Howard, B. V. (1990) Effect of high carbohydrate low saturated fat diet on apoprotein B and triglyceride metabolism in Pima Indians. *J. Clin. Invest.*, **86**, 642–50.

Abraham, R., Riemersma, R. A., Wood, D., Elton, R. and Oliver, M. F. (1989) Adipose fatty acid composition and the risk of serious ventricular arrhythmias in acute myocardial infarction. *Am. J. Cardiol.*, **63**, 269–72.

Abraham, S. and Hillyard, L. A. (1983) Effect of dietary 18-carbon fatty acids on growth of transplantable mammary adenocarcinomas in mice. *J. Natl. Cancer Inst.*, **71**, 601–5.

Ackman, R. G. (1988) Some possible effects on lipid biochemistry of differences in the distribution on glycerol of long chain n-3 fatty acids in the food of marine fish and marine mammals. *Atherosclerosis*, **70**, 171–3.

Adams, L. M., Trout, J. R. and Karmali, R. A. (1990) Effect of n-3 fatty acids on spontaneous and experimental metastasis of rat mammary tumour 13762. *Br. J. Cancer*, **61**, 290–1.

Anonymous. (1990) Fish oils and diabetic microvascular disease. *Lancet*, **1**, 508–9.

Applegate, K. R. and Glomset, J. A. (1986) Computer-based modelling of the conformation and packing properties of docosahexaenoic acid. *J. Lipid Res.*, **27**, 658–80.

Arita, H., Nakano, T. and Hanasaki, K. (1989) Thromboxane A_2: its generation and role in platelet activation. *Prog. Lipid Res.*, **28**, 273–301.

Arntzenius, A. A., Kromhout, D., Barth, J. D. et al. (1985) Diet, lipoproteins, and the progression of coronary atherosclerosis. The Leiden Intervention Trial. *N. Engl. J. Med.*, **312**, 805–11.

Atkinson, P. M., Wheeler, M. C., Mendelsohn, D., Pienaar, N. and Chetty, N. (1987) Effects of a 4-week freshwater fish (trout) diet on platelet aggregation, platelet fatty acids, serum lipids, and coagulation factors. *Am. J. Hematol.*, **24**, 143–9.

Aveldano, M. I. (1987) A novel group of very long chain polyenoic fatty acids in dipolyunsaturated phosphatidylcholines from vertebrate retina. *J. Biol. Chem.*, **262**, 1172–9.

Aveldano, M. I. (1988) Phospholipid species containing long and very long polyenoic fatty acids remain with rhodopsin after hexane extraction of photoreceptor membranes. *Biochem.*, **27**, 1229–39.

Aylsworth, C. F., Welsch, C. W., Kabara, J. J. and Trosko, J. E. (1987) Effects of fatty acids on gap junctional communication: Possible role in tumour promotion by dietary fat. *Lipids*, **22**, 445–54.

Balch, C. M., Doughartz, P. A. and Tilden, A. B. (1982) Excessive prostaglandin E_2 production by suppressor monocytes in head and neck cancer patients. *Ann. Surg.*, **196**, 645–50.

Bamford, J. et al. (1985) Atopic eczema unresponsive to evening primrose oil/linoleic and gamma-linolenic acids. *J. Amer. Acad. Dermatol.*, **13**, 959.

Bandyopadhyay, G. K., Imagawa, W., Wallace, D. and Nandi, S. (1987) Linoleate metabolites enhance the *in vitro* proliferative response of mouse mammary epithelial cells to epidermal growth factor. *J. Biol. Chem.*, **262**, 2750–6.

Bang, H. O., Dyerberg, J. and Hjorne, N. (1976) The composition of food consumed by Greenland Eskimos. *Acta Med. Scand.*, **200**, 69–73.

Bang, H. D., Dyerberg, J., Sinclair, H. H. et al. (1985) Second International Congress on Essential Fatty Acids, Prostaglandins and Leukotrienes, London.

Barcelli, U., Glas-Greenwalt, P. and Pollak, V. E. (1985) Enhancing effect of dietary supplementation with w-3 fatty acids on plasma fibrinolysis in normal subjects. *Thromb. Res.*, **39**, 307–12.

Barker, D. J. P. and Osmond, C. (1986) Infant mortality, childhood nutrition, and ischaemic heart disease in England and Wales. *Lancet*, **i**, 1077–81.

Barker, D. J. P. and Osmond, C. (1987) Inequalities in Health in Britain: specific explanations in three Lancashire towns. *Br. Med. J.*, **294**, 749–52.

Barker, D. J. P., Winter, P. D., Osmond, C., Margetts, B. and Simmonds, S. J. (1989) Weight in infancy and death from heart ischaemic heart disease. *Lancet*, **ii**, 577–80.

Barker, D. J. P., Bull, A. R., Osmond, C. and Simmons, S. J. (1990) Fetal and placental size and risk of hypertension in adult life. *Br. Med. J.*, **301**, 259–62.

Barker, M. E., McClean, S. I., McKenna, P. G., Reid, N. G., Strain, J. J., Thompson, K. A., Williamson, A. P. and Wright, M. E. (1989) *Diet, Lifestyle and Health in Northern Ireland*. A report to the Health Promotion Research Trust. The University of Ulster, Coleraine, Northern Ireland.

Barradas, M. A., Christofides, J. A., Jeremy, J. Y., Mikhailides, D. P., Fry, D. E. and Dandona, P. (1990) The effect of olive oil supplementation on human platelet function, serum-cholesterol-related variables and plasma fibrinogen concentration: a pilot study. *Nutr. Res.*, **10**, 403–11.

Bates, D. (1989) Lipids and multiple sclerosis. *Biochem. Soc. Trans.*, **17**, 289–91.

Bates, D., Cartlidge, N. E. F., French, J. M. et al. (1989) A double blind controlled trial of long chain n-3 polyunsaturated fatty acids in the treatment of multiple sclerosis. *J. Neurol. Neurosurg. Psych.*, **52**, 18–22.

Bauer, K. A. and Rosenberg, R. D. (1987) The pathophysiology of the prethrombotic state in humans: insights gained from studies using markers of haemostatic system activation. *Blood*, **70**, 343–50.

Bazan, N. G. (1989) The supply of omega-3 polyunsaturated fatty acids to photoreceptors and synapses, in Dietary omega-3 and omega-6 Fatty Acids. Biological Effects and Nutritional Essentiality (eds C. Galli and A. P. Simopoulos), Nato ASI Series, Series A: *Life Sciences*, Vol. **171**, Plenum Press, New York and London. pp. 227–39.

Bazan, N. G. (1989) Lipid-derived metabolites as possible retina messengers: arachidonic acid, leukotrienes, eicosanoids and platelet activating factor, in *Extracellular and Intracellular Messengers in the Vertebrate Retina*, Alan. R. Liss Inc., pp. 269–300.

Beardshall, K., Frost, G., Morarji, Y., Domin, J., Bloom, S. R. and Calam, J. (1989) Saturation of fat and cholesytokinin release implications for pancreatic carcinogenesis. *Lancet*, **ii**, 1008–10.

Becker, W. (1989) Comparative uptake in rats and man of omega-3 and omega-6 fatty acids, in Dietary omega-3 and omega-6 fatty acids. Biological Effects and Nutritional Essentiality (eds C. Galli and A. P. Simopoulos) Nato ASI Series. Series A: *Life Sciences*, Vol. **171**, Plenum Press, New York and London, pp. 111–22.

Begin, M. E., Das, U. N., Ells, G. and Horrobin, D. F. (1985a) Selective killing of human cancer cells by polyunsaturated fatty acids. *Prostaglandins Leukotrienes and Medicine*, **19**, 177–86.

Begin, M. E., Ells, G. and Das, U. N. (1985b) Selected fatty acids as possible intermediates for selective cytotoxic activity of anticancer agents involving oxygen radicals. *Anticancer Res.*, **6**, 291–6.

Begin, M. E., Ells, G. and Horrobin, D. F. (1988) Polyunsaturated fatty acid-induced cytotoxicity against tumour cells and its relationship to lipid peroxidation. *J. Natl. Cancer Inst.*, **80**, 188–94.

Belch, J. (1990) Fish oil and rheumatoid arthritis: Does a herring a day keep rheumatologists away? *Ann. Rheum. Dis.*, **49**, 71–2.

Belch, J. J. F., Ansell, D., Madhok, R., O'Dowd, A. and Sturrock, R. D. (1988) Effects of altering dietary fatty acids on requirements for non-steroidal anti-inflammatory drugs in patients with rheumatoid arthritis: a double-blind placebo controlled study. *Ann. Rheum. Dis.*, **47**, 96–104.

Berg Schmidt, E., Ernst, E., Varming, K., Pedersen, J. O. and Dyerberg, J. (1989) The effect of n-3 fatty acids on lipids and haemostasis in patients with type IIa and type IV hyperlipidaemia. *Thromb. Haemostas.*, **62**, 797–801.

Berg Schmidt, E., Varming, K., Ernst, E., Madsen, P. and Dyerberg, J. (1990) Dose-response studies on the effect of n-3 polyunsaturated fatty acids on lipids and haemostasis. *Thromb. Haemostas.*, **63**, 1–5.

Bergan, J. G and Draper, H. H. (1970) Absorption and metabolism of 1–^{14}C-methyl linoleate hydroperoxide. *Lipids*, **5**, 976–82.

Berns, M. A. M., DeVries, J. H. M. and Katan, M. B. (1990) Dietary and other determinants of lipoprotein levels within a population of 315 Dutch males aged 28–29. *Eur. J. Clin. Nutr.*, **44**, 535–44.

Berry, E. M., Hirsch, J., Most, J., McNamara, D. J. and Thornton, J. (1986) The relationship of dietary fat to plasma lipid levels as studied by factor analysis of adipose tissue composition in a free-living population of middle-aged American men. *Am. J. Clin. Nutr.*, **44**, 220–31.

Berry, E. M., Eisenberg, S., Harutz, D., Friedlander, Y., Norman, Y., Kaufmann, N. A. and Stein, Y. (1991) Effects of diets rich in monounsaturated fatty acids on plasma lipoproteins – The Jerusalem Nutrition Study: high MUFAs vs high PUFAs. *Am. J. Clin. Nutr.*, **53**, 899–907.

Beswick, A. D., Fehily, A. M., Sharp, D. S., Renaud, S. and Giddings, J. (1991) Long-term dietary modification and platelet activity. *J. Intern. Med.*, **229**, 511–5.

Bevilacqua, M. P., Pober, J. S., Wheeler, M. E., Cotran, R. S. and Gimbrone, M. A. (1985) Interleukin-1 activation of vascular endothelium. Effects on procoagulant activity and leukocyte adhesion. *Am. J. Pathol.*, **121**, 394–403.

Biagi, P. L. *et al.* (1988) A long-term study in the use of evening primrose oil (Efamol) in atopic children. *Drugs Exptl Clin. Res.*, **14**, 285–90.

Bidard, J-N., Rossi, B., Renaud, J-F. and Lazdunski, M. (1984) A search for an 'ouabain-like' substance from the electric organ of Electrophorus electricus which led to arachidonic acid and related fatty acids. *Biochim. Biophys. Acta*, **769**, 245–52.

Bilheimer, D. W., Ho, Y. K., Brown, M. S *et al.* (1978) Genetics of LDL receptor. Diminished receptor activity in lymphocytes in heterozygotes with familial hypercholesterolaemia. *J. Clin. Invest.*, **61**, 678–96.

Bitman, J., Wood, L., Hamosh, M. *et al.* (1983) Comparison of the lipid composition of breast milk from mothers of term and preterm infants. *Am. J. Clin. Nutr.*, **38**, 300–12.

Bjerve, K. S., Mostad, I. L. and Thoresen, L. (1987) Alpha linolenic acid deficiency in patients on long-term gastric-tube feeding: estimation of linolenic acid and long-chain unsaturated n-3 fatty acid requirement in man. *Am. J. Clin. Nutr.*, **45**, 66–77.

Black, A. K. *et al.* (1990) Pharmacological and clinical effects of Lonapalene, a 5-lipoxygenase inhibitor, in psoriasis. *J. Invest. Dermatol.*, **95**, 50–4.

Blackwell, G. J., Duncombe, W. G., Flower, R. J., Parsons, M. F. and Vane, J. R. (1977) The distribution and metabolism of arachidonic acid in rabbit platelets during aggregation and its modification by drugs. *Br. J. Pharmacol.*, **59**, 353–66.

Blomstrand, R. and Svensson, L. (1974) Studies on phospholipids with particular reference to cardiolipin of rat heart after feeding rapeseed oil. *Lipids*, **9**, 771–80.

Blomstrand, R. and Svensson, L. (1983) The effects of partially hydrogenated marine oils on the mitochondrial function and membrane phospholipid fatty acids in rat heart. *Lipids*, **18**, 151–70.

Bloom, A. L. and Evans, E. P. (1969) Plasma-fibrinogen and the aggregation of platelets by adenosine diphosphate. *Lancet*, **i**, 349–50.

Boberg, M., Vessby, B. and Slinus, I. (1986) Effects of dietary supplementation with n-6 and n-3 long-chain polyunsaturated fatty acids on serum lipoproteins and platelet function in hypertriglyceridaemic patients. *Acta. Med. Scand.*, **220**, 153–60.

Bonaa, K. H., Bjerve, K. S., Straume, B., Gram, I. T. and Thelle, D. (1990) Effect of eicosapentaenoic acid and docosahexaenoic acid on blood pressure in hypertension. A population-based intervention trial from the Tromso Study. *N. Eng. J. Med.*, **322**, 795–801.

Bonanome, A. and Grundy, S. M. (1989) Intestinal absorption of stearic acid after consumption of high fat meals in humans. *J. Nutr.*, **119**, 1556–60.

Bordoni, A. *et al.* (1988) Evening primrose oil (efamol) in the treatment of children with atopic eczema. *Drugs Exptl Clin. Res.*, **14**, 291–7.

Borgeson, C. E., Pardini, L., Pardini, R. S. and Reitz, R. C. (1989) Effects of dietary fish oil on human mammary carcinoma and on lipid-metabolizing enzymes. *Lipids*, **24**, 290–5.

Borgstrom, B. (1980) Importance of phospholipids, pancreatic phospholipase A and fatty acids for the digestion of dietary fat. *Gastroenterol.*, **78**, 954–62.

Born, G. V. R. (1962) Aggregation of blood platelets by adenosine diphosphate and its reversal. *Nature*, **194**, 927–9.

Borschel, M. W., Elkin, R. G., Kirksey, A. *et al.* (1986) Fatty acid composition of mature human milk of Egyptian and American women. *Am. J. Clin. Nutr.*, **44**, 330–5.

Bottino, N. R., Vandenburg, G. A. and Reiser, R. (1967) Resistance of certain long-chain polyunsaturated fatty acids of marine oils to pancreatic lipase hydrolysis. *Lipids*, **2**, 489–93.

Boustani, S. El., Colette, C., Monnier, L., Descamps, B., Crastes de Paulet, A. and Mendy, F. (1987) Enteral absorption in man of eicosapentaenoic acid in different chemical forms. *Lipids*, **22**, 711–4.

Boyer, J. L., Hepler, J. R. and Harden, T. K. (1989) Hormone and growth factor receptor-mediated regulation of phospholipase C. *Trends Pharmacol. Sci.*, **10**, 360–4.

Braden, L. M. and Carroll, K. K. (1986) Dietary polyunsaturated fat in relation to mammary carcinogenesis in rats. *Lipids*, **21**, 285–8.

Brain, S. *et al.* (1984) Release of leukotriene B$_4$-like material in biologically active amounts from the lesional skin of patients with psoriasis. *J. Invest. Dermatol.*, **83**, 70–3.

Brain, S. D. *et al.* (1985) Leukotrienes C$_4$ and D$_4$ in psoriatic skin lesions. *Prostaglandins*, **29**, 611–9.

Brasitus, T. A., Davidson, N. O. and Schachter, D. (1985) Variations in dietary triacylglycerol saturation alter the lipid composition and fluidity of rat intestinal plasma membranes. *Biochim. Biophys. Acta*, **812**, 460–72.

Brenner, R. R. (1989) Factors influencing fatty acid chain elongation and desaturation, in The Role of Fats in Human Nutrition. 2nd edn (eds A. J. Vergroessen and M. Crawford), Academic Press. London, pp. 45–79.

Brenner, R. (1990) Endocrine control of fatty acid desaturation. *Biochem. Soc. Trans.*, **18**, 774.

Brewer, H. B., Grey, R. E., Hoey, J. M. and Fujo, S. S. (1988) Apolipoproteins and lipoproteins in human plasma: an overview. *Clin. Chem.*, **34**, B4–8.

British Nutrition Foundation (1987) *Trans Fatty Acids. Report of the British Nutrition Foundation's Task Force*, The British Nutrition Foundation, London.

British Nutrition Foundation (1988) Are there dietary causes in the causation of cancer? Briefing Paper No. 14.

British Nutrition Foundation (1991) Antoxidant nutrients in health and disease. Briefing Paper No.25.

British Nutrition Foundation (1992) Coronary Heart Disease II – The background factors Briefing Paper. In press.

Brockerhoff, H., Hoyle, R. J. and Huang, P. G. (1966) Positional distribution of fatty acids in the fats of a polar bear and a seal. *Can. J. Biochem.*, **44**, 1519–25.

Broekman, M. J., Handin, R. I., Derksen, A. and Cohen, P. (1976) Distribution of phospholipids, fatty acids, and platelet factor 3 activity among subcellular fractions of human platelets. *Blood*, **47**, 963–71.

Brostoff, J. and Challacombe, S. J. (1987) *Food Allergy and Intolerance*, Baillire Tindall, London.

Brown, G., Albeus, J. J., Fisher, L. D., Schaefer, S. M.,

Lin, J. T., Kaplan, C., Zhao, X. Q., Bisson, B. D., Fitzpatrick, U. F. and Dodge, H. T. (1990) Regression of coronary-artery disease as a result of intensive lipid lowering therapy in men with high levels of apoprotein B. *N. Eng. J. Med.*, **323**, 1289–98.

Brown, A. J. and Roberts, D. C. K. (1991) Moderate fish oil intake improves lipemic response to a standard fat meal. A study in 25 healthy men. *Athero. Thromb.*, **11**, 457–66.

Brox, J. H., Killie, J-E., Gunnes, S. and Nordoy, A. (1981) The effect of cod-liver oil and corn oil on platelets and vessel wall in man. *Throm. Haemostas.*, **46**, 604–11.

Brox, J. H., Killie, J-E., Osterud, B., Holme, S. and Nordoy, A. (1983) Effects of cod-liver oil on platelets and coagulation in familial hypercholesterolaemia (type IIa). *Acta Med. Scand.*, **213**, 137–44.

Bruckdorfer, K. R. (1990) Free radicals, lipid peroxidation and atherosclerosis. *Curr. Op. Lipidol.*, **1**, 529–35.

Brussaard, J. H. and Hautvast, J. G. A. J. (1983) Dietary goals in Europe, in *Dietary Guidelines in Practice*. British Nutrition Foundation, London.

Buchanan, H. M., Preston, S. J., Brooks, P. M. and Buchanan, W. W. (1991) Is diet important in rheumatoid arthritis? *Brit. J. Rheumatol.*, **30**, 125–34.

Buckman, D. K., Chapkin, R. S. and Erickson, K. L. (1990) Modulation of mouse mammary tumour growth and linoleate enhanced metastasis by oleate. *J. Nutr.*, **120**, 148–57.

Bull, A. W., Soullier, B. K., Wilson, P. S., Hayden, M. T. and Nigro, N. D. (1981) Promotion of azoxymethane-induced intestinal cancer by high-fat diets in rats. *Cancer Res.*, **41**, 3700–5.

Bull, H. *et al.* (1990) Proinflammatory mediators induce sustained release of prostaglandin E$_2$ from human dermal microvascular endothelial cells. *Brit. J. Dermatol.*, **22**, 153–64.

Bunting, S., Moncada, S. and Vane, J. R. (1976) The effects of prostaglandin endoperoxides and thromboxane A$_2$ on strips of rabbit coeliac artery and certain other smooth muscle preparations. *Br. J. Pharmacol.*, **57**, 462–63P.

Burr, G. O. and Burr, M. M. (1929) A new deficiency disease produced by rigid exclusion of fat from the diet. *J. Biol. Chem.*, **82**, 345–67.

Burr, G. O. and Burr, M. M. (1930) On the nature and role of the fatty acids essential in nutrition. *J. Biol. Chem.*, **86**, 587.

Burr, M. L. (1989) Fish and the cardiovascular system. *Prog. Food Nutr. Sci.*, **13**, 291–316.

Burr, M. L., Bates, C. J., Fehily, A. M. and St. Leger, A. S. (1981) Plasma cholesterol and blood pressure in vegetarians. *J. Human Nutr.*, **35**, 437–41.

Burr, M. L. and Butland, B. K. (1988) Heart disease in British vegetarians. *Am. J. Clin Nutr.*, **48**, 830–2.

Burr, M. L., Fehily, A. M., Gilbert, J. F., Rogers, S., Holliday, R. M., Sweetnam, P. M., Elwood, P. C. and Deadman, N. M. (1989) Effects of changes in fat, fish and fibre intakes on death and myocardial reinfarction: diet and reinfarction trial (DART). *Lancet*, **ii**, 757–61.

Cahill, P. D., Sarris, G. E., Cooper A. *et al.* (1988) Inhibition of vein graft intimal thickening by eicosapentaenoic acid: reduced thromboxane production without change in lipoprotein levels or low-density lipoprotein receptor density. *J. Vas. Surg.*, **7**, 108–18.

Camp, R. D. R. *et al.* (1983) The identification of hydroxy fatty acids in psoriatic skin. *Prostaglandins*, **26**, 431.

Canessa, M., Adragna, N., Soloman, H. S., Connolly, T. M. and Tosteson, D. C. (1980) Increased sodium-lithium countertransport in red cells from patients with essential hypertension. *N. Eng. J. Med.*, **302**, 772–6.

Cardinal, D. C. and Flower, R. J. (1980) The electronic aggregometer: a novel device for assessing platelet behaviour in blood. *J. Pharmacol. Methods*, **3**, 135–58.

Carey, M. C., Small, D. M. and Bliss, C. M. (1983) Lipid digestion and absorption. *Ann. Rev. Physiol.*, **45**, 651–77.

Carlson, S. E., Rhodes, P. G. and Ferguson, M. G. (1986) Docosahexaenoic acid status of preterm infants at birth and following feeding with human milk formula. *Am. J. Clin. Nutr.*, **44**, 798–804.

Carpenter, K. L. H., Ballantine, J. A., Fussell, B., Enright, J. H. and Mitcheson, A. (1990) Oxidation of cholesteryl linoleate by human monocyte-macrophage *in vitro*. *Atherosclerosis*, **83**, 217–29.

Carroll, K. K. (1980) Lipids and carcinogenesis. *J. Environ. Pathol. Toxicol.*, **3**, 253–71.

Carroll, K. K., Braden, L. M., Bell, J. A. and Kalamegham, R. (1986) Fat and cancer. *Cancer*, **58**, 1818–25.

Carruthers, A. and Melchior, D. L. (1986) How bilayer lipids affect membrane protein activity. *Trends Biochem. Sci.*, **11**, 331–5.

Carson, S. D. and Johnson, D. R. (1990) Consecutive enzyme cascades: complement activation at the cell surface triggers increased tissue factor activity. *Blood*, **76**, 361–7.

Chambaz, J., Pepin, D., Robert, A., Wolf, C. and Bereziat, G. (1983) Protein-stimulated enrichment of human platelet membranes in linoleyl-phosphatidyl-cholines. Effect upon adenylate cyclases and fluidity. *Biochim. Biophys. Acta*, **727**, 313–26.

Chandra, R. K. (1976) Nutrition as a critical determinant in susceptibility to infection. *Wld. Rev. Nutr. Diet.*, **25**, 166–88.

Channing Rogers, R. P. and Levin, J. (1990) A critical reappraisal of the bleeding time. *Sem. Thromb. Haemostas.*, **16**, 1–20.

Charo, I. F., Feinman, R. D. and Detwiler, T. C. (1977) Inter-relationships of platelet aggregation and secretion. *J. Clin. Invest.*, **60**, 866–73.

Chee, K. M., Gong, J. X., Rees, D. M. G., Meydani, M., Ausman, L., Johnson, J., Siguele, E. N. and Schaefer, E. J. (1990) Fatty acids content of marine capsules. *Lipids*, **25**, 523–8.

Chen, I. S., Subramaniam, S., Cassidy, M. M., Sheppard, A. J. and Vahouny, G. V. (1985) Intestinal absorption and lipoprotein transport of n-3 eicosapentaenoic acid. *J. Nutr.*, **115**, 219–25.

Chen, I. S., Hotta, S. S., Ikeda, I., Cassidy, M. M., Sheppard, A. J. and Vahouny, G. V. (1987) Digestion, absorption and effects on cholesterol absorption of menhaden oil, fish oil concentrate and corn oil by rats. *J. Nutr.*, **117**, 1676–80.

Cheng, A., Bustami, M., Norells, M. S., Mitchell, A. G. and Ilsey, C. D. J. (1990) The effect of omega-3 fatty acids on restenosis after coronary angioplasty. *Eur. Heart J.*, **11** (Suppl), 368.

Chiang, B. N., Perlman, L. V., Fulton, M., Ostrander, L. D. and Epstein, F. H. (1970) Predisposing factors to sudden cardiac death in Tecumseh, Michegan: a prospective study. *Circulation*, **41**, 31–7.

Chow, S. C., Sisfontes, L., Bjorkhem, I. and Jondal, M. (1989) Suppression of growth in a leukemic T cell line by n-3 and n-6 polyunsaturated fatty acids. *Lipids*, **24**, 700–4.

Chow, S. L. and Hollander, D. (1978) Arachidonic acid intestinal absorption: mechanism of transport and influence of luminal factors on absorption. *Lipids*, **13**, 768–76.

Chow, S. L. and Hollander, D. (1979) Linoleic acid absorption in the unanesthetized rat: mechanism of transport and influence of luminal factors on absorption. *Lipids*, **14**, 378–85.

Clandinin, M. T., Chappel, J. E., Leong, S., Heim, T., Swyer, P. R. and Chance, G. W. (1980) Intrauterine fatty acid accretion rates in human brain: implications for fatty acid requirements. *Ear. Hum. Dev.*, **4**, 121–9.

Clandinin, M. T., Chappell, J. E. and Heim, T. (1981) Do low weight infants require nutrition with chain elongation-desaturation products of essential fatty acids? *Prog. Lipid Res.*, **20**, 901–4.

Clandinin, M. T. and Chappell, J. E. (1987) Infant formulae. United States Patent no. 4670285.

Clark, R. M., Ferris, A. M., Brown, P. B. *et al.* (1982) Changes in the lipids of human milk from 2 to 16 weeks postpartam. *J. Pediatr. Gastroenterol. Nutr.*, **1**, 311–5.

Clark, W. F., Parbtani, A., Huff, M. W., Reid, B., Holub, B. J. and Falardeau, P. (1989) Omega-3 fatty acid dietary supplementation in systemic lupus erythematosus. *Kidney Int.*, **36**, 653–60.

Cleland, L. G., French, J. K., Betts, W. H., Murphy, G. A. and Elliott, M. J. (1988) Clinical and biochemical effects of dietary fish oil supplements in rheumatoid arthritis. *J. Rheumatol.*, **15**, 1471–5.

Cobiac, L., Nestel, P., Wing, L. M. H. and Howe, P. R. C. (1991) The effects of dietary sodium restriction and fish oil supplements on blood pressure in the elderly. *Clin. Exp. Pharmacol. Physiol.*, **18**, 265–8.

Collins, R., Peto, R., Mac Mahon, S., Hebert, P., Fiebach, N. H., Eberlein, K. A., Godwin, J., Qizilbash, N., Taylor, J. O. and Hennekens, C. H. (1990) Blood pressure, stroke, and coronary heart disease. Part 2, short term reductions in blood pressure: overview of randomised drug trials in their epidemiological context. *Lancet*, **335**, 827–38.

Colman, R. W. (1990) Aggregin: a platelet ADP receptor that mediates activation. *FASEB J.*, **4**, 1425–35.

Committee on Diet, Nutrition and Cancer, Lipids (Fats and Cholesterol) (1982) in *Diet, Nutrition and Cancer*, National Academy Press, Washington DC, pp. 73–105.

Connor, W. E. (1962) The acceleration of thrombus formation by certain fatty acids. *J. Clin. Invest.*, **41**, 1199–205.

Consumers' Association. (1985) *Consumer attitudes to and understanding of nutrition labelling*, British Market Research Bureau Ltd, London.

Cortese, C., Levy, Y., Janus, E. D., Turner, P. R., Rao, S. N., Miller, N. E. and Lewis, B. (1983) Modes of action of lipid-lowering diets in man: studies of apolipoprotein B kinetics in relation to fat. *Eur. J. Clin. Invest.*, **13**, 79–85.

Cottrell, R. C. (1991) Introduction, nutritional aspects of palm oil. *Am. J. Clin. Nutr.*, **53**, 989S–1009S.

Crawford, M. A. and Sinclair, A. J. (1972) Nutritional influences on the evolution of the mammalian brain, in *Lipids, Malnutrition and the Developing Brain*, Ciba Foundation Symposium (eds K. Elliott and J. Knights), Elsevier, Amsterdam, pp. 267–87.

Crawford, M. A., Hassam, A. G., Williams, G. and Whitehouse, W. L. (1976) Essential fatty acids and fetal brain growth. *Lancet*, **i**, 452–3.

Crawford, M. A., Hassam, A. G. and Rivers, J. P. W. (1978) Essential fatty acid requirements in infancy. *Am. J. Clin. Nutr.*, **31**, 2181–5.

Crawford, M. A., Doyle, W. Craft, I. L. and Laurance, B. M. (1986) A comparison of food intakes during pregnancy and birthweight in high and low socioeconomic groups. *Prog. Lipid Res.*, **25**, 249–54.

Crawford, M. A., Doyle, W., Drury, P., Lennon, A., Costeloe, K. and Leighfield, M. (1989) n-6 and n-3 fatty acids during early human development. *J. Int. Med.*, **225**, Suppl. 1, 159–69.

Crawford, M. A., Costeloe, K., Doyle, W., Leighfield, M. J., Lennen, E. A. and Meadows, N. (1990) Potential diagnostic value of the umbilical artery as a definition of neural fatty acid status of the fetus during its growth. *Biochem. Soc. Trans.*, **18**, 761–6.

Crook, S. T., Mattern, M., Sarau, H. M., Winkler, J. D., Balcarek, J., Wong, A. and Bennett, C. F. (1989) The signal transduction system of the leukotriene D_4 receptor. *Trends Pharmacol. Sci.*, **10**, 103–7.

Cunnane, S. C. (1989) Differential retention of long chain fatty acids in liver triglyceride during fasting in the rat: comparison of feeding n-6 and n-3 oils, in *Health Effects of Fish and Fish Oils* (ed. R. K. Chandra), ARTS Biomedical Publishers and Distributors Ltd, St John's, Newfoundland, Canada, pp. 127–141.

Curb, J. D. and Reed, D. M. (1985) Fish consumption and mortality from coronary heart disease. *N. Engl. J. Med.*, **313**, 820–1.

Cutler, M. G. and Hayward, M. A. (1974) Effect of lipid peroxides on fat absorption and folic acid status in the rat. *Nutr. Metab.*, **16**, 87–93.

Dart, A. M., Riemersma, R. A. and Oliver, M. F. (1989) Effects of MaxEPA on serum lipids in hypercholesterolaemic subjects. *Atherosclerosis*, **80**, 119–24.

Das, U. N., Begin, M. E., Ells, G., Huang, Y. S. and Horrobin, D. F. (1987) Polyunsaturated fatty acids augment free radical generation in tumour cells in vitro. *Biochem. Biophys. Res. Commun.*, **145**, 15–24.

Davies, H. R., Bridenshire, R. T., Vesselinovitch, D. and Wissler, R. W. (1987) Fish oil inhibits development of atherosclerosis in rhesus monkeys. *Arteriosclerosis*, **7**, 441–8.

Davies, P. A and Stewart, A. L. (1975) Low birth weight infants: neurological sequelae and later intelligence. *Br. Med. Bull.*, **31**, 85–91.

Dayton, S., Pearce, M. L., Hashimoto, S., Dixon, W. J. and Tomiyasu, U. (1969) A controlled clinical trial of a diet high in unsaturated fat in preventing complications of atherosclerotic complications. *Circulation*, **60**, (suppl. II), 1–63.

Dayton, S., Hashimoto, S. and Wollman, J. (1977) Effect of high oleic and high linoleic safflower oils on mammary tumours induced in rats by 7, 12-dimethylbenz (alpha) anthracene. *J. Nutr.*, **107**, 1353–60.

Dehmer, G. J., Popman, J. J., Van den Berg, E. K. *et al.* (1988) Reduction in the rate of early restenosis after coronary angioplasty by a diet supplemented with n-3 fatty acids. *N. Engl. J. Med.*, **319**, 733–40.

Denman, A. M. (1987) Allergy and joint complaints, in *Allergy: An International Textbook* (eds M. H. Lessof, T. H. Lee, D. M. Kemeny), John Wiley, Chichester, pp. 565–76.

Department of Health and Social Security (1980) Diet and cardiovascular disease. *Report on health and social subjects No.18.* Committee on Medical Aspects of Food Policy, HMSO, London.

Department of Health and Social Security (1984) Diet and Cardiovascular Disease. Report of the Panel on diet in relation to cardiovascular disease. *Report on Health and Social Subjects 28.* HMSO, London.

Department of Health and Social Security. (1989) Present day practice in infant feeding: third report. *Report on Health and Social Subjects 32.* HMSO, London.

Department of Health (1991) Dietary Reference Values for Food Energy and Nutrients for the United Kingdom. Committee on Medical Aspects of Food Policy. *Report on Health and Social Subjects 41.* HMSO, London.

Despres, J. P. (1991) Obesity and lipid metabolism: relevance of body fat distribution. *Curr. Op. Lipidol.*, **2**, 5–15.

Di Minno, G., Silver, M. J. and Murphy, S. (1982) Stored human platelets retain full aggregation potential in response to pairs of aggregating agents. *Blood*, **59**, 563–8.

Di Minno, G., Silver, M. J. and Murphy, S. (1983) Monitoring the entry of new platelets into the circulation after ingestion of aspirin. *Blood*, **61**, 1081–5.

Dippenaar, N., Booyens, J., Fabbri, D. and Katzeff, I. E. (1982) The reversibility of cancer: evidence that malignancy in melanoma cells is gamma-linoleic acid deficiency-dependent. *S. Afr. Med. J.*, **62**, 505–9.

Dobbing, J. (1972) Vulnerable periods of brain development, in *Lipids, Malnutrition and the Developing Brain*, Ciba Foundation Symposium (eds K. Elliott and J. Knights), Elsevier, Amsterdam, pp. 1–7.

Dolecek, T. A., Grandits, G., Caggiula, A. W. and Tillotson, J. (1989) *Dietary omega-3 fatty acids and mortality in the multiple risk factor intervention trial.* Proc. Second World Congress Prev. Cardiol., USA, Abstract No. 255.

Downing, D. T. (1976) Mammalian waxes, in *Chemistry and Biochemistry of Natural Waxes* (ed P. E. Kolattukudy), Elsevier, New York, pp. 17–48.

Downing, D. T. *et al.* (1986) Essential fatty acids and acne. *J. Amer. Acad. Dermatol.*, **14**, 221–5.

Doyle, W. Crawford, M. A., Laurance, B. M. and Drury, P. (1982) Dietary survey during pregnancy in a low socioeconomic group. *Hum. Nutr. App. Nutr.*, **36A**, 95–106.

Doyle, W., Crawford, M. A., Wynn, A. H. and Wynn. S. W. (1989) Maternal nutrient intake and birthweight. *J. Hum. Nutr. Diet.*, **62**, 407–14.

Doyle, W., Crawford, M. A., Wynn, A. H. and Wynn, S. W. (1990) The association between maternal diet and birth dimensions. *J. Nutr. Med.*, **61**, 9–17.

Dratz, E. A. and Deese, A. J. (1986) The role of docosahexaenoic acid (22:6 n-3) in biological membranes: examples from photoreceptors and model membrane bilayers, in *Health Effects of Polyunsaturated Fatty Acids in Seafoods* (eds A. P. Simopoulos, R. R. Kifer and R. E. Martin), Academic Press, New York, pp. 319–51.

Dreon, D. M., Vranizan, K. M., Krauss, R. M., Austin, M. A. and Wood, P. D. (1990) The effects of polyunsaturated fat vs monounsaturated fat on plasma lipoproteins. *N. Engl. J. Med.*, **323**, 439–45.

Driss, F., Vericel, E., Lagarde, M., Dechavanne, M. and Darcet, P. (1984) Inhibition of platelet aggregation and thromboxane synthesis after intake of small amounts of eicosapentaenoic acid. *Thromb. Res.*, **36**, 389–96.

Dunn, H. G. (1986) *Sequelae of Low Birth Weight: The Vancouver Study*, Blackwell Scientific Publications, Oxford.

Dunn, J. T., Silverberg, M. and Kaplan, A. P. (1982) The cleavage and formation of activated human Hageman factor by autodigestion and by kallikrein. *J. Biol. Chem.*, **257**, 1779–84.

Dworkin, R. H., Bates, D., Millar, J. H. D. and Paty, D. W. (1984) Linoleic acid and multiple sclerosis: a reanalysis of three double blind trials. *Neurol.*, **34**, 1441–5.

Dyerberg, J., Bang, H. O., Stoffersen, E., Moncada, S. and Vane, J. R. (1978) Eicosapentaenoic acid and prevention of thrombosis and atherosclerosis. *Lancet*, ii, 117–9.

Dyerberg, J. and Bang, H. O. (1979) Haemostatic function and platelet polyunsaturated fatty acids in Eskimos. *Lancet*, ii, 433–5.

Ederer, F., Leren, P., Turpeiren, O. and Frandt, I. D. (1971) Cancer among men on cholesterol-lowering diets. Experience from five clinical trials. *Lancet*, i, 203–5.

Edwards-Webb, J. D. and Gurr, M. I. (1988) The influence of dietary fats on the chemical composition and physical properties of biological membranes. *Nutr. Res.*, 8, 1297–1305.

Elias, P. M. *et al.* (1980) The permeability barrier in essential fatty acid deficiency; evidence of a direct role for linoleic acid in barrier function. *J. Invest. Dermatol.*, 74, 230–3.

Elwood, P. C., Renaud, S., Sharp, D. S., Beswick, A. D., O'Brien, J. R. and Yarnell, J. W. G. (1991) Ischaemic heart disease and platelet aggregation: The Caerphilly Collaborative Heart Disease study. *Circulation*, 83, 38–44.

Emeis, J. J., van Houwelingen, A. C., van den Hoogen, C. M. and Hornstra, G. (1989) A moderate fish intake increases plasminogen activator inhibitor type-1 in human volunteers. *Blood*, 74, 233–7.

Endres, S., Kelley, V. E. and Dinarello, C. A. (1987) Effects of omega-3 fatty acids on the *in vitro* production of human interleukin-1. *J. Leukocyte Biol.*, 42, 617.

Endres, S., Ghorbani, R., Kelley, V. E. *et al.* (1989) The effect of dietary supplementation with n-3 polyunsaturated fatty acids on the synthesis of interleukin-1 and tumor necrosis factor by mononuclear cells. *N. Engl. J. Med.*, 320, 265–71.

Enig, M. G., Munna, R. J. and Keeney, M. (1978) Dietary fat and cancer trends – a critique. *Fed. Proc.*, 37, 2215–8.

Erickson, K. L., Schlanger, D. S., Adams, D. A., Fregeau, D. R. and Stern, J. S. (1984) Influence of dietary fatty acid concentration and geometric configuration on murine mammary tumorigenesis and experimental metastasis. *J. Nutr.*, 114, 1834–42.

Esterbauer, H., Dieber-Rotheneder, M., Waeg, G., Striegi, G., Ashby, A., Sattler, W. and Jurgens, G. (1990) Biochemical, structural and functional properties of oxidized low-density lipoproteins. *Chem. Res. Toxicol.*, 3, 77–92.

Ettinger, W. H., Dysko, R. C. and Clarkson, T. B. (1989) Prednisolone increases low density lipoprotein in Cynomogus monkeys fed saturated fat and cholesterol. *Arteriosclerosis*, 9, 848–55.

European Atherosclerosis Society (1987) Strategies for the prevention of coronary heart disease: a policy statement of the European Atherosclerosis Society. *Eur. Heart J.*, 8, 77–88.

European Community (1990) Nutrition Labelling for foodstuffs (90/496/EEC), *Off. J. Eur. Comm.*, 6/10/90.

Evans, R. W. and Tinoco, J. (1978) Monolayers of sterols and phosphatidylcholines containing a 20-carbon chain. *Chem. Phys. Lipids*, 22, 207–20.

Fady, C., Reisser, R., Lagadec, P., Pelletier, H., Olsson, N. O. and Jeannin, J. F. (1988) *In vivo* and *in vitro* effects of fish containing diets on colon tumour cells in rats. *Anticancer Res.*, 8, 225–8.

FAO/WHO (1978) *Report of an Expert Consultation, The Role of Dietary Fats and Oils in Human Nutrition*, FAO, Rome.

Fehily, A. M., Phillips, K. M. and Sweetnam, P. M. (1984) A weighed dietary survey of men in Caerphilly, South Wales. *Hum. Nut. Appl. Nutr.*, 38A, 270–6.

Fehily, A. M., Yarnell, J. W. G. and Butland, B. K. (1987) Diet and ischaemic heart disease in the Caerphilly study. *Hum. Nutr. Appl. Nutr.*, 41A, 319–26.

Fehily, A. M., Yarnell, J. W. G., Bolton, C. H. and Butland, B. K. (1988) Dietary determinants of plasma lipids and lipoproteins: the Caerphilly Study. *Eur. J. Clin. Nutr.*, 42, 405–13.

Fehily, A. M., Barker, M. E., Thomson, M., Yarnell, J. W. G., Holliday, R. M., Thompson, K. A., Elton, R., Bainton, D. and Baker, I. A. (1990) The diets of men in four areas of the UK: The Caerphilly, Northern Ireland, Edinburgh and Speedwell studies. *Eur. J. Clin. Nutr.*, 44, 813–7.

Feldman, E. B., Russell, B. S., Schnare, F. H., Miles, B. C., Doyle, E. A. and Morettio-Rojas, I. (1979) Effects of tristearin, triolein and safflower oil diets on cholesterol balance in rats. *J. Nutr.*, 109, 2226–36.

Field, C. J. and Clandinin, M. T. (1984) Modulation of adipose tissue fat composition by diet: a review. *Nutr. Res.*, 4, 743–55.

Filer L. J, Mattson F. M. and Fomon, S. J. (1969) Triglyceride configuration and fat absorption by the human infant. *J. Nutr.*, 99, 293–8

Fincham, N. *et al.* (1985) Synthesis of arachidonate lipoxygenase products of epidermal cells. *J. Invest. Dermatol.*, 84, 447.

Finley, D. A., Lonnerdal, B., Dewey, K. G. and Grivetti, L. E. (1985) Breast milk composition: fat content and fatty acid composition in vegetarians and non-vegetarians. *Am. J. Clin. Nutr.*, 41, 787–800.

Fisher, M., Levine, P. H., Weiner, B., Ockene, I. S., Johnson, B., Johnson, M. H., Natale, A. M., Vaudreuil, C. H. and Hoogasian, J. (1986) The effect of vegetarian diets on plasma lipid and platelet levels. *Arch. Int. Med.*, 146, 1193–7.

Fisher, S. and Weber, P. C. (1986) The prostacyclin/thromboxane balance is favourably shifted in Greenland Eskimos. *Prostaglandins*, 32, 235–41.

Fleischman, A. I., Justice, D., Bierenbaum, M. L., Stier, A. and Sullivan, A. (1975) Beneficial effects of increased dietary linoleate upon *in vivo* platelet function in men. *J. Nutr.*, 105, 1286–90.

Flores, C. A., Brannon, P. M., Wells, M. A., Morrill, M. and Koldovsky, O. (1990) Effect of diet on triolein absorption in weanling rats. *Am. J. Phys.*, 258, G38-G44.

Food Advisory Committee (1989) *Use of nutrition claims in food labelling and advertising*, Ministry of Agriculture, Fisheries and Food, London.

Fox, P. L. and DiCorleto, P. E. (1988) Fish oils inhibit endothelial cell production of platelet-derived growth factor-like protein. *Science*, 241, 453–6.

Frankel, T. L. (1980) The inability of the cat to desaturate. Thesis, University of Cambridge.

Franzen, D., Hopp, H. W., Gunther, H. *et al.* (1990) Prospective, randomized and double-blinded trial about the effect of fish oil on the incidence of restenosis following PTCA and on coronary artery disease progression. *Eur. Heart J.*, 11 (Suppl.), 367.

Fraumeni, J. F. (1975) Cancers of the pancreas and biliary tract; epidemiological considerations. *Cancer. Res.*, 35, 3437–46.

Freedman, L. S., Clifford, C. and Messina, M. (1990) Analysis of dietary fat, calories, body weight and the development of mammary tumours in rats and mice: A review. *Cancer Res.*, 50, 5710–9.

Frick, M. H., Elo, O., Haapa, K., Heinonen, O. P.,

Heinsalmi, P., Helo, P., Huttunen, J. K., Kaitaniemi, P., Koskinen, P., Manninen, V., Maempaa, H., Malkonen, M., Manttari, M., Norola, S., Pasternack, A., Pikkarainen, J. M., Romo, M., Sjoblom, T. and Nikkila, E. A. (1987) Helsinki Heart Study: Primary prevention trial with gemfibrozil in middle aged men with dyslipidemia. *N. Engl. J. Med.*, **317**, 1237–45.

Friday, K. E., Childs, N. T., Tsunehara, C. H., Fujimoto, W. Y., Bierman, E. L. and Ensinck, J. W. (1989) Elevated plasma glucose and lowered triglyceride levels from omega-3 fatty acid supplementation in Type II diabetes. *Diabetes Care*, **12**, 276–81.

Friedman, Z. and Frolich, J. C. (1979) Essential fatty acids and the major urinary metabolites of the E prostaglandins in thriving infants and infants receiving parenteral fat emulsions. *Pediatr. Res.*, **13**, 932–6.

Frisch, R. E. (1977) In *Anorexia Nervosa*. (ed. R. Vigersky,), Raven Press, New York.

Frojmovic, M. M., Milton, J. G. and Gear, A. R. L. (1989) Platelet aggregation measured *in vitro* by microscopic and electronic particle counting. *Methods Enzymol.*, **169**, 134–49.

Furchgott, R. F. and Zawadzki, J. V. (1980) The obligatory role of endothelial cells in the relaxation of arterial smooth muscle by acetylcholine. *Nature*, **288**, 373–6.

Galli, C., Galli, G., Spagnuolo, C *et al.* (1977) In *Function and Biosynthesis of Lipids* (eds N. G. Bazan, R. R. Brenner, and N. M. Giusto) Plenum Press, New York. pp. 561–73.

Galli, C. and Socini, A. (1983) Dietary lipids in pre- and post-natal development. In *Dietary Fats and Health* (eds. E. G. Perkins and W. J. Visek), Proc. Am. Oil. Chem. Soc. Conf., **16**, 278–301.

Gans, R. O. B., Bilo, H. J. G., Weersink, E. G. L., Rauwerda, J. A., Fonk, T., Popp-Snijders, C. and Donker, A. J. M. (1990) Fish oil supplementation in patients with stable claudication. *Am. J. Surgery*, **160**, 490–5.

Garcia-Palmieri, M. R., Sorlie, P., Tillotson, J., Costas R., Cordero, E. and Rodriguez, M. (1980) Relationship of dietary intake to subsequent coronary heart disease incidence: the Puerto Rico heart health programme. *Am. J. Clin. Nutr.*, **33**, 1818–27.

Garg, A., Bonanome, A., Grundy, S. M., Zhang, Z. J. and Unger, R. H. (1988) Comparison of a high-carbohydrate diet with a high monounsaturated fat diet in patients with non-insulin dependent diabetes mellitus. *N. Eng. J. Med.*, **319**, 829–34.

Garton, G. A. (1985) Aspects of chemistry and biochemistry of branched chain fatty acids (Hilditch Memorial Lecture, 1984) *Chem. Ind.*, **9**, 295–300.

Gerkens, J. F., Williams, A. and Branch, R. A. (1981) Effect of precursors of the 1,2 and 3 series prostaglandins on renin release and renal blood flow in the dog. *Prostaglandins*, **22**, 513–5.

Gibson, N. A. and Kneebone, G. M. (1981) Fatty acid composition of human milk colostrum and mature breast milk. *Am. J. Clin. Nutr.*, **34**, 252–7.

Ginsberg, H. N., Barr, S. L., Gilbert, A. *et al.* (1990) Reduction of plasma total cholesterol levels in normal men on an American Heart Association Step 1 diet or Step 2 diet with added monounsaturated fat. *N. Engl. J. Med.*, **322**, 574–9.

Giroud, A. (1970) *Nutrition of the embryo*, Springfield, Thomas.

Glauber, H., Wallace, P., Griver, K. and Brechtel, G. (1988) Adverse metabolic effect of omega-3 fatty acids in non-insulin-dependent diabetes mellitus. *Ann. Intern. Med.*, **108**, 663–8.

Glueck, C. J., Mattson, F. H. and Jandacek, R. J. (1979) The lowering of plasma cholesterol by sucrose polyester in subjects consuming diets with 800, 300 or less than 50 mg cholesterol per day. *Am. J. Clin. Nutr.*, **32**, 1636–44.

Godfried, S. L., Combs, G. R., Saroka, J. M. and Dillingham, L. A. (1989) Potentiation of atherosclerotic lesions in rabbits by a high dietary level of vitamin E. *Br. J. Nutr.*, **61**, 607–17.

Goodnight, S. H., Harris, W. S. and Connor, W. E. (1981) The effects of dietary w3 fatty acids on platelet composition and function in man: a prospective, controlled study. *Blood*, **58**, 880–5.

Gordon, T., Kagan, A., Garcia-Palmieri, M., Kannel, W. B., Zukel, W. J., Tillotson, J., Sorlie, P. and Hjortland, M. (1981) Diet and its relation to coronary heart disease and death in three populations. *Circulation*, **63**, 500–15.

Grant, H. W., Palmer, K. R., Kelly, R. W., Wilson, N. H. and Misiewicz, J. J. (1988) Dietary linoleic acid, gastric acid and prostaglandin secretion. *Gastroenterol.*, **94**, 955–9.

Greaves, M. W. and McDonald Gibson, W. (1972) Prostaglandin biosynthesis by human skin and its inhibition by corticosteroids. *Br. J. Pharmacol.*, **46**, 172–5.

Greaves, M. W. (1988) Inflammation and mediators. *Br. J. Dermatol.*, **119**, 419–26.

Gregory, J., Foster, K., Tyler, H. and Wiseman, M. (1990) *The Dietary and Nutritional Survey of British Adults*, HMSO, London.

Gries, A., Malle, H. and Kostner, G. M. (1990) Influence of dietary fish oils on plasma Lp(a) levels. *Throm. Res.*, **58**, 667–8.

Grigg, L. E., Kay, T. W. H., Valentine, P. A. *et al.* (1989) Determinants of restenosis and lack of effect of dietary supplementation with eicosapentaenoic acid on the incidence of coronary artery restenosis with angioplasty. *J. Am. Coll. Cardiol.*, **13**, 665–72.

Grundy, S. M. (1986) Comparison of monounsaturated fatty acids and carbohydrates for plasma total cholesterol lowering. *N. Engl. J. Med.*, **314**, 745–8.

Grundy, S. M. (1989) Monounsaturated fatty acids and cholesterol metabolism: implications for dietary recommendations. *J. Nutr.*, **119**, 529–33.

Grundy, S. M. (1990) *Trans* monounsaturated fatty acids and some cholesterol levels. *N. Engl. J. Med.*, **323**, 480.

Grundy, S. M. (1991) Multifactorial aetiology of hypercholesterolaemia: implications for prevention of coronary heart disease. *Athero. Thromb.*, **11**, 1619–35.

Grundy, S. M. and Denke, M. A. (1990) Dietary influences on serum lipids and lipoproteins. *J. Lipid Res.*, **31**, 1149–72.

Gudbjarnason, S. (1989) Dynamics of n-3 and n-6 fatty acids in phospholipids of heart muscle. *J. Inter. Med.*, **225** (Suppl. 1), 117–28.

Gudbjarnason, S., Emilsson, A. and Gudmundsdottir, A. (1983) Fatty acid composition of phospholipids of heart muscle in relation to age, dietary fat and stress, in *Arterial Pollution* (eds H. Peeters, G. A. Gresham and R. Paoletti), Plenum, New York, pp. 115–24.

Gunstone, F. D., Harwood, J. L. and Padley, F. B. (eds) (1986) *The Lipid Handbook*, Chapman and Hall, London.

Gurr, M. I. (1981) Review of the progress of dairy science: human and artificial milks for infant feeding. *J. Dairy Res.*, **48**, 519–54.

Gurr, M. I. (1983) The nutritional significance of lipids, in

Developments in Dairy Chemistry – 2 Lipids (ed. P. F. Fox), Applied Science Publishers, London.

Gurr, M. I. (1984) Agricultural aspects of lipids in *The Lipid Handbook* (eds F. D. Gunstone, J. L. Harwood and F. B. Padley), Chapman and Hall London.

Gurr, M. I. (1988) Lipids: products of industrial hydrogenation, oxidation and heating, in *Nutritional and Toxicological Aspects of Food Processing* (eds. R. Walker and E. Quattrucci), Taylor and Francis, London.

Gurr, M. I. (1991) *Polyunsaturates. Their role in health and nutrition.* The Butter Council, Sevenoaks.

Gurr, M. I. and Harwood, J. L. (1991) *Lipid Biochemistry. An Introduction.* 4th edn, Chapman and Hall, London.

Hagberg, B., Hagberg, G. and Zetterstrom, R. (1989) Decreasing perinatal mortality – increase in cerebral palsy morbidity. *Acta Pediatr. Scand.*, **78**, 664–70.

Haglund, O., Wallin, R., Luostarinen, R. and Saldeen, T. (1990) Effect of a new fluid fish oil concentrate, ESKIMO-3, on triglyceride, cholesterol, fibrinogen and blood pressure. *J. Intern. Med.*, **227**, 347–53.

Hague, T. A. (1988) Effects of unsaturated fatty acids on cell membrane functions. *Scand. J. Clin. Lab. Invest.*, **48**, 381–8.

Haines, A. P., Chakrabarti, R., Fisher, D., Meade, T. W., North, W. R. S. and Stirling, Y. (1980) Haemostatic variables in vegetarians and non-vegetarians. *Thromb. Res.*, **19**, 139–48.

Haines, A. P., Sanders, T. A. B., Imeson, J. D., Mahler, R. F., Martin, J., Mistry, M., Vickers, M. and Wallace, P. G. (1986) Effects of a fish oil supplement on platelet function, haemostatic variables and albuminuria in insulin-dependent diabetics. *Thromb. Res.*, **43**, 643–55.

Hajjar, K. A., Hamel, N. M., Harpel, P. C. and Nachman, R. L. (1987) Binding of tissue plasminogen activator to cultured human endothelial cells. *J. Clin. Invest.*, **809**, 1712–9.

Hall, B. (1979) The uniformity of human milk. *Am. J. Clin. Nutr.*, **32**, 304–12.

Hamazaki, T., Urakaze, M., Makuta, M., Ozawa, A., Soda, Y., Tatsume, H., Yano, S. and Kumagai, A. (1987) Intake of different eicosapentaenoic acid-containing lipids and fatty acid pattern of plasma lipids in the rat. *Lipids*, **22**, 994–8.

Hamberg, M. (1980) Transformation of 5,8,11,14,17-eicosapentaenoic acid in human platelets. *Biochim. Biophys. Acta*, **618**, 389–98.

Hammerstrom, S. *et al.* (1975) Increased concentrations of non-esterified arachidonic acid 12L-hydroxy-5, 8, 10, 14-eicosatetraenoic acid prostaglandin E_2 and prostaglandin F_2 in epidermis of psoriasis. *Proc. Natl Acad. Sci. USA*, **72**, 5130–4.

Hanasaki, K., Nagasaki, T. and Arita, H. (1989) Characterization of platelet thromboxane A_2/prostaglandin H_2 receptor by a novel thromboxane receptor antagonist, [^{3}H] S-145. *Biochim. Pharmacol.*, **38**, 2007–17.

Hansen, A. E. (1933) Serum lipid changes and therapeutic effects of various oils in infantile eczema. *Proc. Soc. Exp. Biol. Med.*, **31**, 160–1.

Hansen, A. E. *et al.* (1958) Essential fatty acids in infant nutrition. III Clinical manifestation of linoleic acid deficiency. *J. Nutr.*, **66**, 565–76.

Hansen, A. E., Wiese, H. F., Boelsche, A. N., Haggard, M. E., Adam, D. J. D. and Davies, H. (1963) Role of linoleic acid in infant nutrition, clinical and chemical study of 428 infants fed on milk mixtures varying in kind and amount of fat. *Pediatrics*, **31**, 171–92.

Hansen, H. S. (1989) Linoleic acid and epidermal water

barrier in *Dietary omega-3 and omega-6 Fatty Acids. Biological Effects and Nutritional Essentiality.* (eds C. Galli and A. P. Simopoulos), Nato ASI Series A: Life Sciences, Vol. **171**, Plenum Press, New York and London, pp. 333–41.

Hansen, I. B., Friis-Hansen, B. and Clausen, J. (1969) The fatty acid composition of umbilical cord serum, infant serum and maternal serum and its relation to diet. *Zeits. Ernahrung.*, **9**, 352–63.

Hansen, J. B., Olsen, J. O., Wilsgard, L. and Osterud, B. (1989) Effects of dietary supplementation with cod-liver oil on monocyte thromboplastin synthesis, coagulation and fibrinolysis. *J. Intern. Med.*, **225**, (suppl. 1), 133–9.

Harmon, J. T. and Jamieson, G. A. (1986) Activation of platelets by alpha-thrombin is a receptor-mediated event. *J. Biol. Chem.*, **261**, 15928–33.

Harris, W. S. and Connor, W. E. (1980) The effects of salmon oil upon plasma lipids, lipoproteins, and triglyceride clearance. *Trans. Ass. Am. Phys.*, **93**, 148–55.

Harris, W. S., Connor, W. E., McMurray, M. P. (1983) The comparative reductions of the plasma lipids and lipoproteins by dietary polyunsaturated fats: salmon oil vs vegetable oils. *Metabolism*, **32**, 179–84.

Harris, W. S., Connor, W. E. and Lindsey, S. (1984) Will dietary w-3 fatty acids change the composition of human milk? *Am. J. Clin. Nutr.*, **40**, 780–5.

Harris, W. S., Zucker, M. L. and Dujovne, C. A. (1988) w3 fatty acids in hypertriglyceridaemic patients: triglycerides vs methyl esters. *Am. J. Clin. Nutr.*, **48**, 992–7.

Harris, W. S. (1989) Fish oils and plasma lipid and lipoprotein metabolism in humans: a critical review. *J. Lipid. Res.*, **30**, 785–807.

Harris, W. S., Connor, W. E., Illingworth, D. R., Rothrock, D. W. and Foster, D. M. (1990) Effects of fish oil on VLDL triglyceride kinetics in humans. *J. Lipid. Res.*, **31**, 1549–58.

Hartog, J. M. (1989) Dietary fish oil and experimental atherosclerosis. Doctoral Thesis. Erasmus University, Rotterdam.

Harzer, G., Haug, M., Dieterich, I. *et al.* (1983) Changing patterns of human milk lipids in the course of lactation and during the day. *Am. J. Clin. Nutr.*, **37**, 612–21.

Hassam, A. G. and Crawford, M. A. (1976) The differential incorporation of labelled linoleic, gamma-linolenic, di-homogamma-linolenic and arachidonic acids into the developing rat brain. *J. Neurochem.*, **27**, 967–8.

Hassid, A. and Dunn, M. J. (1980) Microsomal prostaglandin biosynthesis of human kidneys. *J. Biol. Chem.*, **255**, 2472–5.

Hatzakis, H., Sanduja, S. K. and Wu, K. K. (1987) Stimulation of endothelial cell production of tissue plasminogen activator and eicosanoids by phorbol ester. *Blood*, **70**, (suppl 1), 404a.

Hayes, K. C., Pronczuk, A., Lindsey, S. L. and Diersen-Schade, D. (1991) Dietary saturated fatty C12:0, 14:0, 16:0) differ in their impact on plasma cholesterol and lipoproteins in non human primates. *Am. J. Clin. Nutr.*, **53**, 491–9.

Health and Welfare Canada (1990) *Nutrition recommendations.* The report of the scientific review committee, Ministry of Supply and Services, Canada.

Health Education Authority (1991) *Enjoy healthy eating*, Health Education Authority, London.

Heine, R. J., Mulder, C., Popp-Snijders, C., van der Meer, J., van der Veen, E. A. (1989) Linoleic acid enriched diet: long-term effects of serum lipoprotein and apolipoprotein concentrations and insulin sensitivity in non-insulin dependent diabetic patients. *Am. J. Clin. Nutr.*, **49**, 448–56.

Heine, R. J. and Schouten, J. A. (1989) The role of fat in the treatment of diabetes mellitus, in *The Role of Fat in Human Nutrition* (eds A. J. Vergroessen and M. Crawford), Academic Press, London.

Hendra, T. J., Britton, M. E., Roper, D. R., Wagaine-Twabwe, D., Jeremy, J. Y., Dandona, P., Haines, A. P. and Yudkin, J. S. (1990) Effects of fish oil supplements in NIDDM subjects. Controlled study. *Diabetes Care*, **13**, 821–9.

Herrmann, W., Biermann, J., Lindhofer, H. G. and Kostner, G. (1989) [Modification of the atherogenic risk factor Lp(a) by supplementary fish oil administration in patients with moderate physical training]. Beeinflussung des atherogenen Risikofaktors Lp(a) durch supplementare Fischolaufnahme bei patienten mit moderatem physischem Training. *Med. Klin.*, **84**, 429–33.

Heyden, S. (1974) Polyunsaturated fatty acids and colon cancer. *Nutr. Med.*, **17**, 321–6.

Hillyard, L. A. and Abrahams, S. (1979) Effect of dietary polyunsaturated fatty acids on growth of mammary adenocarcinomas in mice and rats. *Cancer Res.*, **39**, 4430–7.

Hirai, A., Terano, T., Hamazaki, T., Sajiki, J., Kondo, S., Ozawa, A., Fugita, T., Miyamoto, T., Tamura, Y. and Kumagai, A. (1982) The effects of the oral administration of fish oil concentrate on the release and the metabolism of [^{14}C] arachidonic acid and [^{14}C] eicosapentaenoic acid by human platelets. *Thromb. Res.*, **28**, 285–98.

Hirayama, T. (1978) Epidemiology of breast cancer with special reference to the role of diet. *Prev. Med.*, **7**, 173–95.

Hjermann, I., Velve-Byre, K., Holme, I. and Leren, P. (1981) Effect of diet and smoking intervention on incidence of CHD: report from the Oslo Study Group of a randomized trial in healthy men. *Lancet*, **ii**, 1303–10.

Hjermann, I., Holme, I. and Leren, P. (1986) Oslo study diet and anti-smoking trial. Results after 102 months. *Am. J. Med.*, **80** (suppl 2A), 7–11.

Hock, C. E., Holahan, M. A. and Reibel, D. K. (1987) Effect of dietary fish oil on myocardial phospholipids and myocardial ischemic damage. *Am. J. Physiol.*, **252**, H554–60.

Holland, B., Welch, P. A., Unwin, I. D., Buss, D. H., Paul, A. A. and Southgate, D. A. T. (1991) *McCance and Widdowson's The Composition of Foods*, 5th edn, Royal Society of Chemistry, Cambridge.

Holman, R. T. (1970) Biochemical activities of and requirement for polyunsaturated fatty acids. *Prog. Chem. Fats Lipids*, **9**, 607–87.

Holman, R. T. (1986) Control of polyunsaturated fatty acids in tissue lipids. *J. Am. Coll. Nutr.*, **5**, 183–211.

Holman, R. T. and Johnson, S. (1981) Changes in essential fatty acid profile of serum phospholipids in human disease. *Prog. Lipid Res.*, **20**, 67–73.

Holman, R. T., Johnstone, S. B. and Hatch, T. F. (1982) A case of human linoleic acid deficiency involving neurological abnormalities. *Am. J. Clin. Nutr.*, **35**, 617–23.

Holroyde, C. P., Skutches, C. L. and Reichard, G. A. (1988) Effect of dietary enrichment with n-3 polyunsaturated fatty acid (PUFA) in metastatic breast cancer. *Proc. Soc. Clin. Oncol.*, **7**, 42.

Hornstra, G., Lewis, B., Chait, A., Turpeinen, O., Karvonen, M. J. and Vergroesen, A. J. (1973) Influence of dietary fat on platelet function in men. *Lancet*, **i**, 1155–7.

Hornstra, G. (1988) Dietary lipids and cardiovascular disease: effects of palm oil. *Oleagineux*, **43**, 75–81.

Hornstra, G. (1992) The effect of dietary lipids on blood coagulation and fibrinolysis (in press).

Hornstra, G., van Houwelingen, A. C., Kivito, G. A. A., Fischer, S. and Vedelhoven, W. (1990) Influence of dietary fish on eicosanoid metabolism in man. *Prostaglandins*, **40**, 311–29.

Horrobin, D. F. (1980) The reversibility of cancer: the relevance of cyclic AMP, calcium, essential fatty acids and prostaglandin E_1. *Med. Hypotheses*, **6**, 469–86.

Horrobin, D. F. (1981) The possible roles of prostaglandin E1 and of essential fatty acids in mania, depression and alcoholism. *Prog. Lipid Res.*, **20**, 539–41.

Horrobin, D. F. (1990) Gamma linolenic acid. *Rev. Con. Pharmacother.*, **1**, 1–41.

Hostmark, A. T., Bjerkedal, T., Kierulf, P., Flaten, H. and Ulshagen, K. (1988) Fish oil and plasma fibrinogen. *Br. Med. J.*, **297**, 180–1.

Houtsmuller, A. J., Zalin, K. J. and Henkes, H. E. (1980) Favourable influence of linoleic acid on the progression of diabetic micro and macroangiopathy. *Nutr. Metab.*, **24** (supp.1), 105–12.

Howard-Williams, J., Patel, P., Jelfs, R., Carter, R. D., Awdry, P., Bron, A., Mann, J. I. and Hockaday, T. D. R. (1985) Polyunsaturated fatty acids and diabetic retinopathy. *Brit. J. Ophth.*, **69**, 15–8.

Hubbard, N. E. and Erickson, K. L. (1987a) Influence of dietary fats on cell populations of line 168 mouse mammary tumours: a morphometric and ultrastructural study. *Cancer Lett.*, **35**, 281–94.

Hubbard, N. E. and Erickson, K. L. (1987b) Enhancement of metastasis from a transplantable mouse mammary tumour by dietary linolenic acid. *Cancer Res.*, **47**, 6171–5.

Hubbard, N. E., Chapkin, R. S. and Erickson, K. L. (1988) Inhibition of growth and linoleate-enhanced metastasis of a transplantable mammary tumour by indomethacin. *Cancer Lett.*, **43**, 111–20.

Hubbard, N. E. and Erickson, K. L. (1989) Effect of dietary linoleic acid level on lodgement, proliferation and survival of mammary tumour metastases. *Cancer Lett.*, **44**, 117–25.

Huff, M. W. and Telford, D. E. (1989) Dietary fish oil increase conversion of very low density lipoprotein apoprotein B to low density lipoprotein. *Arteriosclerosis*, **9**, 58–66.

Hunt, S. C., Wu, L. L., Hopkins, P. N. *et al.* (1989) Apolipoprotein, low density lipoprotein subfraction and insulin associations with familial combined hyperlipaemia. Study of Utak patients with Familial Dyslipidaemic hypertension. *Arteriosclerosis*, **9**, 335–44.

Hurley, L. S. (1979) Nutritional deficiencies and excesses, in *Handbook of Teratology*, Plenum Press, New York. pp. 261–308.

Hytten, P. E. and Leitch, I. (1971) *The Physiology of Human Pregnancy*, Blackwell Scientific Publications, Oxford.

Iacono, J. M., Puska, P., Dougherty, R. M., Pietinen, P., Vartiainen, E., Leino, U., Mutanen, M. and Moisio, S. (1983) Effect of dietary fat on blood pressure in a rural Finnish population. *Am. J. Clin. Nutr.*, **38**, 860–7.

Ibrahim, J. B. T. and McNamara, D. J. (1988) Cholesterol homeostasis in guinea pigs fed saturated and polyunsaturated fat diets. *Biochim. Biophys. Acta*, **963**, 109–18.

Illingworth, D. R., Sundberg, E. E., Becker, N., Connor, W. E. and Alaupovic, P. (1981) Influence of saturated, monounsaturated and w6-PUFA on LDL metabolism in man. *Arteriosclerosis*, **1**, 380.

Imagawa, W., Bandyopadhyay, G. K., Wallace, D. and Nandi, S. (1989) Phospholipids containing polyunsaturated fatty acyl groups are mitogenic for normal mouse epithelial cells in serum-free primary cell culture. *Proc. Natl Acad. Sci. USA*, **86**, 4122–6.

Innerarity, T. L., Mahley, R. W., Weisgraber, K. H., Bersot, T. P., Krauss, R. M., Vega, G. L., Grundy, S. M., Friedl, W., Davignon, J. and McCarthy, B. J. (1990) Familial defective apolipoprotein B-100: a mutation of apolipoprotein B that causes hypercholesterolemia. *J. Lipid Res.*, **31**, 1337–49.

Innis, S. M. and Kuhulein, H. V. (1988) Long chain n-3 fatty acids in breast milk of Inuit women consuming traditional foods. Early Hum. Dev.,

Insel, P. A., Nirenberg, P., Turnbull, J. and Shattil, S. J. (1978) Relationships between membrane cholesterol, alpha-adrenergic receptors, and platelet function. *Biochem.*, **17**, 5269–74.

Ip, C. (1980) Ability of dietary fat to overcome the resistance of mature female rats to 7, 12-dimethylbenz(a)anthracene-induced mammary tumorigenesis. *Cancer Res.*, **40**, 2785–9.

Ip, C., Carter, C. A. and Ip, M. M. (1985) Requirement of essential fatty acid for mammary tumorigenesis in the rat. *Cancer Res.*, **45**, 1997–2001.

Ivy, A. C., Nelson, D. and Bucher, G. (1941) The standardisation of certain factors in the cutaneous 'venostasis' bleeding time technique. *J. Lab. Clin. Med.*, **26**, 1812–22.

James, W. P. T. (1988) *Healthy nutrition: preventing nutrition-related diseases in Europe*. WHO Regional Publications, European Series, No.24 WHO.

Jandacek, R. J. (1977) US Patent 4 005 195.

Jandacek, R. H., Mattson, F. H., McNeedy, S., Gallon, L., Yanker, R. and Glueck, C. J. (1980) Effect of sucrose polyester on faecal steroid excretion by 24 normal men. *Am. J. Clin Nutr.*, **33**, 251–9.

Jandacek, R. J., Hollenbach, E. J., Holcombe, B. N., Kuehlthan, C. M., Peters, J. C. and Taulbee, J. O. (1991) Reduced storage of dietary eicosapentaenoic and docosahexaenoic acids in the weaning rat. *J. Nutr. Biochem.*, **2**, 142–9.

Jansson, L., Akesson, B., Holmberg, L. (1981) Vitamin E and fatty acid composition of human milk. *Am. J. Clin. Nutr.*, **34**, 8–13.

Jantti, J., Nikkari, T., Solakivi, T., Vapaatalo, H. and Isomaki, H. (1989) Evening primrose oil in rheumatoid arthritis; changes in serum lipids and fatty acids. *Ann. Rheum. Dis.*, **48**, 124–7.

Jensen, T., Stender, S., Goldstein, K., Holmer, G. and Deckert, T. (1989) Partial normalization by dietary cod-liver oil of increased microvascular albumin leakage in patients with insulin-dependent diabetes and albuminuria. *N. Engl. J. Med.*, **321**, 1572–7.

Jessup, S. J. *et al.* (1970) Biosynthesis and release of prostaglandins on hormonal treatment of frog skin and their effect on ion transport. *Fed. Proc.*, **29**, 387.

Jilial, I., Vega, G. L. and Grundy, S. M. (1990) Physiological levels of ascorbate inhibit the oxidative modification of low density lipoprotein. *Atherosclerosis*, **82**, 185–190.

Johnson, A. R. and Davenport, J. B. (1971) *Biochemistry and Methodology of Lipids*, John Wiley and Sons New York.

Johnson, M., Carey, F. and McMillan, R. M. (1983) Alternative pathways of arachidonate metabolism: prostaglandins, thromboxane and leukotrienes, in *Es-*

says in Biochemistry (eds P. N. Campbell and R. D. Marshall), Academic Press, pp. 40–141.

Jones, P. J. H., Pencharz, P. B. and Clandinin, M. T. (1985) Whole body oxidation of dietary fatty acids: implications for energy utilisation. *Am. J. Clin. Nutr.*, **42**, 769–77.

Jos, A. F., den Kamp, O., Roelofsen, B. and Van Deenen, L. L. M. (1985) Structural and dynamic aspects of phosphatidylcholine in the human erythrocyte membrane. *Trends Biochem. Sci.*, **10**, 320–3.

Kannel, W. B., Neaton, J. B., Wentworth, D., Thomas, H. E., Stamler, J., Hulley, S. B. and Kjelsberg, M. D. (1986) Overall and coronary heart disease mortality rates in relation to major risk factors in 325,348 men screened MRFIT. *Am. Heart J.*, **112**, 825–36.

Karmali, R. A., Marsh, J. and Fuchs, C. (1984) Effect of omega-3 fatty acids on growth of a rat mammary tumor. *J. Natl Cancer Inst.*, **73**, 457–61.

Karmali, R. A., Marsh, J., Fuchs, C., Hare, W. and Crawford, M. (1985) Effects of dietary enrichment with gamma-linoleic acid upon growth of the R3230AC mammary adenocarcinoma. *J. Nutr. Growth Cancer*, **21**, 41–51.

Karmali, R. A., Reichel, P., Cohen, L. A., Terano, T., Hirai, A., Tamura, Y. and Yoshida, S. (1987) The effects of dietary n-3 fatty acids on the DU-145 transplantable human prostatic tumor. *Anticancer Res.*, **7**, 1173–80.

Kasim, S. E., Stern, B., Khilnani, S., McLin, P., Baciorowski, S., Jen, K-L. C. (1988) Effects of omega-3 fish oils on lipid metabolism, glycaemic control and blood pressure in Type II diabetic patients. *J. Clin. Endocrinol. Metab.*, **67**, 1–5.

Katz, E. B. and Boylan, E. S. (1987) Stimulatory effect of high polyunsaturated fat diets on lung metastasis from the 13762 mammary adenocarcinoma in female retired breeder rats. *J. Natl Cancer. Inst.*, **79**, 351–8.

Kesaniemi, Y. A., Belz, W. F. and Grundy, S. M. (1985) Comparison of clofibrate and caloric restriction on kinetics of very low density lipoprotein triglycerides. *Arteriosclerosis*, **5**, 153–61.

Kestin, M., Clifton, P., Belling, G. B. and Nestel, P. J. (1990) n-3 fatty acids of marine origin lowers systolic blood pressure and triglycerides but raise LDL cholesterol compared with n-3 and n-6 fatty acids from plants. *Am. J. Clin. Nutr.*, **51**, 1028–34.

Keys, A. (1970) Coronary heart disease in seven countries. *Circulation*, **41**, 1–211.

Keys, A. (1980) *Seven Countries: A multivariate analysis of death and coronary heart disease*, Harvard University Press, London.

Kinsella, J. E. (1990) Possible mechanisms underlying the effects of dietary n-3 polyunsaturated fatty acids. *Omega-3 News*, **V**, 1–5.

Knapp, H. R., Reilly, I. A. G., Alessandrini, P. and FitzGerald, G. A. (1986) *In vivo* indexes of platelet and vascular function during fish-oil administration in patients with atherosclerosis. *N. Engl. J. Med.*, **314**, 937–42.

Knapp, H. R. and Fitzgerald, G. A. (1989) The antihypertensive effects of fish oil. A controlled study of polyunsaturated fatty acid supplements in essential hypertension. *N. Eng. J. Med.*, **320**, 1037–43.

Koletzko, B., Mrotzek, M. and Bremer, H. J. (1988) Fatty acid composition of mature milk in Germany. *Am. J. Clin. Nutr.*, **47**, 954–9.

Koletzko, B. and Braunn, M. (1991) Arachidonic acid and early human growth: is there a relation? *Am. Nutr. Metab.*, **35**, 128–31.

Kollmorgen, G. M., King, M. M. and Kosanke, S. D. (1983) Influence of dietary fat and indomethacin on the growth of transplantable mammary tumors in rats. *Cancer Res.*, **43**, 4714–9.

Kort, W. J., Weijma, I. M., Bijma, A. M., van Schakwijk, W. P., Vergroesen, A. J. and Westbrock, D. L. (1987) Omega-3 fatty acids inhibiting the growth of a transplantable rat mammary adenocarcinoma. *J. Natl. Cancer Inst.*, **79**, 593–9.

Kragbulle, K. *et al.* (1986) Dermis-derived 15-hydroxy-eicosatetraenoic acid inhibits epidermal 12-lipoxygenase activity. *J. Invest. Dermatol.*, **87**, 494–8.

Kremer, J. M., Lawrence, D. A., Jubiz, W., DiGiacomo, R., Rynes, K., Bartholomew, L. E. and Sherman, M. (1990) Dietary fish oil and olive oil supplementation in patients with rheumatoid arthritis; clinical and immunological effects. *Arth. Rheumat.*, **33**, 810–20.

Kristensen, P., Larsson, L. I., Nielsen, L. S., Grondahl-Hansen, J., Andreasen, P. A. and Dano, K. (1984) Human endothelial cells contain one type of plasminogen activator. *FEBS Lett.*, **168**, 33–7.

Kritchevsky, D. (1988) Atherosclerosis, in *Comparative Nutrition* (ed K. Blaxter and I. MacDonald), John Libbey, London, pp. 185–98.

Kritchevsky, D., Davidson, L. M., Weight, M., Kriek, N. P. J. and du Plessis, J. P. (1984) Effect of *trans* unsaturated fats on experimental atherosclerosis in vervet monkeys. *Atherosclerosis*, **51**, 123–33.

Kritchevsky, D., Weber, M. M., Nuck, C. L. and Klurfeld, D. M. (1986) Calories, fat and cancer. *Lipids*, **21**, 272–4.

Kromhout, D. and Coulander., C. L. (1984) Diet, prevalence and 10-year mortality from coronary heart disease in 871 middle aged men. The Zutphen Study. *Am. J. Epidemiol.*, **119**, 733–41.

Kromhout, D., Bosschieter, E. B. and Coulander., C. L. (1985) The inverse relation between fish consumption and 20-year mortality from coronary heart disease. *N. Engl. J. Med.*, **312**, 1205–9.

Kromhout, D., Arntzenius, A. C., Kempen-Voogd, N., Kempen, H. J., Barth, J. D., van der Voort, H. A. and van der Velde, E. A. (1987) Long-term effects of a linoleic acid-enriched diet, changes in body weight and alcohol consumption on serum total and HDL-cholesterol. *Atherosclerosis*, **66**, 99–105.

Kurzrok, R. and Lieb, C. C. (1930) Biochemical studies of human semen II. The action of semen on the human uterus. *Proc. Soc. Exp. Biol. Med.*, **28**, 268–72.

Kushi, L. H., Lew, R. A., Stare, F. J., Ellison, C. R., El Lozy, M., Bourke, G., Daly, L., Graham, I., Hickey, N., Mulcahy, R. and Kevaney, J. (1985) Diet and 20 year mortality from coronary heart disease: the Ireland-Boston diet-heart study. *N. Engl. J. Med.*, **312**, 811–8.

Kuusi, T., Ehnholm, C., Huttunen, J. K., Kostiainen, E., Pietenen, P., Leino, U., Uusitalo, U., Nikkari, T., Iacono, J. M. and Puska, P. (1985) Concentration and composition of serum lipoproteins during a low-fat diet at two levels of polyunsaturated fat. *J. Lipid Res.*, **26**, 360–7.

Kyle, D. J. (1992) Microbial omega-3 containing fats and oils for food use. *Adv. App. Biotechnol.* (in press).

Lachmann, P. J. and Peters, D. K. (1982) *Clinical Aspects of Immunology*, 4th edn, Blackwell Scientific Publications, Oxford.

Lambers, J. W. J., Cammenga, M., Konig, B. W., Mertens, K., Pannekoek, H. and van Mourik, J. A. (1987) Activation of human endothelial cell-type plasminogen activator inhibitor (PAI-1) by negatively charged phospholipids. *J. Biol. Chem.*, **262**, 17492–6.

Lands, W. E. M. (1991) Dose-response relationships for omega-3/omega-6 effects, in *Health Effects of Omega-3 Polyunsaturated Fatty Acids in Sea Foods* (eds A. P. Simopoulos, R. R. Kifer, R. E. Martin and S. M. Barlow), Karger, Basel, pp. 177–94.

Lands, W. E. M., Culp, B. R., Hirai, A. and Gorman, R. (1985) Relationship of thromboxane generation to the aggregation of platelets from humans: effects of eicosapentaenoic acid. *Prostaglandins*, **30**, 819–25.

Landymore, R. W., Kinley, C. E., Cooper, J. H. *et al.* (1985) Cod-liver oil in the prevention of intimal hyperplasia in autologous vein grafts used for arterial bypass. *J. Thorac. Card. Surg.*, **89**, 351–7.

Lanson, M., Bougnoux, P., Beeson, P., Lansac, J., Hubert, B., Couet, C. and Le Floch, O. (1990) n-6 polyunsaturated fatty acids in human breast carcinoma phosphatidylethanolamine and early relapse. *Br. J. Cancer*, **61**, 776–8.

Lapidus, L., Anderson, H., Bengtsson, C. and Bosaeus, I. (1986) Dietary habits in relation to incidence of cardiovascular disease and death in women: a 21-year follow-up of participants in the population study of women in Gothenburg, Sweden. *Am. J. Clin. Nutr.*, **44**, 444–8.

Laustiola, K., Salo, M. K. and Metsa-Ketela, T. (1986) Altered physiological responsiveness and decreased cyclic AMP levels in rat atria after dietary cod liver oil supplementation and its possible association with an increased membrane phospholipid n-3/n-6 fatty acid ratio. *Biochim. Biophys. Acta*, **889**, 95–102.

Lawson, L. D., Hill, E. G. and Holman, R. T. (1985) Dietary fats containing concentrates of *cis* or *trans* octadeconates and the patterns of polyunsaturated fatty acids on liver phosphatidylcholine and phosphatidyl ethanolamine. *Lipids*, **20**, 262–7.

Lawson, L. D. and Hughes, B. G. (1988) Human absorption of fish oil fatty acids as triacylglycerols, free acids or ethyl esters. *Biochem. Biophys. Res. Comm.*, **152**, 328–35.

Leaf, A. and Weber, P. C. (1988) Cardiovascular effects of n-3 fatty acids. *N. Engl. J. Med.*, **318**, 549–57.

Leat, W. M. F. (1989) Dietary linolenic acid and tissue function in rodents, in *Dietary omega-3 and omega-6 Fatty Acids. Biological Effects and Nutritional Essentiality* (ed C. Galli and A. P. Simopoulos), Nato ASI Series. Series A: Life Sciences, Vol. 171, Plenum Press, New York and London, pp. 219–26.

Lee, T. H. *et al.* (1984) Effects of exogenous arachidonic eicosapentaenoic and docosahexaenoic acids on the generation of 5-lipoxygenase pathway products by ionophore activated human neutrophils. *J. Clin. Invest.*, **74**, 1922–33.

Lee, T. H., Hoover, R. L., Williams, J. D., Sperling, R. I., Ravalese, J. R. III., Spur, B. W., Robinson, D. R., Corey, E. J., Lewis, R. A. and Austen, K. F. (1985) Effect of dietary enrichment with eicosapentaenoic and docosahexaenoic acids on *in vitro* neutrophil and monocyte leukotriene generation and neutrophil function. *N. Engl. J. Med.*, **312**, 1217–24.

Lees, A. B. and Karel, C. D. (1990) *Omega-3 Fatty Acids in Health and Disease*, Marcel Dekker, New York.

Lefer, A. M., Ogletree, M. L., Smith, J. B., Silver, M. J., Nicolson, K. C., Barnette, W. E., and Gasic, G. P. (1978) Prostacyclin: a potentially valuable agent for preserving myocardial tissue in acute myocardial ischaemia. *Science*, **200**, 52–4.

Leren, P. (1970) The Oslo diet-heart study. Eleven-year report. *Circulation*, **42**, 935–42.

Leslie, C. A., Gonnermen, W. A., Ullman, M. D., Hayes,

K. C., Franzblau, C. and Cathcart, E. S. (1985) Dietary fish oil modulates macrophage fatty acids and decreases arthritis susceptibility in mice. *J. Exp. Med.*, **162**, 1336–49.

Leth-Espensen, P., Stender, S., Ravn, H. and Kjeldsen, K. (1988) Anti-atherogenic effect of olive and corn oils in cholesterol-fed rabbits with the same plasma cholesterol levels. *Arteriosclerosis*, **8**, 281–7.

Levenson, D. J., Simmons, C. E. and Brenner, B. M. (1982) Arachidonic acid metabolism, prostaglandins and the kidney. *Am. J. Med.*, **72**, 354–74.

Levin, E. G., Marzec, U., Anderson, I. and Harker, L. A. (1984) Thrombin stimulates tissue plasminogen activator release from cultured human endothelial cells. *J. Clin. Invest.*, **74**, 1988–95.

Lewis, J. C. and Taylor, R. G. (1989) Effects of varying dietary fatty acid ratios on plasma lipids and platelet function in the African green monkey. *Atherosclerosis*, **77**, 167–74.

Leyton, J., Drury, P. J. and Crawford, M. A. (1987) Differential oxidation of saturated and unsaturated fatty acids *in vivo* in the rat. *Br. J. Nutr.*, **57**, 383–93.

Li, X. and Steiner, M. (1990) Fish oil: a potent inhibitor of platelet adhesiveness. *Blood*, **76**, 938–45.

Lin, D. S. and Connor, W. E. (1990) Are the n-3 fatty acids from dietary fish oil deposited in the triglyceride stores of adipose tissue? *Amer. J. Clin. Nutr.*, **51**, 535–9.

Lippiello, L., Fienhold, M. and Grandjean, C. (1990) Metabolic and ultra-structural changes in articular cartilage of rats fed dietary supplements of omega-3 fatty acids. *Arth. Rheumat.*, **33**, 1029–36.

Lisanti, M. P. and Rodriguez-Boulan, E. (1990) Glyco-phospholipid membrane anchoring provides clues to mechanisms of protein sorting in polarised epithelial cells. *Trends Biochem. Sci.*, **15**, 113–8.

Liu, C. C., Carlson, S. E., Rhodes, P. G. *et al.* (1986) Increase in plasma phospholipid docosahexaenoic and eicosapentaenoic acids as a reflection of their intake and mode of administration. *Pediatr. Res.*, **22**, 292.

Loiacono, R. E., Kudo, K., Dusting, G. J. and Sinclair, A. J. (1987) Effects of arachidonate deficiency on endothelium-dependent vascular reactions. *Clin. Exp. Pharmacol. Physiol.*, **14**, 149–54.

Lollar, P., Knutson, G. L. and Fass, D. N. (1984) Stabilization of thrombin-activated porcine factor VIII:C by factor IXa and phospholipid. *Blood*, **63**, 1303–8.

Lorenz, R., Spengler, U., Fischer, S., Duhm, J. and Weber, P. C. (1983) Platelet function, thromboxane formation and blood pressure control during supplementation of the Western diet with cod liver oil. *Circulation*, **67**, 504–11.

Lovell, C. *et al.* (1981) Treatment of atopic eczema with evening primrose oil. *Lancet*, **i**, 278.

Low, M. G., Ferguson, M. A. J., Futerman, A. H. and Silman. I. (1986) Covalently attached phosphatidyl-inositol as a hydrophobic anchor. *Trends Biochem. Sci.*, **11**, 212–5.

Lucas, A., Morley, R. Cole, T. J., Gore, S. M., Lucas, P. J., Crowle, P., Pearse, R., Boon, A. J. and Powell, R. (1990) Early diet in preterm babies and developmental status at 18 months. *Lancet*, **i**, 1477–81.

Lucas, A., Morley, R., Cole, T. J., Lister, G. and Leeson-Payne, C. (1992) Breast milk and subsequent intelligence quotient in children born pre-term. *Lancet*, **339**, 261–4.

Lynch, D. V. and Thompson, G. A. (1984) Re-tailored lipid molecular species: a tactical mechanism for modulating membrane properties. *Trends Biochem. Sci.*, **9**, 442–5.

Mac Mahon, S., Peto, R., Cutler, J., Collins, R., Sorlie, P., Neaton, J., Abbott, R., Godwin, J., Dyer, A. and Stamler, J. (1990) Blood pressure, stroke, and coronary heart disease: prospective observational studies corrected for the regression dilution bias. *Lancet*, **335**, 765–74.

MacFarlane, R. G., Trevan, J. W. and Attwood, A. M. P. (1941) Participation of a fat soluble substance in coagulation of the blood. *J. Physiol.*, **99**, 7–8P.

Mahley, R. W., Innerarity, T. L., Weisgraber, K. H. and Oh, S. Y. (1979) Altered metabolism (*in vivo* and *in vitro*) of plasma lipoproteins after selective chemical modification of lysine residues of the apoproteins. *J. Clin. Invest.*, **64**, 743–50.

Mahley, R. W. (1982) Atherogenic hyperlipoproteinaemia. The cellular and molecular biology of plasma lipoproteins altered by dietary fat and cholesterol. *Med. Clin. N. Am.*, **66**, 375–402.

Manku, M. S. *et al.* (1984) Essential fatty acids in plasma phospholipids of patients with atopic eczema. *Br. J. Dermatol.*, **110**, 643–8.

Manning, A. S. and Hearse, D. J. (1984) Reperfusion-induced arrhythmias: mechanisms and prevention. *J. Mol. Cell. Cardiol.*, **16**, 497–518.

Margetts, B. M., Beilin, L. J., Armstrong, B. K., Rouse, I. L., Vandongen, R., Croft, K. D. and McMurchie, E. J. (1985) Blood pressure and dietary polyunsaturated and saturated fats: a controlled trial. *Clin. Sci.*, **69**, 165–75.

Markmann, P., Sandstrom, B. and Jespersen, J. (1990) Effects of total fat content and fatty acid composition in diet on factor VII coagulant activity and blood lipids. *Atherosclerosis*, **80**, 227–33.

Marlow, N. and Chiswick, M. L. (1985) Neurodevelopmental outcome in extremely low birthweight survivors. *Rec. Adv. Perinatal Med.*, **2**, 181–205.

Martinez, M. (1989) Dietary polyunsaturated fatty acids in relation to neural development in humans, in *Dietary n-3 and n-6 fatty acids* (eds C. Galli and A. P. Simopoulos), NATO ASI series, vol **171**, 123–33.

Masuda, J. and Ross, R. (1990a) Atherogenesis during low level hypercholesterolaemia in the non human primate. I. Fatty streak formation. *Arteriosclerosis*, **10**, 164–77.

Masuda, J. and Ross, R. (1990b) Atherogenesis during low level hypercholesterolaemia in the non human primate. II. Fatty streak conversion to fibrous plaque. *Arteriosclerosis*, **10**, 164–77.

Masuzawa, Y., Sugiura, T., Sprecher, H. and Waku, K. (1989) Selective acyl transfer in the reacylation of brain glycerophospholipids. Comparison of three acylation systems for 1-alk-1'-enylglycero-3-phosphoethanolamine, 1-acylglycero-3-phosphoethanolamine and 1-acylglycero-3-phosphocholine in rat brain microsomes. *Biochim. Biophys. Acta*, **1005**, 1–12.

Mattson, F. H. and Volpenheim, R. A. (1969) Relative rates of hydrolysis by rat pancreatic lipase of esters of C2-C18 fatty acids with C1-C18 primary n-alcohols. *J. Lipid Res.*, **10**, 271–6.

Mattson, F. H. and Nolen, G. A. (1972a) Absorbability by rats of compounds containing 1–8 ester groups. *J. Nutr.*, **102**, 1171–76.

Mattson, F. H. and Nolen, G. A. (1972b) Rate and extent of absorption of the fatty acids of fully esterified glycerol, erythritol, xylitol and sucrose as measured in thoracic duct cannulated rats. *J. Nutr.*, **102**, 1177–80.

Mattson, F. H. and Volpenheim, R. A. (1972) Hydrolysis of fully esterified alcohols containing from one to eight hydroxyl groups by the lipolytic enzymes of rat pancreatic juice. *J. Lipid Res.*, **13**, 325–8.

Mattson, F. H., Hollenback, E. J. and Kligman, A. M. (1975) Effect of hydrogenated fat on the plasma total cholesterol and triglyceride levels of man. *Am. J. Clin. Nutr.*, **28**, 726–31.

Mattson, F. H. and Hollenbach, E. J. (1979) The effect of a non-absorbable fat, sucrose polyester, on the metabolism of vitamin A by the rat. *J. Nutr.*, **109**, 1688–93.

Mattson, F. H. and Grundy, S. M. (1986) Comparison of dietary saturated, monounsaturated and polyunsaturated fatty acids on plasma lipids and lipoproteins in man. *J. Lipid Res.*, **26**, 194–202.

Maurice, P. D. L. *et al.* (1987) The effects of dietary supplementation with fish oil in patients with psoriasis. *Br. J. Dermatol.*, **117**, 599–606.

McDonald, B. E., Gerrard, J. M., Bruce, V. M. and Corner, E. J. (1989) Comparison of the effect of canola oil and sunflower oil on plasma lipids and lipoproteins and on *in vivo* thromboxane A$_2$ and prostacyclin production in healthy young men. *Am. J. Clin. Nutr.*, **50**, 1382–8.

McGee, D. L., Reed, D. M., Yano, K., Kagan, A. and Tillotson, J. (1984) Ten-year incidence of coronary heart disease in the Honolulu heart programme: relationship to nutrient intake. *Am. J. Epidemiol.*, **119**, 667–76.

McKeigue, P. M., Marmot, M. G., Adelstein, A. M., Hunt, S. P., Shipley, M. J., Butler, S. M., Riemersma, R. A. and Turner, P. R. (1985) Diet and risk factors for coronary heart disease in Asians in Northwest London. *Lancet*, **ii**, 1086–90.

McKeigue, P. M., Marmot, M. G., Syndercombe Court, Y. D., Cottier, D. E., Rahman, S. and Riemersma, R. A. (1988) Diabetes, hyperinsulinaemia, and coronary risk fctors in Bangladeshis in East London. *Br. Heart. J.*, **60**, 390–6.

McKeigue, P. M., Adelstein, A. M., Marmot, M. G *et al.* (1989) Diet and fecal steroid profile in a South Asian population with a low colon cancer rate. *Am. J. Clin. Nutr.*, **50**, 151–4.

McLennan, P. L., Abeywardena, M. Y. and Charnock, J. S. (1990) Reversal of the arrhythmogenic effects of long term saturated fatty acid intake by dietary n-3 and n-6 polyunsaturated fatty acids. *Am. J. Clin. Nutr.*, **51**, 53–8.

Meade, T. W., Vickers, M. V., Thompson, S. G. and Seghatchian, M. J. (1985a) The effect of physiological levels of fibrinogen on platelet aggregation. *Thromb. Res.*, **38**, 527–34.

Meade, T. W., Vickers, M. V., Thompson, S. G., Stirling, Y., Haines, A. P. and Miller, G. J. (1985b) Epidemiological characteristics of platelet aggregability. *Br. Med. J.*, **290**, 428–32.

Meade, T. W., Mellows, S., Brozovic, M., Miller, G. J., Chakrabarti, R. R., North, W. R. S., Haines, A. P., Stirling, Y., Imeson, J. D. and Thompson, S. G. (1986) Haemostatic function and ischaemic heart disease: principal results of the Northwick Park Heart Study. *Lancet*, **ii**, 533–7.

Medalie, J. H., Papier, C. M., Goldbouert, B. and Herman, J. B. (1975) Major factors in the development of diabetes mellitus in 10,000 men. *Arch. Intern. Med.*, **135**, 811–8.

Medicines Bill (1968) HMSO.

Mehrabian, M., Peter, J. B., Barnard, R. J. and Lusis, A. J. (1990) Dietary regulation of fibrinolytic factors. *Atherosclerosis*, **84**, 25–32.

Mehta, J., Lawson, D. and Saldeen, T. (1988) Reduction in plasminogen activator inhibitor-1 (PAI-1) with omega-3 polyunsaturated fatty acid (PUFA) intake. *Am. Heart J.*, **116**, 1201–6.

Mellies, M. J., Ishikawa, T. T., Gartside, P. S. *et al.* (1979) Effects of varying maternal dietary fatty acids in lactating women and their infants. *Am. J. Clin. Nutr.*, **32**, 299–303.

Mellies, M. J., Jandacek, R. J., Taulbee, J. D., Tewkesbury, M. B., Lamkin, G., Baehler, L., King, P., Boggs, D., Goldman, S., Gouge, A., Tsang, R. and Glueck, C. J. (1983) A double-blind, placebo-controlled study of sucrose polyester in hypercholesterolaemic outpatients. *Am. J. Clin. Nutr.*, **37**, 339–46.

Mendelsohn, D. and Mendelsohn, J. (1989) Effect of polyunsaturated fat on regression of atheroma in the non-human primate. *S. Afr. Med. J.*, **76**, 371–3.

Mensink, R. P. and Katan, M. B. (1987) Effect of monounsaturated fatty acids versus complex carbohydrates on high density lipoproteins in healthy men and women. *Lancet*, **i**, 122–5.

Mensink, R. P., Janssen, M. C. and Katan, M. B. (1988) Effect of blood pressure of two diets differing in total fat but not in saturated and polyunsaturated fatty acids in healthy volunteers. *Am. J. Clin. Nutr.*, **47**, 976–80.

Mensink, R. P., Groot, M. J. M., Van den Broek, L. T., Severijne-Nobels, A. P., Demacker, P. N. M. and Katan, M. B. (1989) Effects of monounsaturated fatty acids versus complex carbohydrates and serum lipoproteins and apoproteins in healthy men and women. *Metabolism*, **38** (Suppl 2), 172–8.

Mensink, R. P. and Katan, M. B. (1989) Effect of a diet enriched with monounsaturated or polyunsaturated fatty acids on levels of low density and high density lipoprotein cholesterol in healthy men and women. *N. Engl. J. Med.*, **321**, 436–41.

Mensink, R. P. and Katan, M. B. (1990) Effects of dietary *trans* fatty acids on high density and low density lipoprotein levels in health subjects. *N. Engl. J. Med.*, **323**, 439–44.

Miettinen, T. A., Naukkarinen, V., Hattunen, J. K., Mattial, S. and Kumlin, T. (1982) Fatty acid composition of serum lipids predicts myocardial infarction. *Br. Med. J.*, **285**, 993–6.

Milian, M. G. (1901) Influence de la peau sur la coagulabilitee du sang. *C. R. Soc. Biol.* (Paris), **53**, 576–8.

Millar, J. H. D., Zilkha, K. J., Langman, M. J. S., Wright, H. P., Smith, A. D., Belin, J. and Thompson, R. H. S. (1973) Double blind trial of linoleate supplementation of the diet in multiple sclerosis. *Br. Med. J.*, **i**, 765–8.

Miller, G. J., Martin, J. C., Webster, J., Wilkes, H. C., Miller, N. E., Wilkinson, W. H. and Meade, T. W. (1986) Association between dietary fat intake and plasma factor VII coagulant activity – a predictor of cardiovascular mortality. *Atherosclerosis*, **60**, 269–77.

Miller, G. J., Kotecha, S., Wilkinson, W. H., Stirling, Y., Sanders, T. A. B., Broadhurst, A., Allison, J. and Meade, T. W. (1988) Dietary and other characteristics relevant for coronary heart disease in men of Indian, West Indian and European descent in London. *Atherosclerosis*, **61**, 63–72.

Miller, G. J., Martin, J. C., Mitropoulos, K. A., Reeves, B. E. A., Thompson, R. L., Meade, T. W., Cooper, J. A. and Cruickshank, J. K. (1991) Plasma factor VII is activated by post-prandial triglyceridaemia, irrespective of dietary fat composition. *Atherosclerosis*, **86**, 163–71.

Miller, J. P., Heath, I. D., Choraria, S. K. *et al.* (1988)

Triglyceride lowering effect of MaxEPA fish lipid concentrate: a multicentre placebo controlled double blind study. *Clin. Chem.*, **178**, 215–60.

Miller, M. J. S., Pinto, A. and Mullane, K. M. (1987) Impaired endothelium-dependent relaxations in rabbits subjected to aortic coarctation hypertension. *Hypertension*, **10**, 164–70.

Mills, D. C. B., Figures, W. R., Scearce, L. M., Stewart, G. J., Colman, R. F. and Colman, R. W. (1985) Two mechanisms for inhibition of ADP-induced platelet shape change by 5-p-fluorosulfonylbenzoyl adenosine. Conversion of adenosine, and covalent modification at an ADP binding site distinct from that which inhibits adenylate cyclase. *J. Biol. Chem.*, **260**, 8078–83.

Milner, M. R., Gallino, R. A., Leffingwell, A. et al. (1989) Usefulness of fish oil supplements in preventing clinical evidence of restenosis after percutaneous transluminal coronary angioplasty. *Am. J. Cardiol.*, **64**, 394–9.

Ministry of Agriculture, Fisheries and Food (1959–1990) *Household Food Consumption and Expenditure*, Annual reports of the National Food Survey, HMSO, London.

Ministry of Agriculture, Fisheries and Food. (1988) *Guidelines on nutrition labelling*, London.

Ministry of Agriculture, Fisheries and Food, Food Advisory Committee (1989) *Use of nutrition claims in food labelling and advertising*, HMSO, London.

Ministry of Agriculture, Fisheries and Food (1990) *Eight guidelines for a healthy diet*.

Ministry of Agriculture, Fisheries and Food (1990) *Food labelling survey: England and Wales*, HMSO, London.

Ministry of Agriculture, Fisheries and Food and Department of Health. (1991) *Dietary Supplements and Health Foods*. Report of the Working Group, MAFF Publications, London.

Minoura, T., Takata, T., Sakaguchi, M., Takada, H., Yamamura, M., Koshire, I. and Yamamoto, M. (1988) Effect of dietary eicosapentaenoic acid on azoxymethane-induced colon carcinogenesis in rats. *Cancer Res.*, **48**, 4790–4.

Mishira, T., Watanbe, H., Arake, H. and Nakao, M. (1985) Epidemiological study of prostatic cancer by matched-pair analysis. *The Prostate*, **6**, 423–6.

Mitropoulos, K. A., Martin, J. C., Reeves, B. E. A. and Esnouf, M. P. (1989) The activation of the contact phase of coagulation by physiologic surfaces in plasma: the effect of large negatively charged liposomal vesicles. *Blood*, **73**, 1525–33.

Mohrhauer, H. and Holman, R. T. (1963a) The effect of dose levels of essential fatty acids upon fatty acid composition of the rat liver. *J. Lipid. Res.*, **4**, 151–9.

Mohrhauer, H. and Holman, R. T. (1963b) The effect of dietary essential fatty acids upon composition of polyunsaturated fatty acids in depot fat and erythrocytes of the rat. *J. Lipid. Res.*, **4**, 346–50.

Molgaard, J., von Schenck, H. V., Lassvik, C. L. et al. (1990) Effect of fish oil treatment on plasma lipoproteins in type III hyperlipoproteinaemia. *Atherosclerosis*, **81**, 1–9.

Moncada, S., Gryglewski, R., Bunting, S. and Vane, J. R. (1976) An enzyme isolated from arteries transforms prostaglandin endoperoxides to an unstable substance that inhibits platelet aggregation. *Nature*, **263**, 663–5.

Moncada, S., Higgs, E. A. and Vane, J. R. (1977) Human arterial and venous tissue generate prostacyclin (prostaglandin X), a potent inhibitor of platelet aggregation. *Lancet*, **i**, 18–21.

Morinelli, T. A., Niewiarowski, S., Daniel, J. L. and Smith, J. B. (1987) Receptor-mediated effects of a PGH$_2$ analogue (U46619) on human platelets. *Am. J. Physiol.*, **253**, H1035–43.

Morinelli, T. A., Niewiarowski, S., Kornacki, E., Figures, W. R., Wachtfogel, Y. and Colman, R. W. (1983) Platelet aggregation and exposure to fibrinogen receptors by prostaglandin endoperoxide analogues. *Blood*, **61**, 41–9.

Moroi, M., Jung, S. M., Okuma, M. and Shinmyozu, K. (1989) A patient with platelets deficient in glycoprotein VI that lacks both collagen induced aggregation and adhesion. *J. Clin. Invest.*, **84**, 1440–5.

Morris, J. N., Marr, J. W. and Clayton, D. G. (1977) Diet and heart: a postscript. *Br. Med. J.*, **ii**, 1307–14.

Morse, P. F., Horrobin, D. F., Manku, M. S., Stewart, J. C. M., Allen, R., Littlewood, S., Wright, S., Burton, J., Gould, D. J., Holt, P. J., Jansen, C. T., Mattila, L., Meigel, W., Dettke, T., Wexler, D., Guenther, L., Bordon, A. and Patriz, A. (1989) Meta-analysis of placebo-controlled studies of the efficacy of Epogam in the treatment of atopic eczema. Relationship between plasma essential fatty acids changes and clinical response. *Br. J. Dermatol.*, **121**, 75–90.

Mortensen, J. Z., Schmidt, E. B., Nielsen, A. H. and Dyerberg, J. (1983) The effect of n-6 and n-3 polyunsaturated fatty acids on hemostasis, blood lipids and blood pressure. *Thromb. Haemost.*, **50**, 543–6.

MRC Research Committee (1968) Controlled trial of soyabean oil in myocardial infarction. *Lancet*, **ii**, 693–700.

MRC Vitamin Study Research Group (1991) Prevention of neural tube defects: results of the MRC vitamin study. *Lancet*, **338**, 131–7.

Muller, A. D., v. Houwellingen, A. C., van Dam-Mieras, M. C. E., Bas, B. M. and Hornstra, G. (1989) Effects of a moderate fish intake on haemostatic parameters in healthy males. *Thromb. Haemostas.*, **61**, 468–73.

Multiple Risk Factor Intervention Trial Research Group (1982) Multiple risk factor intervention trial. *J.A.M.A.*, **248**, 1465–77.

Murray, R. K., Granner, D. K., Mayes, P. A. and Rodwell, V. W. (1988) *Harper's Biochemistry*. 21st edn, Appleton and Lange, Prentice Hall, USA.

Naismith, D. J. (1973) Kwashiorkor in western Nigeria: a study of traditional weaning foods, with particular reference to energy and linoleic acid. *Br. J. Nutr.*, **30**, 567–76.

Naismith, D. J., Deeprose, S. P., Supramaniam, G. and Williams, M. J. H. (1978) Reappraisal of linoleic acid requirement of the young infant, with particular regard to use of modified cows' milk formulae. *Arch. Dis. Child.*, **53**, 845–9.

Narisawa, T., Magadia, N. E., Weisburger, J. H. and Wynder, E. L. (1974) Promoting effect of bile acids on colon carcinogenesis after intrarectal instillation of N-methyl-N^1-nitro-N-nitrosoguanadine in rats. *J. Natl. Cancer Inst.*, **53**, 1093–7.

National Advisory Committee on Nutrition Education (1983) *Proposals for nutritional guidelines for health education in Britain*, Health Education Council, London.

National Research Council (1989) *Recommended Daily Allowances*, 10th edn, Food and Nutrition Board, Nut. Acad. Sci. USA.

Needleman, P., Moncada, S., Bunting, S., Vane, J. R., Hamberg, M. and Samuelsson, B. (1976) Identification of an enzyme in platelet microsomes which generates thromboxane A$_2$ from prostaglandin endoperoxides. *Nature*, **261**, 558–60.

Needleman, P., Raz, A., Minkes, M. S., Ferrendelli, J. A. and Sprecher, H. (1979) Triene prostaglandins: prostacyclin and thromboxane biosynthesis and unique biological properties. *Proc. Natl Acad. Sci. USA*, **76**, 944–8.

Needleman, P., Whitaker, M. O., Wyche, A., Watters, K., Sprecher, H. and Raz, A. (1980) Manipulation of platelet aggregation by prostaglandins and their fatty acid precursors: pharmacological basis for a therapeutic approach. *Prostaglandins*, **19**, 165–81.

Neilsen, N. H. and Hansen, J. P. (1980) Breast cancer in Greenland – selected epidemiological, clinical and histological features. *J. Cancer Res. Clin. Oncol.*, **98**, 287–99.

Nestel, P. J. (1991) Review: fish oil and cardiac function, in Health effects of w3 polyunsaturated fatty acids in seafoods, (eds A. D. Simopoulos, R. R. Kifer, R. E. Martin, and S. M. Barlow), *World Rev. Nutr. Diet.*, **66**, Basel, Karger, pp. 268–77.

Nestel, P. J., Conner, W. E., Reardon, M. F., Conner, S., Wary, S. and Boston, R. (1984) Suppression by diets rich in fish oil of very low density lipoprotein production in man. *J. Clin. Invest.*, **74**, 82–9.

Neuringer, M. and Connor, W. E. (1989) Omega-3 fatty acids in the retina, in *Dietary omega-3 and omega-6 fatty acids. Biological Effects and Nutritional Essentiality* (eds C. Galli and A. P. Simopoulos) Nato ASI Series, Series A: Life Sciences, Vol. **171**. Plenum Press, New York and London, pp. 177–90.

Neuringer, M., Anderson, G. J. and Connor, W. E. (1988) The essentiality of n-3 fatty acids for the development and function of the retina and brain. *Ann. Rev. Nutr.*, **8**, 517–41.

Nichols, A. B., Ravenscroft, C., Lamphiear, D. E. and Ostrander, L. D. (1976) Daily nutritional intake and serum lipid levels. The Tecumseh Study. *Am. J. Clin. Nutr.*, **29**, 1384–92.

Nieuwenhuis, H. K., Akkerman, J. W. N., Houdijk, W. P. M. and Sixma, J. J. (1985) Human blood platelets showing no response to collagen fail to express surface glycoprotein 1a. *Nature*, **318**, 470–2.

Nishizuka, Y. (1986) Studies and perspectives of protein kinase C. *Science*, **233**, 305–12.

Nordoy, A. and Rodset, J. M. (1971) The influences of dietary fats on platelets in man. *Acta Med. Scand.*, **190**, 27–4.

Norell, S., Ahlbom, A., Feychting, M. and Pedersen, N. L. (1986) Fish consumption and mortality from coronary heart disease. *Br. Med. J.*, **293**, 426.

Norris, P. G., Jones, C. J. H. and Weston, M. J. (1986) Effect of dietary supplementation with fish oil on systolic blood pressure in mild essential hypertension. *Br. Med. J.*, **293**, 104–5.

Norum, K. R., Christiansen, E. N., Christophersen, B. O. and Bremer, J. (1989) Metabolic and nutritional aspects of long-chain fatty acids of marine origin, in *The Role of Fats in Human Nutrition*, 2nd edn (eds A. J. Vergroessen and M. Crawford), Academic Press, London, pp. 118–49.

Novotny, W. F., Girard, T. J., Miletich, J. P. and Broze, G. J. (1989) Purification and characterization of the lipoprotein-associated coagulation inhibitor from human plasma. *J. Biol. Chem.*, **264**, 18832–7.

O'Connell, M. L., Canipari, R. and Strickland, S. (1987) Hormonal regulation of tissue plasminogen activator secretion and mRNA levels in rat granulosa cells. *J. Biol. Chem.*, **262**, 2339–44.

O'Connor, T. P., Roebuck, B. D., Peterson, F. and Campbell, T. C. (1985) Effect of dietary intake of fish oil and fish protein on the development of L-azaserine-induced preneoplastic lesions in the rat pancreas. *J. Natl Cancer Inst.*, **75**, 959–62.

Odeleye, O. and Watson, R. R. (1991) Health implications of the n-3 fatty acids. *Am. J. Clin. Nutr.*, **53**, 177–8.

Office of Population Censuses and Surveys, (1982–1990) *Mortality statistics: cause*. Review of the Registrar General on deaths by cause, sex and age in England and Wales, HMSO, London: HMSO.

Olegard, R. and Svennerholm, L. (1970) Fatty acid composition of plasma and red cell phosphoglycerides in full term infants and their mothers. *Acta Paediat. Scand.*, **59**, 637–47.

Ongari, M. A., Ritter, J. M., Orchard, M., Waddell, K. A., Blair, I. A. and Lewis, P. J. (1984) Correlation of prostacyclin synthesis by human umbilical artery with status of essential fatty acid. *Am. J. Obstet. Gynaecol.*, **149**, 455–60.

Osmundsen, H. and Hovik, R. (1988) Beta oxidation of polyunsaturated fatty acids. *Biochem. Soc. Trans.*, **16**, 420–3.

Osterud, B., Bouma, B. N. and Griffin, J. H. (1978) Human blood coagulation factor IX. Purification, properties and mechanisms of activation by activated factor XI. *J. Biol. Chem.*, **253**, 5946–51.

Owens, M. R. and Cave, W. T. (1990) Dietary fish lipids do not diminish platelet adhesion to subendothelium. *Br. J. Haematol.*, **75**, 82–5.

Pace-Asciak, C. and Wolfe, L. S. (1971) A novel prostaglandin derivative formed from arachidonic acid by rat stomach homogenates. *Biochem.*, **10**, 3657–64.

Packard, C. J., Munro, A., Lorimer, A. R., Gotto, A. M. and Shepherd, J. (1986) Metabolism of apolipoprotein B in large triglyceride-rich very low density lipoprotein of normal and hypertriglyceridemic subjects. *J. Clin. Invest.*, **75**, 2178–92.

Palmer, R. M., Ferrige, A. G. and Moncada, S. (1987) Nitric oxide release accounts for the biological activity of enothelium-derived relaxing factor. *Nature*, **327**, 524–6.

Panza, J. A., Quyyumi, A. A., Brush, J. E. and Epstein, S. E. (1990) Abnormal endothelium-dependent vascular relaxation in patients with essential hypertension. *N. Eng. J. Med.*, **323**, 22–7.

Parks, J. S., Kaduck-Sawyer, J., Bullock, B. C. and Rudel, L. L. (1990) Effect of dietary fish oil on coronary artery and aortic atherosclerosis in African green monkeys. *Arteriosclerosis*, **10**, 1102–12.

Parm, J., Horton, C. E., Mencia-Huerta, J. M., House, F., Eiser, N. M., Clark, T. J. H., Spur, B. W. and Lee, T. H. (1988) Effect of dietary supplementation with fish oil lipids on mild asthma. *Thorax*, **43**, 84–92.

Parratt, J. R. and Coker, S. J. (1985) Arachidonic acid cascade and the generation of ischaemia- and reperfusion-induced ventricular arrhythmias. *J. Cardiovasc. Pharmacol.*, **7** (Suppl. 5), S65–70.

Parthasarathy, S., Quinn, M. T., Schwenke, D. C., Carew, T. E. and Steinberg, D. (1990) Low density lipoprotein rich in oleic acid is protected against oxidative modification: implications for the dietary prevention of atherosclerosis. *Proc. Natl Acad. Sci. USA*, **87**, 3894–8.

Patsch, J. R. and Gotto, A. M. (1987) Metabolism of high density lipoproteins, in *Plasma Lipoproteins* (ed A. M. Gotto), Elsevier, Amsterdam, pp. 221–59.

Paty, D. W., Cousin, H. K., Read, S. and Adlakha, K. (1978) Linoleic acid in multiple sclerosis: failure to show any therapeutic benefit. *Acta Neurol. Scand.*, **58**, 53–8.

Payan, D. G., Wong, M. Y. S. and Chernov-Rogan, T. (1986) Alterations in human leukocyte function induced by ingestion of eicosapentaenoic acid. *J. Clin. Immunol.*, **6**, 402–10.

Pearce, M. L. and Dayton, S. (1971) Incidence of cancer in men on a diet higher in polyunsaturated fat. *Lancet*, **i**, 464–8.

Pharoah, P. O. D., Cooke, T., Cooke, R. W. I. and Rosenbloom, L. (1990) Birthweight specific trends in cerebral palsy. *Arch. Dis. Child.*, **65**, 602–6.

Phillipson, B. E., Rothrock, D. W., Connor, W. E. *et al.* (1985) Reduction of plasma lipids, lipoproteins and apoproteins by dietary fish oils in patients with hypertriglyceridaemia. *N. Engl. J. Med.*, **319**, 1210–6.

Picado, C., Castillo, J. A., Schinca, N., Pujades, M., Ordinas, A., Coronas, A. and Agusti-Vidal, A. (1988) Effects of a fish oil enriched diet on aspirin intolerant asthmatic patients: a pilot study. *Thorax*, **43**, 93–7.

Pixley, F., Wilson, D., McPherson, K. and Mann, J. (1985) Effect of vegetarianism on development of gallstones in women. *Br. Med. J.*, **291**, 11–2.

Plow, E. F., Pierschbacher, M. D., Ruoslahti, E., Marguerie, G. A. and Ginsberg, M. H. (1985) The effect of Arg-Gly-Asp-containing peptides on fibrinogen and von Willebrand factor binding to platelets. *Proc. Natl. Acad. Sci. USA*, **82**, 8057–61.

Plow, E. F. and Ginsberg, M. H. (1989) Cellular adhesion: GP IIb–IIIa as a prototypic adhesion receptor. *Prog. Haemostas. Thromb.*, **9**, 117–56.

Plummer, N. A. *et al.* (1978) Prostaglandins E_2, F_2 and arachidonic acid levels in irradiated and unirradiated skin of psoriatic patients receiving PUFA treatment. *Clin. Exp. Dermatol.*, **3**, 367–9.

PNUN (1989) *Nordishe Naringsrekomendationer Andra Uplagen*, The National Food Agency, Sweden.

Poole, J. C. F. (1955) The effect of certain fatty acids on the coagulation of plasma *in vitro*. *Br. J. Exp. Pathol.*, **36**, 248–53.

Pooling Project Research Group (1978) Relationship of blood pressure, serum cholesterol, smoking habit, relative weight and ECG abnormalities incidence of major coronary events: final report of the Pooling Project. *J. Chron. Dis.*, **31**, 201–306.

Popp-Snijders, C., Schouten, J. A., Heine, R. J., van der Meer, J., van der Veen, E. A. (1987) Dietary supplementation of omega-3 polyunsaturated fatty acids improves insulin sensitivity in non-insulin-dependent diabetes. *Diab. Res.*, **4**, 141–7.

Prescott, S. M. (1984) The effect of eicosapentaenoic acid on leukotriene production by human neutrophils. *J. Biol. Chem.*, **259**, 7615–21.

Press, M. *et al.* (1974a) Correction of essential fatty acid deficiency in man by cutaneous application of sunflower seed oil. *Lancet*, **i**, 597.

Press, M. *et al.* (1974b) Diagnosis and treatment of essential fatty acid deficiency in man. *Br. Med. J.*, **2**, 247–50.

Prickett, J. D., Trentham, D. E. and Robinson, D. R. (1984) Dietary fish oil augments the induction of arthritis in rats immunized with type II collagen. *J. Immunol.*, **132**, 725–9.

Prottey, C. (1977) Investigation of functions of essential fatty acids in the skin. *Br. J. Dermatol.*, **97**, 29–38.

Pudelkewicz, C., Seufert, J. and Holman, R. T. (1968) Requirements of the female rat for linoleic and linolenic acids. *J. Nutr.*, **94**, 138–46.

Pullman-Mooar, S., Laposata, M., Lem, D., Holman, R. T., Leventhal, L. D., Demarco, D. and Zurier, R. B. (1990) Alterations of the cellular fatty acid profile and the production of eicosanoids in human monocytes by gamma-linolenic acid. *Arth. Rheumat.*, **33**, 1526–30.

Puska, P., Iacona, J. M., Nissinen, A., Korhonen, H. J., Vartiainen, E., Pietinen, P., Dougherty, R., Leino, U., Mutanen, M., Moisio, S. and Huttunen, J. (1983) Controlled, randomised trial of the effect of dietary fat on blood pressure. *Lancet*, **i**, 1–5.

Putnam, J. C., Carlson, S. E., DeVoe, P. W. *et al.* (1982) The effect of variations in dietary fatty acids on the fatty acid composition of erythrocyte phosphotidyl choline and phosphotidyl ethanolamine in human infants. *Am. J. Clin. Nutr.*, **36**, 106–14.

Quinn, D., Shirai, K. and Jackson, R. L. (1982) Lipoprotein lipase: mechanism of action and role in lipoprotein metabolism. *Prog. Lipid Res.*, **22**, 35–78.

Quinn, M. T., Partharsarathy, S., Fong, L. C. and Steinberg, D. (1987) Oxidatively modified LDL: a potential role in recruitment and retention of monocyte/macrophage in atherogenesis. *Proc. Natl Acad. Sci.*, **84**, 2995–7.

Radack, K., Deck, C. and Huster, G. (1989) Dietary supplementation with low-dose fish oils lowers fibrinogen levels: a randomized, double-blind controlled study. *Ann. Intern. Med.*, **111**, 757–8.

Radack, K. L., Dock, C. C. and Huder, G. A. (1990) n-3 fatty acids effects on lipids, lipoproteins and apoproteins in very low doses: results of a randomised controlled trial in hypertriglyceridaemic subjects. *Am. J. Clin. Nutr.*, **51**, 599–605.

Radcliffe, R., Bagdasarian, A., Colman, R. and Nemerson, Y. (1977) Activation of bovine factor VII by Hageman factor fragments. *Blood*, **50**, 611–7.

Radomski, M. W., Palmer, R. M. and Moncada, S. (1987) The anti-aggregatory properties of vascular endothelium: interactions between prostacyclin and nitric oxide. *Br. J. Pharmacol.*, **92**, 639–46.

Ramesha, C. S., Gronke, R. S., Sivarajan, M. and Lands, W. E. M. (1985) Metabolic products of arachidonic acid in rats. *Prostaglandins*, **29**, 991–1008.

Ramsay, L. E., Yeo, W. W. and Jackson, P. R. (1991) Dietary reduction of serum cholesterol concentration : time to think again. *Br. Med. J.*, **303**, 953–7.

Reaven, G. M. (1988a) Dietary therapy for non-insulin dependent diabetes. *N. Eng. J. Med.*, **319**, 862–4.

Reaven, G. M. (1988b) Banting Lecture. Role of insulin resistance in human disease. *Diabetes*, **37**, 1595–1607.

Reddy, B. S. (1983) Tumor promotion in colon carcinogenesis, in *Mechanisms of Tumor Promotion* (ed T. J. Slaga), Vol. 1, CRC Press Inc., Boca Raton FL, pp. 107–29

Reddy, B. S. and Maeura, Y. (1984) Tumor promotion by dietary fat in azoxymethane-induced colon carcinogenesis in female F344 rats: influence of amount and sources of dietary fat. *J. Natl Cancer Inst.*, **72**, 745–50.

Reddy, B. S., Tanaka, T. and Simi, B. (1985) Effect of different levels of dietary *trans* fat or corn oil on azoxymethane-induced colon carcinogenesis in F344 rats. *J. Natl Cancer Inst.*, **75**, 791–8.

Reddy, B. S. and Maruyama, (1986) Effect of dietary fish oil on azoxymethane-induced colon carcinogenesis in male F344 rats. *Cancer Res.*, **46**, 3367–70.

Redgrave, T. A. (1983) Formation and metabolism of chylomicrons. *Int. Rev. Physiol.*, **28**, 103–30.

Regan, J. W., Nakata, H., De Marinis, R. M., Caron, M. G. and Lefkowitz, R. J. (1986) Purification and characterization of the human platelet alpha$_2$-adrenergic receptor. *J. Biol. Chem.*, **261**, 3894–900.

Reich, R., Royce, L. and Martin, G. R. (1989) Eicosapentaenoic acid reduces the invasive and metastatic activities of malignant tumor cells. *Biochem. Biophys. Res. Comm.*, **160**, 559–64.

Reiss, G. J., Boucher, T. M., Sipperly, M. E. *et al.* (1989) Randomised trial of fish oil for prevention of restenosis after coronary angioplasty. *Lancet*, **ii**, 177–81.

Renaud, S., Dumont, E., Godsey, F., Suplisson, A. and Thevenon, C. (1978) Platelet functions in relation to dietary fats in farmers from two regions in France. *Thromb. Haemostas.*, **40**, 518–31.

Renaud, S., Godsey, F., Dumont, E., Thevenon, C., Ortchanian, E. and Martin, J. L. (1986) Influence of long-term diet modification on platelet function and composition in Moselle farmers. *Am. J. Clin. Nutr.*, **43**, 136–50.

Reue, K. L., Warden, C. H. and Lusis, A. J. (1990) Animal models for genetic lipoprotein disorders. *Curr. Op. Lipidol.*, **1**, 143–50.

Revack, S. D., Cochrane, C. G. and Griffin, J. H. (1977) Multiple forms of active Hageman factor (HF) (coagulation factor XII) produced during contact activation. *Fed. Proc.*, **36**, 329.

Revack, S. D., Cochrane, C. G., Bonma, B. N. and Griffin, J. H. (1978) Surface and fluid phase activities of two forms of activated Hagemen factor produced during contact activation of plasma. *J. Exp. Med.*, **147**, 719–29.

Reynier, M. O., Crotte, C., Chautan, P., Lafont, H. and Gerolami, A. (1989) Influence of unsaturated oils on intestinal absorption of cholesterol. *J. Nutr. Res.*, **9**, 663–71.

Rich, S., Miller, J. F., Charous, S., Davis, H. R., Shanks, P., Glasgov, S. and Lands, W. E. M. (1989) Development of atherosclerosis in genetically hyperlipidaemic rabbits during chronic fish oil ingestion. *Arteriosclerosis*, **9**, 189–94.

Riella, M. C. *et al.* (1975) Essential fatty acid deficiency in human adults during total parenteral nutrition. *Ann. Int. Med.*, **83**, 786–9.

Riemersma, R. A., Wood, D. A., Butler, S., Elton, R. A., Oliver, M., Salo, M., Nikkari, T., Vartiainen, E., Puska, P., Gey, F., Rubba, P., Mancini, M. and Fidanza, F. (1986) Linoleic acid content in adipose tissue and coronary heart disease. *Br. Med. J.*, **292**, 1423–7.

Riemersma, R. A., Sargent, C. A., Saman, S., Rebergen, S. A. and Abraham, R. (1988) Dietary fatty acids and ischaemic arrhythmias. *Lancet*, **ii**, 285–6.

Riemersma, R. A. and Sargent, C. A. (1989) Dietary fish oil and ischaemic arrhythmias. *J. Intern. Med.*, **225** (Suppl.1), 111–6.

Rittenhouse-Simmons, S. (1979) Production of diglyceride from phosphatidylinositol in activated human platelets. *J. Clin. Invest.*, **63**, 580–7.

Ritter, J. M. and Taylor, G. W. (1988) Editorial. Fish oil in asthma. *Thorax*, **43**, 81–3.

Rivers, J. P. W. and Frankel, T. L. (1981) Essential fatty acid deficiency. *Br. Med. Bull.*, **37**, 59–64.

Roebuck, B. D., Longnecker, D. S., Baumgartner, K. J. and Thron, C. D. (1985) Carcinogen-induced lesions in the rat pancreas: effects of varying levels of essential fatty acid. *Cancer Res.*, **45**, 5252–6.

Rogers, A. E. and Wetsel, W. C. (1981) Mammary carcinogenesis in rats fed different amounts and types of fat. *Cancer Res.*, **41**, 3735–7.

Rogers, S., James, K. S., Butland, B. K., Etherington, M. D., O'Brien, J. R. and Jones, J. G. (1987) Effects of a fish oil supplement on serum lipids, blood pressure, bleeding time, haemostatic and rheological variables.

A double blind randomised controlled trial in healthy volunteers. *Atherosclerosis*, **63**, 137–43.

Roitt, I. M. (1988) *Essential Immunology*, 6th edn, Blackwell Scientific Publications, Oxford.

Rose, G. A., Thomson, W. B. and Williams, R. T. (1965) Corn oil in treatment of ischaemic heart disease. *Br. Med. J.*, **i**, 1531–3.

Rose, D. P. and Connolly, J. M. (1989) Stimulation of growth of human breast cancer cell lines in culture by linoleic acid. *Biochem. Biophys. Res. Comm.*, **164**, 277–83.

Rosenthal, M. D., Garcia, M. C., Jones, M. R. and Sprecher, H. (1991) Retroconversion and delta-4 desaturation of docosatetraenoate [22:4(n-6)] and docosapentaenoate [22:5 (n-3)] by human cells in culture. *Biochim. Biophys. Acta*, **1083**, 29–36.

Ross, R. (1986) The pathogenesis of atherosclerosis – an update. *N. Engl. J. Med.*, **314**, 488–500.

Roth, M., Lewit-Bentley, A., Michel, H., Deisenhofer, J., Huber, R. and Oesterfelt, D. (1989) Detergent structure in crystals of a bacterial photosynthetic reaction centre. *Nature*, **340**, 659–62.

Rouse, I. L., Beilin, L. J., Armstrong, B. K., and Vandongen, R. (1983) Blood pressure lowering effect of a vegetarian diet: controlled diet in normotensive subjects. *Lancet*, **i**, 5–10.

Royal College of Physicians of London and the British Cardiac Society (1976) Prevention of coronary heart disease. *J. Roy. Coll. Phys.*, **10**, 1–63.

Royal College of Physicians of London (1983) Obesity. *J. Roy. Coll. Phys.*, **17**, 3–58.

Ruberman, W., Weinblatt, E., Goldberg, J. D., Frank, C. W., and Shapiro, S. (1977) Ventricular premature beats and mortality after myocardial infarction. *N. Engl. J. Med.*, **297**, 750–7.

Sachs, F. M., Rosner, B., and Kass, E. H. (1974) Blood pressure in vegetarians. *Am. J. Epidemiol.*, **100**, 390–8.

Sakai, K., Hayashi, M., Kawashima, S. and Akanuma, H. (1989) Calcium-induced localization of calcium-activated neutral proteinase on plasma membranes. *Biochim. Biophys. Acta*, **985**, 51–4.

Salesse, R. and Garnier, J. (1984) Adenylate cyclase and membrane fluidity. The repressor hypothesis. *Mol. Cellul. Biochem.*, **60**, 17–31.

Sanders, T. A. B., Ellis, F. R. and Dickerson, J. W. T. (1978) Studies of vegans: the fatty acid composition of plasma choline phosphoglycerides, erythrocytes, adipose tissue, and breast milk and some indicators of susceptibility to ischaemic heart disease in vegans and omnivore controls. *Am. J. Clin. Nutr.*, **31**, 805–13.

Sanders, T. A. B and Naismith, D. J. (1979) A comparison of the influence of breast-feeding and bottle-feeding on the fatty acid composition of erythrocytes. *Br. J. Nutr.*, **41**, 619–23.

Sanders, T. A. B., Vickers, M. and Haines, A. P. (1981) Effect on blood lipids and haemostasis of a supplement of cod-liver oil rich in eicosapentaenoic and docosahexaenoic acids in healthy young men. *Clin. Sci.*, **61**, 317–24.

Sanders, T. A. B. and Hochland, M. C. (1983) A comparison of the influence on plasma lipids and platelet function of supplements of w3 and w6 polyunsaturated fatty acids. *Br. J. Nutr.*, **50**, 521–9.

Sanders, T. A. B. and Roshanai, F. (1983) The influence of different types of w3 polyunsaturated fatty acids on blood lipids and platelet function in healthy volunteers. *Clin. Sci.*, **63**, 91–9.

Sanders, T. A. B. and Roshanai, F. (1984) Assessment

of fatty acid intakes in vegans and omnivores. *Hum. Nutr. App. Nutr.*, **38A**, 345–54.

Sanders, T. A. B., Sullivan, D. R., Reeve, J. and Thompson, G. R. (1985) Triglyceride-lowering effect of marine polyunsaturates in patients with hypertriglyceridaemia. *Arteriosclerosis*, **5**, 459–65.

Sanders, T. A. B., Hinds, A. and Perreira, C. C. (1989) Influence of n-3 fatty acids on blood lipids in normal subjects. *J. Int. Med.*, **225** (Suppl 1), 99–104.

Sanders, T. A. B. and Reddy, S. (1992) The influence of a vegetarian diet on the fatty acid composition of breast milk and the essential fatty acid status of the infant. *J. Pediatrics* (in press).

Santoli, D., Phillips, P. D., Colt, T. L. and Zurier, R. B. (1990) Suppression of interleukin-2 dependent human T cell growth by E-series prostaglandins (PGE) and their precursor fatty acids: evidence for a PGE-independent mechanism of inhibition by the fatty acids. *J. Clin. Invest.*, **85**, 424–32.

Sarris, G. E., Fan, J. I., Sokoloff, M. H. *et al.* (1989) Mechanisms responsible for inhibition of vein-graft arteriosclerosis by fish oil. *Circulation*, **80**, I-109-I-123.

Sauer, L. A. and Dauchy, R. T. (1988) Identification of linoleic and arachidonic acids as the factors in hyperlipaemic blood that increase [³H]thymidine incorporation in hepatoma 7288CTC perfused *in situ*. *Cancer Res.*, **48**, 3106–11.

Saynor, R., Verel, D. and Gillot, T. (1984) The long-term effect of dietary supplementation with fish lipid concentrate on serum lipids, bleeding time, platelets and angina. *Atherosclerosis*, **50**, 3–10.

Saynor, R. and Gillott, T. (1988) Fish oil and plasma fibrinogen. *Br. Med. J.*, **297**, 1196.

Schaeffer, B. E. and Zadunaisky, J. A. (1979) Stimulation of chloride transport by fatty acids in corneal epithelium and relation to changes in membrane fluidity. *Biochim. Biophys. Acta*, **556**, 131–43.

Schaeffer, O. (1959) Medical observations on Canadian Eskimos. *Can. Med. Ass. J.*, **81**, 386–93.

Schectman, G., Kaul, S. and Kissbah, A. H. (1989a) Heterogeneity of low density lipoprotein responses to fish oil supplementation in hypertriglyceridemic subjects. *Arteriosclerosis*, **9**, 345–454.

Schectman, G., Kaul, S., Cherayil, G. D. *et al.* (1989b) Can the hypotriglyceridemic effect of fish oil concentrate be sustained. *Ann. Intern. Med.*, **110**, 346–52.

Schmidt, E. B., Varming, K., Ernst, E., Madsen, P. and Dyerberg, J. (1990) Dose-response studies on the effect of n-3 polyunsaturated fatty acids on lipids and haemostasis. *Thromb. Haemostas.*, **63**, 1–5.

Schonfeld, G. (1990) The genetic dyslipoproteinemias – nosology update 1990. *Atherosclerosis*, **81**, 81–93.

Scott, J. (1989) Lipoprotein (a): thrombogenesis linked to atherogenesis at last? *Nature*, **341**, 22.

Scott, J. (1991) Lipoprotein(a) *Br. Med. J.*, **303**, 663–4.

Seed, M., Hoppichler, F., Reaveley, D., McCarthy, S., Thompson, G. R., Boerwinkle, E. and Utermann, G. (1990) Relation of serum lipoprotein (a) concentration and apolipoprotein A phenotype to coronary heart disease in patients with familial hypercholesterolaemia. *N. Engl. J. Med.*, **322**, 1494–9.

Segrest, J. P. and Albers, J. J. (1986) The lipoproteins. Vol **128**: Separation, structure and molecular biology. Vol **129**. Characterization cell biology and metabolism, in *Methods in Enzymology*, Academic Press, New York.

Selenskas, S. L., Ip, M. M. and Ip, C. (1984) Similarity between *trans* fat and saturated fat in the modification of rat mammary carcinogenesis. *Cancer Res.*, **41**,

1321–6.

Seligsohn, U., Osterud, B., Brown, S. F., Griffin, J. H. and Rapaport, S. I. (1979) Activation of human factor VII in plasma and in purified systems. Roles of activated factor IX, kallikrein and activated factor XII. *J. Clin. Invest.*, **64**, 1056–65.

Seyberth, H. W., Oelz, O., Kennedy, T., Sweetmen, B. J., Danon, A., Frolich, J. C., Heimberg, M. and Oates, J. A. (1975) Increased arachidonate in lipids after administration to men; effects on prostacyclin biosynthesis. *Clin. Pharmacol. Therapeut.*, **18**, 521–9.

Seymour, A. A., Davis, J. O., Freeman, R. H., De Forrest, J. M., Rowe, B. P. and Wiliams, G. M. (1979) Renin release from kidneys stimulated with PGI_2 and PGD_2. *Am. J. Physiol.*, **237**, F285–90.

Sharp, D. S., Beswick, A. D., O'Brien, J. R., Renaud, S., Yarnell, J. W. G. and Elwood, P. C. (1990) The association of platelet and red cell count with platelet impedence changes in whole blood and light scattering changes in platelet rich plasma; evidence from the Caerphilly Collaborative Heart Disease Study. *Thromb. Haemostas.*, **64**, 211–5.

Shattil, S. J., Anaya-Galindo, R., Bennett, J., Colman, R. W. and Cooper, R. A. (1975) Platelet hypersensitivity induced by cholesterol incorporation. *J. Clin. Invest.*, **55**, 636–43.

Shattil, S. J. and Cooper, R. A. (1976) Membrane microviscosity and human platelet function. *Biochem*, **15**, 4832–7.

Shekelle, R. B., MacMillan Shryock, A., Paul, O., Lepper, M., Stamler, J., Liu, S. and Raynor, W. J. (1981) Diet, serum cholesterol and death from coronary heart disease. The Western Electric Study. *N. Engl. J. Med.*, **304**, 65–70.

Shekelle, R. B., Paul, O., MacMillan Shryock, A. and Stamler, J. (1985) Fish consumption and mortality from coronary heart disease. *N. Engl. J. Med.*, **313**, 820.

Shimokawa, H., Lam, J. Y. T., Chesebro, J. H., Bowie, E. J. W. and Vanhoutte, P. M. (1987) Effects of dietary supplementation with cod liver oil on endothelium-dependent responses in porcine coronary arteries. *Circulation*, **76**, 898–905.

Shimokawa, H. and Vanhoute, P. M. (1988) Dietary cod-liver oil improves endothelium-dependent responses in hypercholesterolaemia and atherosclerotic porcine coronary arteries. *Circulation*, **78**, 1421–30.

Shurtleff, D. (1974) Some characteristics related to the incidence of cardiovascular disease and death: Framingham Study, 18 year follow-up, in *The Framingham Study: an epidemiological investigation of cardiovascular disease*, Publication No (NIH) 74–599. US Dept of Health, Education and Welfare, Section 30. Washington DC.

Siegel, I., Liu, T. L., Yaghoubzadeh, E., Keskey, T. S. and Gleicher, N. (1987) Cytotoxic effects of free fatty acids on ascites tumor cells. *J. Natl. Cancer Inst.*, **78**, 271–7.

Siess, W., Roth, P., Scherer, B., Kurzmann, I., Bohlig, B. and Weber, P. C. (1980) Platelet-membrane fatty acids, platelet aggregation and thromboxane formation during a mackerel diet. *Lancet*, **i**, 441–4.

Simons, L. A., Dwyer, T., Simons, J., Bernstein, L., Lock, P., Poonia, N. S., Balasubramanian, S., Baren, D., Brunsen, J., Morgan, J. and Roy, P. (1987) Chylomicron and chylomicron remnants in coronary artery disease : a case-controlled study. *Atherosclerosis*, **65**, 181–9.

Simpson, H. C. R., Barker, K., Carter, R. D., Cassels, E. and Mann, J. I. (1982) Low dietary intake of linoleic acid predisposes to myocardial infarction. *Br. Med. J.*,

285, 683–4.

Sinclair, A. J. and Crawford, M. A. (1972) The accumulation of arachidonate and docosahexaenoate in the developing rat brain. *J. Neurochem.*, **19**, 1753–8.

Sinclair, A. J. and Crawford, M. A. (1973) The effect of a low fat maternal diet on neonatal rats. *Br. J. Nutr.*, **29**, 127–37.

Sinclair, A. J. (1975) The incorporation of radioactive polyunsaturated fatty acids into the liver and brain of the developing rat. *Lipids*, **10**, 175–84.

Sinclair, H. M. (1953) The diet of Canadian Indians and Eskimos. *Proc. Nutr. Soc.*, **12**, 69–82.

Sinensky, M., Pinkerton, F., Sutherland, E. and Simon, F. R. (1979) Rate limitation of (Na$^+$ and K$^+$)-stimulated adenosinetriphosphatase by membrane acyl chain ordering. *Proc. Natl Acad. Sci. USA*, **76**, 4893–7.

Singer, S. J. and Nicholson, G. L. (1972) The fluid mosaic model of the structure of cell membranes. *Science*, **175**, 720–31.

Singer, P., Berger, I., Luck, K., Taube, C., Naumann, E. and Godicke, W. (1986) Long-term effect of mackerel diet on blood pressure, serum lipids and thromboxane formation in patients with mild essential hypertension. *Atherosclerosis*, **62**, 259–65.

Singer, P., Jaeger, W., Berger, I., Barleben, H., Wirth, M., Richter-Heinrich, E., Voigt, S. and Godicke, W. (1990) Effects of dietary oleic, linoleic and alpha linolenic acids on blood pressure, serum lipids, lipoproteins and the formation of eicosanoid precursors in patients with mild essential hypertension. *J. Human Hypertension*, **4**, 227–33.

Singer, P., Wirth, M. and Berger, I. (1990) A possible contribution of decrease in free fatty acids to low serum triglyceride levels after diets supplemented with n-6 and n-3 polyunsaturated fatty acids. *Atherosclerosis*, **83**, 167–75.

Singh, A., Balint, J. A., Edmonds, R. H. and Rodgers, J. B. (1972) Adaptive changes of the rat small intestine in response to a high fat diet. *Biochim. Biophys. Acta*, **260**, 708–15.

Sinha, A. K., Shattil, S. J. and Colman, R. W. (1977) Cyclic AMP metabolism in cholesterol-rich platelets. *J. Biol. Chem.*, **252**, 3310–14.

Sinha, A. K., Scharschmidt, L. A., Neuwirth, R., Holthofer, H., Gibbons, N., Arbeeny, C. M. and Schlondorff, D. (1990) Effects of fish oil on glomerular function in rats with diabetes mellitus. *J. Lipid. Res.*, **31**, 1219–28.

Sirtori, C. R., Tremoli, E., Gatti, E., Montanari, G., Sirtori, M., Colli, S., Gianfranceschi, G., Maderna, P., Dentone, C. Z., Testolin, G. and Galli, C. (1986) Controlled evaluation of fat intake in the Mediterranean diet: comparative activities of olive oil and corn oil on plasma lipids and platelets in high-risk patients. *Am. J. Clin. Nutr.*, **44**, 635–42.

Sivakoff, M., Pure, E., Hsueh, W. and Needleman, P. (1979) Prostaglandins and the heart. *Fed. Proc.*, **38**, 78–2.

Skorve, J., Asiedu, D., Rustan, A. C., Drevon, C. A., Shurbaji, A. L. and Berge, R. K. (1990) Regulation of fatty acid oxidation and triglyceride and phospholipid metabolism by hypolipidemic sulphur-substituted fatty acid analogues. *J. Lipid. Res.*, **31**, 1627–36.

Skuladottir, G., Benediktsdottir, E., Hardarson, T., Hallgrimsson, J., Oddsson, G., Sigfusson, N. and Gudbjarnason, S. (1988) Arachidonic acid level of non-esterified fatty acids and phospholipids in serum and heart muscle of patients with fatal myocardial infarction. *Acta Med. Scand.*, **223**, 233–8.

Small, D. M. (1988) Progression and regression of atherosclerotic lesions. Insights from lipid physical biochemistry. *Arteriosclerosis*, **8**, 103–29.

Smith, D. L., Willis, A. L., Nguyen, N. *et al.* (1989) Eskimo plasma constituents dihomo gamma linolenic acid and eicosapentaenoic acid and docosahexaenoic acid inhibit the release of atherogenic mitogens. *Lipids*, **24**, 70–8.

Smithells, R. W., Sheppard, S., Schorah, C. J., Seller, M. J., Nevin, N. C., Harris, R., Read, A. P. and Fielding, D. W. (1981) Apparent prevention of neural tube defects by periconceptional vitamin supplementation. *Arch. Dis. Child.*, **56**, 911–18.

Snowdon, D. A. (1988) Animal product consumption and mortality because of all causes combined, coronary heart disease, stroke, diabetes and cancer in Seventh-Day Adventists. *Am. J. Clin. Nutr.*, **48**, 739–48.

Soderhjelm, J., Wiese, H. F. and Holman, R. T. (1970) The role of polyunsaturated acids in human nutrition and metabolism. *Prog. Chem. Fats Lipids*, **9**, 555.

Sorci-Thomas, M., Wilson, M. D., Johnson, F. L., Williams, D. L. and Rudel, L. L. (1989) Studies on the expression of genes encoding apoproteins B100 and B48 and the low density lipoprotein receptors in non human primates. Comparison of dietary fat and cholesterol. *J. Biol. Chem.*, **264**, 9039–45.

Sorisky, A. and Robbins, D. C. (1989) Fish oil and diabetes. *Diabetes Care*, **12**, 302–4.

Spady, D. K. and Dietschy, J. M. (1989) Interaction of aging and dietary fat in the regulation of low density lipoprotein transport in the hamster. *J. Lipid. Res.*, **30**, 559–69.

Specker, B. L., Wey, H. E. and Miller, D. (1987) Differences in fatty acid composition of human milk in vegetarian and non-vegetarian women: long-term effect of diet. *J. Ped. Gastr. Nutr.*, **6**, 764–8.

Sperling, R. I., Robin, J. L., Kylander, K. A., Lee, T. H., Lewis, R. A. and Austen, K. F. (1987a) The effect of n-3 polyunsaturated fatty acids on the generation of platelet-activating factor-A by human monocytes. *J. Immunol.*, **138**, 4186–91.

Sperling, R. I., Weinblatt, M. E., Robin, J. L., Ravalese, J., Hoover, R. L., House, F., Coblyn, J. S., Fraser, P. A., Spur, B. W., Robinson, D. R. Lewis, R. A. and Austen, K. F. (1987b) Effects of dietary supplementation with marine fish oil on leukocyte lipid mediator generation and function. *Arth. Rheum.*, **30**, 987–8.

Sperling, R. I. (1989) Diet therapy in rheumatoid arthritis. *Curr. Op. Rheumatol.*, **1**, 33–8.

Sprecher, H. (1989) (n-3) and (n-6) fatty acid metabolism, in *Dietary omega-3 and omega-6 Fatty Acids. Biological Effects and Nutritional Essentiality* (ed C. Galli and A. P. Simopoulos), Nato ASI Series A: Life Sciences, Vol. **171**, Plenum Press, New York and London, pp. 69–79.

Sprecher, H. (1991) Enzyme activities affecting tissue lipid fatty acid composition, in *Health Effects of Omega-3 Polyunsaturated Fatty Acids in Sea Foods* (eds A. P. Simopoulos, R. R. Kifer, R. E. Martin and S. M. Barlow), Karger, Basel, pp. 166–76.

Sprengers, E. D. and Kluft, C. (1987) Plasminogen activator inhibitors. *Blood*, **69**, 381–7.

Stacpoole, P. W., Alig, J., Ammon, L. and Crockett, S. E. (1989) Dose-response effects on dietary marine oils on carbohydrate and lipid metabolism in normal subjects and patients with hypertriglyceridaemia. *Metabolism*, **38**, 946–56.

Stammers, J. P., Hull, D., Abraham, R. and McFadyen, I. R. (1989) High arachidonic acid levels in the cord blood of mothers on vegetarian diets. *Br. J. Nutr.*, **61**, 89.

Statutory Instrument No. 1305 (1984) *The Food Labelling Regulations 1984*, HMSO, London.

Stefansson, V. (1960) *Cancer: Disease of Civilization?*, Hill and Wang, New York.

Steinberg, D., Parthasarathy, S., Carew, T. E., Khoo, J. C. and Witzum, J. L. (1989) Beyond cholesterol. Modifications of low density lipoprotein that increase its atherogenicity. *N. Engl. J. Med.*, **320**, 915–23.

Stern, D. M., Bank, I., Nawroth, P. P., Cassimeris, J., Kisiel, W., Fenton, J. W., Dinarello, C., Chess, L. and Jaffe, E. A. (1985) Self-regulation of procoagulant events in the endothelial cell surface. *J. Exp. Med.*, **162**, 1223–35.

Stewart, M. E. *et al.* (1978) The fatty acids of human sebaceous gland phosphatidyl choline. *Biochem. Biophys. Acta.*, **529**, 380–6.

Storlein, L. H., Borkman, M., Jenkins, A. B. and Campbell, L. V. (1992) Diet and *in vivo* insulin action: of rats and man. *Diabetes, Nutr. Meta.* (in press).

Streb, H., Irvine, R. F., Berridge, M. J. and Schulz, I. (1983) Release of Ca^{2+} from a nonmitochondrial intracellular store in pancreatic acinar cells by inositol-1,4,5-triphosphate. *Nature*, **306**, 67–9.

Strittmatter, P., Thiede, M. A., Hackett, C. S. and Ozols, J. (1988) Bacterial synthesis of active stearoyl CoA desaturase lacking the 26-residue amino terminal acid sequence. *J. Biol. Chem.*, **263**, 2532–5.

Stuart, M. J., Gerrard, J. M. and White, J. G. (1980) Effect of cholesterol on production of thromboxane B_2 by platelets *in vitro*. *N. Engl. J. Med.*, **302**, 6–10.

Stubbs, C. D. and Smith, A. D. (1990) Essential fatty acids in membrane: physical properties and function. *Biochem. Soc. Trans.*, **18**, 779–81.

Sturdevant, R. A. L., Pearce, M. L. and Dayton, S. (1973) Increased prevalence of cholelithiasis in men ingesting a serum-cholesterol-lowering diet. *N. Engl. J. Med.*, **288**, 24–7.

Sullivan, D. R., Sanders, T. A. B., Trayner, I. M. and Thompson, G. R. (1986) Paradoxical elevation of LDL apoprotein B levels in hypertriglyceridaemic patients and normal subjects ingesting fish oil. *Atherosclerosis*, **61**, 129–34.

Surawicz, C. M., Saunders, D. R., Sillery, J. and Rubin, C. E. (1981) Linoleate transport by human jejunum: presumptive evidence for portal transport at low absorption rates. *Am. J. Physiol.*, **240**, G157–62.

Swank, R. L. and Dugan, B. B. (1990) Effect of low saturated fat diet in early and late cases of multiple sclerosis. *Lancet*, **336**, 37–9.

Takimoto, G., Galang, J., Lee, G. K. and Bradlow, B. A. (1989) Plasma fibrinolytic activity after ingestion of omega-3 fatty acids in human subjects. *Thromb. Res.*, **54**, 573–82.

Tan, S. Y., Bravo, E. and Mulrow, P. J. (1978) Impaired renal prostaglandin E_2 biosynthesis in human hypertensive states. *Prostaglandins Med.*, **1**, 76–85.

Tate, G. A., Mandell, B. F., Karmali, R. A., Laposata, M., Baker, D. G., Schumacher, H. R. and Zurier, R. B. (1988) Suppression of monosodium urate crystal induced acute inflammation by diets enriched with gamma linolenic acid and eicosapentaenoic acid. *Arth. Rheum.*, **31**, 1543–51.

Tate, G., Mandell, B. F., Laposate, M., Ohliger, D., Baker, D. G., Schumacher, H. R. and Zurier, R. B. (1989) Suppression of acute and chronic inflammation by dietary gamma-linolenic acid. *J. Rheumatol.*, **16**, 729–33.

Telang, N. T., Basu, A., Kurihara, H., Osborne, M. P. and Modatz, M. J. (1988) Modulation in the expression of murine mammary tumor virus, *ras* proto-oncogene and of alveolar hyperplasia by fatty acids in mouse mammary explant cultures. *Anticancer Res.*, **8**, 971–6.

Tempel, H. van der., Tulleken, J. E., Limbuerg, P. C., Muskiet, F. A. J. and Rijswijk, M. H. van. (1990) Effects of fish oil supplementation in rheumatoid arthritis. *Ann. Rheum. Dis.*, **49**, 76–80.

Terano, T., Hirai, A., Hamazaki, T., Kobayashi, S., Fujita, T., Tamura, Y. and Kumagai, A. (1983) Effect of oral administration of highly purified eicosapentaenoic acid on platelet function, blood viscosity and red cell deformability in healthy human subjects. *Atherosclerosis*, **46**, 321–31.

Thiede, M. A., Ozols, J. and Strittmatter, P. (1986) Construction and sequence of cDNA from rat liver stearyl coenzyme A desaturase. *J. Biol. Chem.*, **261**, 13230–5.

Thomas, L. H., Olpin, S. O., Scott, R. G. and Wilkins, M. P. (1987) Coronary heart disease and the composition of adipose tissue taken at biopsy. *Hum. Nutr. Food Sci. Nutr.*, **41F**, 167–72.

Thompson, B. J. and Smith, S. (1985) Biosynthesis of fatty acids by lactating human breast epithelial cells. An evaluation of the contribution to the overall composition of human milk. *Paediatr. Res.*, **19**, 139–43.

Thompson, G. R. (1990) *A Handbook of Hyperlipidaemia*, Current Science Ltd, London.

Thompson, S. G. and Vickers, M. V. (1985) Methods in dose response platelet aggregometry. *Thromb. Haemostas.*, **53**, 216–8.

Thomson, A. B. R., Keelan, M., Garg, M. L. and Clandinin, M. T. (1989a) Influence of dietary fat composition on intestinal absorption in the rat. *Lipids*, **24**, 494–501.

Thomson, A. B. R., Keelan, M., Garg, M. L. and Clandinin, M. T, (1989b) Intestinal aspects of lipid absorption: in review. *Can. J. Physiol. Pharmacol.*, **67**, 179–91.

Thomson, M., Logan, R. L., Sharman, M., Lockerbie, L., Riemersma, R. A. and Oliver, M. F. (1982) Dietary survey in 40 year old Edinburgh men. *Hum. Nut. Appl. Nutr.*, **36A**, 272–80.

Thomson, M., Fulton, M., Wood, D. A., Brown, S., Elton, R. A., Birtwhistle, A. and Oliver, M. F. (1985) A comparison of the nutrient intake of some Scotsmen with dietary recommendations. *Hum. Nut. Appl. Nutr.*, **39A**, 443–55.

Thorngren, M. and Gustafson, A. (1981) Effects of 11-week increase in dietary eicosapentaenoic acid on bleeding time, lipids and platelet aggregation. *Lancet*, **ii**, 1190–3.

Thorngren, M., Shafi, S. and Born, G. V. R. (1984) Delay in primary haemostasis produced by a fish diet without change in local thromboxane A_2. *Br. J. Haematol.*, **58**, 567–78.

Tichelaar, H. Y. (1990) Eicosapentaenoic acid composition of different fish oil concentrates. *Lancet*, **ii**, 1450.

Tikkanen, M. J., Hullenen, J. K., Ehnholm, C. and Pielenen, P. (1990) Apolipoprotein E_4 homozygosity to serum cholesterol elevation during high fat diet. *Arteriosclerosis*, **10**, 285–8.

Tinoco, J. (1982) Dietary requirements and functions of alpha-linolenic acid in animals. *Prog. Lipid Res.*, **21**, 1–46.

Tisdale, M. J. and Dhesi, J. K. (1990) Inhibition of weight loss by n-3 fatty acids in an experimental cachexia model. *Cancer Res.*, **50**, 5022–6.

Tisdale, M. J. and Beck, S. A. (1991) Inhibition of tumour-induced lipolysis *in vivo* by eicosapentaenoic acid. *Biochem. Pharmacol.*, **41**, 103–7.

Trebitsch, R. (1907) Die Krankenheiten der Eskimos in Westgronland. *Wien. Klin. Wochenschr.*, **20**, 1404–8.

Tsai, M. H., Yu, C. L., Wei, F. S. and Stacey, D. W. (1989) The effect of GTPase activating protein upon rats is inhibited by mitogenically responsive lipids. *Science*, **243**, 522–6.

Tulleken, J. E., Limburg, P. C., Muskiet, F. A. J. and Rijswijk, M. H. van (1990) Vitamin E status during dietary fish oil supplementation in rheumatoid arthritis. *Arth. Rheumat.*, **33**, 1416–9.

Turner, J. D., Ngoc-anh, Le. and Brown, W. V. (1981) Effect of changing dietary fat saturation on low-density lipoprotein metabolism in man. *Am. J. Physiol.*, **241**, E57-E63.

Turner, M. and Gray, J. (1982) *Implementation of dietary guidelines. Obstacles and opportunities.* British Nutrition Foundation, London.

Turpeinen, U., Karvonen, M. J., Pekkarinen, M., Miettinen, M., Elosuo, R. and Paavilainen, E. (1979) Dietary prevention of coronary heart disease: the Finnish mental hospital study. *Int. J. Epidemiol.*, **8**, 99–118.

Uauy, R., Treen, M. and Hoffman, D. (1989) Essential fatty acid metabolism and requirements during development. *Sem. Perinatol.*, **13**, 118–30.

Uauy, R. D., Birch, D. G., Birch, E. E., Tyson, J. E. and Hoffman, D. R. (1990) Effect of dietary omega-3 fatty acids on retinol function of very low-birthweight neonates. *Ped. Res.*, **28**, 485–92.

Ulbricht, T. L. V. and Southgate, D. A. T. (1991) Coronary heart disease: seven dietary risk factors. *Lancet*, **338**, 985–92.

Ura, H., Makino, T., Ito, S., Tsutsumi, M., Kinigasa, T., Kamano, T., Ichimaya, H. and Konishi, Y. (1986) Combined effects of cholecystectomy and lithocholic acid on pancreatic carcinogenesis. *Cancer Res.*, **46**, 4782–6.

Vallance, P., Collier, J. and Moncada, S. (1989) Effects of endothelium-derived nitric oxide on peripheral arteriolar tone in man. *Lancet*, **ii**, 997–1000.

Vallot, A., Bernard, A. and Carlier, H. (1985) Influence of the diet on the portal and lymph transport of decanoic acid in rats. Simultaneous study of its mucosal catabolism. *Com. Biochem. Physiol.*, **82A**, 693–9.

Van Amelsvoort, J. M. M., Van Stratum, P., Krall, J. H., Lussenburg, R. N. and Houtsmuller, U. M. T. (1989) Effects of varying the carbohydrate fat ratio in a hot lunch on postprandial variables in male volunteers. *Br. J. Nutr.*, **61**, 267–83.

van der Merwe, C. F. (1984) The reversibility of cancer. *S. Afr. Med. J.*, **65**, 712.

van der Merwe, C. F., Booyens, J., Joubert, M. F. and van der Merwe, C. A. (1990) The effect of gamma-linolenic acid , an *in vitro* cytostatic substance contained in evening primrose oil, on primary liver cancer. A double-blind placebo controlled trial. *Prostaglandins, Leukotrienes and Essential Fatty Acids*, **40**, 199–202.

Vandongen, R., Mori, T., Coddle, J. P., Stanton, K. G. and Maserei, J. R. L. (1988) Hypercholesterolaemic effect of fish oil in insulin-dependent diabetic patients. *Med. J. Austr.*, **148**, 141–3.

Van Dorp, D. H. (1971) Recent developments in the biosynthesis and analysis of prostaglandins. *Ann. N.Y. Acad. Sci.*, **180**, 181–99.

Van Houwelingen, R., Nordoy, A., Van der Beek, E., Houtsmuller, U., de Metz, M., and Hornstra, G. (1987) Effect of a moderate fish intake on blood pressure, bleeding times, hematology and clinical chemistry in healthy males. *Am. J. Clin. Nutr.*, **46**, 424–36.

Van Houwelingen, A. C., Hornstra, G., Kromhout, D. and de Lezenne Coulander, C. (1989) Habitual fish consumption, fatty acids of serum phospholipids and platelet function. *Atherosclerosis*, **75**, 157–65.

Van Meer, G. (1988) How epithelia grease their microvilli. *Trends Biochem. Sci.*, **13**, 242–3.

Van Meer, G. (1989) Lipid traffic in animal cells. *Ann. Rev. Cell Biol.*, **5**, 247–75.

Velardo, B., Lagarde, M., Guichardant, M. and Dechavanne, M. (1982) Decrease in platelet activity after intake of small amounts of eicosapentaenoic acid in diabetes. *Thromb. Haemostas.*, **48**, 344.

Vessby, B. and Boberg, M. (1990) Supplementation with n-3 fatty acids may impair glucose homeostasis in patients with non-insulin-dependent diabetes mellitus. *J. Intern. Med.*, **228**, 165–71.

Vickers, M. V. and Thompson, S. G. (1985) Sources of variability in dose response platelet aggregometry. *Thromb. Haemostas.*, **53**, 219–20.

Villar, J. and Rivers, J. (1988) Nutritional supplementation during two consecutive pregnancies and the interim lactation period: effect on birth weight. *Pediatrics*, **81**, 51–7.

Vitale, J. J. and Broitman, S. A. (1981) Lipids and immune function. *Cancer Res.*, **41**, 3706–10.

Vollset, S. E., Heuch, I. and Bjelke, E. (1985) Fish consumption and mortality from coronary heart disease. *N. Engl. J. Med.*, **313**, 821.

Von Euler, U. S. (1934) *Arch. Exptl Pathol. Pharmakol.*, **175**, 78.

von Schacky, C., Fischer, S. and Weber, P. C. (1985) Long-term effects of dietary marine w-3 fatty acids upon plasma and cellular lipids, platelet function, and eicosanoid formation in humans. *J. Clin. Invest.*, **76**, 1626–31.

Vueri, E., Kuira, K., Makinen, S. M. *et al.* (1982) Maternal diet and fatty acid pattern of breast milk. *Acta. Paedriatr. Scand.*, **71**, 959–63.

Wade, A. E., Harley, W. and Bunce, O. R. (1982) The effects of dietary corn oil on the metabolism and mutagenic activation of N-nitrosodimethylamine (DMN) by hepatic microsomes from male and female rats. *Mutation Res.*, **102**, 113–21.

Wahlqvist, M. L., Lo, C. S. and Myers, K. A. (1989) Fish intake and arterial wall characteristics in healthy people and diabetic patients. *Lancet*, **ii**, 944–6.

Walker, A. F. and Wood, C. (1991) *Cholesterol, its role in health and nutrition*, The Butter Council, Sevenoaks.

Watts, G. F., Ahmed, W., Quiney, J., Houlston, R., Jackson, P., Iles, C. and Lewis, B. (1988) Effective lipid lowering diets including lean meat. *Br. Med. J.*, **296**, 235–7.

Weber, P. C., Larsson, C., Anggard, E., Hamberg, M., Corey, E. J., Nicolaou, K. C., and Samuelsson, B. (1976) Stimulation of renin release from rabbit renal cortex by arachidonic acid and prostaglandin endoperoxides. *Circ. Res.*, **39**, 868–74.

Weintraub, M. S., Zechner, R., Brown, A. *et al.* (1988) Dietary polyunsaturated fats of the n-6 and n-3 series reduce postprandial lipoprotein levels. Chronic and acute effects of fat saturation on postprandial lipoprotein metabolism. *J. Clin. Invest.*, **82**, 1884–93.

Welsch, C. W., House, J. L., Herr, B. L., Eliasberg, S. J. and Welsch, M. A. (1990) Enhancement of mammary carcinogenesis by high levels of dietary fat: A phenomenon dependent on *ad libitum* feeding. *J. Natl Cancer Inst.*, **82**, 1615–20.

Welsch, K. W. and Aylsworth, C. F. (1983) Enhancement of murine mammary tumorigenesis by feeding high levels of dietary fat: a hormonal mechanism? *J. Natl Cancer Inst.*, **70**, 215–21.

Wencel-Drake, J. D., Plow, E. G., Kunicki, T. J., Woods, V. L., Keller, D. M. and Ginsberg, M. H. (1986) Localization of internal pools of membrane glycoproteins involved in platelet adhesive responses. *Am. J. Pathol.*, **124**, 324–34.

Wertz, P. W. and Downing, D. T. (1982) Glycolipids in mammalian epidermis: structure and function in the water barrier. *Science*, **217**, 1261–2.

Wertz, P. W. *et al.* (1983) Effect of essential fatty acid deficiency on the epidermal sphingo lipids of the rat. *Biochim. Biophys. Acta*, **753**, 350–5.

Westberg, G., Tarkowski, A. and Svalander, C. (1989) Effect of eicosapentaenoic acid rich menhaden oil and maxEPA on the autoimmune disease of MRL/1 mice. *Int. Arch. Aller. Appl. Immunol.*, **88**, 454–61.

Westberg, G. and Tarkowski, A. (1990) Effect of MaxEPA in patients with SLE. *Scand. J. Rheumatol.*, **19**, 137–43.

WHO European Collaborative Group. (1986) European collaborative trial of multifactorial prevention of coronary heart disease: final report on the six year results. *Lancet*, **i**, 869–72.

Widdowson, E. M. (1970) The harmony of growth. *Lancet*, **i**, 901.

Wiener, B. H., Ockene, I. S., Levine, P. H. *et al.* (1986) Inhibition of atherosclerosis by cod-liver oil in a hyperlipidaemic swine model. *N. Engl. J. Med.*, **315**, 841–6.

Wiklund, O., Angelen, B., Olofsson, S. O., Eriksson, M., Fager, G., Berglund, I. and Bondjers, G. (1990) Apolipoprotein (a) and ischaemic heart disease in familial hypercholesterolemia. *Lancet*, **335**, 1360–3.

Wilhelmsen, L., Berglund, G., Elmfeldt, D., Tibblin, G., Wedel, H., Pennert, K., Vedin, A., Wilhelmsson, C. and Werko, L. (1986) The multifactor primary prevention trial in Goteborg, Sweden. *Eur. Heart J.*, **7**, 279–88.

Willett, W. C., Stampfer, M. J., Colditz, G. A., Rosner, B. A., Hennekens, C. H. and Speizer, F. E. (1987) Dietary fat and the risk of breast cancer. *New Eng. J. Med.*, **316**, 22–8.

Willett, W. C., Stampfer, M. J., Colditz, G. A., Rosner, B. A. and Speizer, F. E. (1990) Relation of meat, fat and fibre intake to the risk of colon cancer in a prospective study among women. *N. Engl. J. Med.*, **323**, 1664–72.

Wilt, T. J., Lofgren, R. P., Nichol, K. L. *et al.* (1989) Fish oil supplementation does not lower plasma total cholesterol in men with hypercholesterolemia. *Ann. Intern. Med.*, **111**, 900–5.

Wissler, R. W. and Vesselinovitch, D. (1988) Brief overview of the mounting evidence that atherosclerosis is both preventable and reversible. *J. Clin. Apheresis*, **4**, 52–8.

Wong, S. H., Fisher, E. A. and Marsh, J. B. (1989) Effects of eicosapentaenoic acid and docosahexaenoic acids on apoprotein B mRNA and secretion of very low density lipoprotein in HepG2 cells. *Arteriosclerosis*, **9**, 836–41.

Wood, D. A., Butler, S., Riemersma, R. A., Thomson, M., Oliver, M. F., Fulton, M., Birthwhistle, A. and Elton, R. (1984) Adipose tissue and platelet fatty acids and coronary heart disease in Scottish men. *Lancet*, **ii**, 117–21.

Wood, D. A., Butler, S., MacIntyre, C., Riemersma, R. A., Thomson, M., Elton, R. A. and Oliver, M. F. (1987) Linoleic and eicosapentaenoic acids in adipose tissue and platelets and risk of coronary heart disease. *Lancet*, **i**, 117–83.

Wood, P. D., Stefanick, M. L., Williams, P. T. and Haskell, W. L. (1991) The effects on plasma lipoproteins of a prudent weight-reducing diet, with or without exercise, in overweight men and women. *N. Engl. J. Med.*, **325**, 461–6.

Woodhill, J. M., Palmer, A. J., Leelarthaepin, B., McGilchrist, C. and Blacket, R. B. (1978) Low fat low cholesterol diet in secondary prevention of coronary heart disease. *Adv. Exp. Med. Biol.*, **109**, 317–30.

World Health Organisation (1982). *Prevention of coronary heart disease*. Technical report No. 678.

World Health Organisation (1990). *Diet, nutrition and the prevention of chronic diseases*, Technical report No. 797.

Wright, S. and Burton, J. L. (1982) Oral evening primrose seed oil improves atopic eczema. *Lancet*, **i**, 1120–2.

Wynn, M. and Wynn, A. (1981) *The prevention of handicap of early pregnancy origin*, The Foundation for Education and Research in Child Bearing, London.

Yamamoto, N., Saitoh, M., Moriuchi, A., Nomura, M. and Okuyama, H. (1987) Effect of dietary alpha-linolenate/linoleate balance on brain lipid composition and learning ability in rats. *J. Lipid Res.*, **28**, 144–51.

Yamanaka, W. K., Clemans, G. W. and Hutchinson, M. L. (1981) Essential fatty acids deficiency in humans. *Prog. Lipid Res.*, **19**, 187–215.

Yarnell, J. W. G., Elwood, P. C. and Renaud, S. (1987) Platelet function and ischaemic heart disease in the Caerphilly Study: preliminary results. *Biblthca. Nutr. Dieta.*, **40**, 19–27.

Young, M. R. I. and Young, M. E. (1989) Effects of fish oil and corn oil diets on prostaglandin-dependent myelopoiesis-associated immune suppressor mechanisms of mice bearing metastatic Lewis lung carcinoma tumors. *Cancer Res.*, **49**, 1931–6.

Zhu, Y. P., Su, Z. W. and Li, C. H. (1989) Growth inhibitory effects of oleic acid, linoleic acid and their methyl esters on transplanted tumors in mice. *J. Natl Cancer Inst.*, **81**, 1302–6.

Ziboh, V. A. *et al.* (1986) Effects of dietary supplementation of fish oil on neutrophil and epidermal fatty acids. *Arch. Dermatol.*, **122**, 1277–82.

Ziboh, V. A. (1989) Epidermal lipogenesis: Essential fatty acids and lipid inhibitors, in *Pharmacology of the Skin I* (eds Greaves M. W., Shuster, S.), Handbook of Experimental Pharmacology, **87**, Springer Verlag, Berlin, pp. 59–68.

Zilversmit, D. B. (1979) Atherogenesis: a postprandial phenomenon. *Circulation*, **60**, 473–85.

INDEX